MATHEMATICS
SOCIETY AND CURRICULA

MATHEMATICS
SOCIETY AND CURRICULA

H. B. GRIFFITHS
Professor of Mathematics, Southampton University

A. G. HOWSON
Senior Lecturer in Mathematics
Southampton University

CAMBRIDGE UNIVERSITY PRESS

To our parents

Published by the Syndics of the Cambridge University Press
Bentley House, 200 Euston Road, London NW1 2DB
American Branch: 32 East 57th Street, New York, N.Y. 10022

© Cambridge University Press 1974

Library of Congress Catalogue Card Number: 73–84320

ISB NUMBERS
0 521 20287 6 (casebound)
0 521 09892 0 (paperback)

First published 1974

Printed in Great Britain by
William Clowes & Sons, Limited
London, Beccles and Colchester

CONTENTS

CONTENTS

vi

CONTENTS

CONTENTS

PART 6 THE CURRICULUM IN THE SMALL

viii

CONTENTS

PART 7 A CLOSER LOOK AT EXAMINATIONS

PREFACE

Although this book is now intended for a wider audience, it is important to say that it grew from a course in 'Mathematical Curriculum Studies' given by the authors in Southampton since 1966. Consequently it owes much to our students, whose essays and questions introduced us to new ideas and attitudes. Colleagues in Southampton and elsewhere have also helped to educate us, by discussion and by allowing us to observe their ways of working.

The structure of the book can be inferred by looking at the table of contents, but Chapter 1 provides an introduction to our philosophy and conventions. Because our subject is as yet so unformed that we can only roughly summarise it at the beginning, we conclude the book with an Epilogue, which continues Chapter 1 but uses the language developed in the body of the text.

We hope that readers will not be alarmed by our unwillingness to confine ourselves to mathematics. If they are, we hope that they may nevertheless be persuaded to contemplate some of the questions we consider, if only because our pleasure in so doing might be infectious.

Southampton, January 1973

H. B. GRIFFITHS
A. G. HOWSON

Category	−2000	−500	−400	−300	−200	0	1300	1400	1500	1600	1700	1750	'76	'89	1800
Associations/Projects											Royal Society				
Individuals			Plato	Aristotle			Arab & Jewish translators	Renaissance artists			Jesuits				Rousseau
Administration											Scottish Act (1696)				
Schools — British							Universities { Oxford / Cambridge		Grammar Schools / Scottish Universities	Dissenting Academies					Private Tuition / Public Schools
Schools — Foreign							Bologna / Paris			Harvard	Yale				
Examinations	Apprenticeship of priests and scribes					Written (China)	Oral (West)				Math. Tripos (Written)		East India Co. Exams		
Conflict, etc.						Rome	Christianity / Cathedrals / Islam		Reformation / English Civil War / Navigation				American War of Independence	French Revolution	Industrial Rev.
Mathematicians	Babylon, Egypt	Pythagoras / Zeno	Eudoxus	Euclid	Archimedes				Copernicus / Kepler / Galileo / Newton / Leibniz				Euler		Laplace / Gauss

Time Scale: −2000 ... −500 −400 −300 −200 0 1300 1400 1500 1600 1700 ... 1750 ... '76 '89 180[0]

WARNING. This chart is a skeleton to help the reader to organise

A timeline chart (read with time running left to right, categories as rows).

Category	~1800	'28–'38	1850–'58	'70–'71	1900	'02–'09	'14–18	'33	'39–45	1950	'57–'60	'64–'65
Mathematicians	Lagrange, Fourier	Galois, Abel	Cauchy, Riemann	Boole, Sylvester, Cayley, Maxwell, Cantor, Poincaré	Hilbert, Russell	Einstein		Pontrjagin, Turing	von Neumann		Thom	
World Events	Steam	Waterloo, Railways	Great Exhibition, Crimean War, US Civil War, Franco-Prussian War			Aircraft	World War I, Russian Revolution	Hitler	World War II, Computers, Radar, Chinese Revolution		Korean War, Sputnik	'Third World' and Vietnam War
Examinations	Written BA at Oxford		London Matriculation	India Civil Service, Examining Boards, Home Civil Service		End of Exams for Mandarins (China)		SSEC		GCE Exams		CSE Exams, Schools Council
Grandes Ecoles / Foreign	Grandes Ecoles		Prussian Secondary Education, Tech. Hochschulen, Specialised Research		Rise of State Univs. in USA			Institute for Advanced Study (Princeton)				
Universities / Schools	Dame Schools, Bible Schools	London University, Durham University, Church Training Colleges, Elementary Schools	Owens College, Manchester		Manchester University	Birmingham University	Redbrick Universities				Comprehensive Schools begin, Plate-Glass Universities	Open University
Education Acts / Reports		First public funds for English educn., HMI's start	Clarendon and Taunton Commissions	Forster Act (Compulsory Elementary Education)		Secondary Schools Act, Hadow Report			Spens Report, Butler Act (Secondary Schools)		Newsom Report, Plowden Report	James Report
Educators	Froebel	T. Arnold	Kaye-Shuttleworth, M. Arnold, Darwin		F. Klein	Perry, Godfrey & Siddons, Hardy attacks, Tripos			Influence of N. Bourbaki		Piaget	
Mathematical Societies	Analytical Society (Babbage et al.)		London Math. Society	AIGT formed	Mathematical Association						ATM, UICSM, SMSG	SMP, MME, Nuffield, etc., Entebbe, etc.

Date axis: 1800 | '28 '38 | 1850 '58 | '70 '71 | 1900 | '02 '09 '14–18 | '33 '39–45 | 1950 | '57 '60 | '64 | '65

...king. For details and corrective emphases, consult the text.

ACKNOWLEDGEMENTS

Thanks are due to the following people and institutions for permission to include examination questions and examination papers: Bristol University, Exeter University, Hull University, Southampton University, Cambridge Colleges Joint Examinations, Associated Examining Board, Joint Matriculation Board, University of London School Examinations Department, Oxford and Cambridge Schools Examinations Board, Southern Universities Joint Board, Metropolitan Regional Examinations Board, Southern Regional Examinations Board, the Headmaster of Abbey Wood School and the Headmaster of Winchester College.

The authors and publisher also wish to thank the following for permission to quote extracts or to include diagrams; the School Mathematics Project for an extract 'On the "O" Level Course' and 'The "O" Level Syllabus' from *SMP: The First Ten Years*, edited by Bryan Thwaites; the Councils and Education Press Ltd, the National Council of Educational Technology and Dr M. R. Eraut for a diagram from *Continuing Mathematics*; Professor G. Matthews for a diagram outline of the St Dunstan's Syllabus, and Appleton-Century-Crofts/New Century for an extract from *Theories of Learning*, Second Edition, by E. R. Hilgard; Her Majesty's Stationery Office for an extract from *Schools Council Working Paper No. 14 – Mathematics for the Majority*, and Houghton-Mifflin and Mr Paul St Anthony for a diagram from *Goals for the Education of Elementary School Teachers*.

The authors are most grateful to the staff of the Cambridge University Press for their encouragement during the writing of the book and the skill and care they have shown during its printing.

1

What is Mathematical Education?

1. CHANGE, AND THOSE WHO IMPLEMENT IT

Great changes are taking place in schools and universities throughout the world today. Old courses are being replaced by new ones: and mathematics courses, especially, are changing rapidly. These developments have resulted from a reappraisal of the content and purposes of mathematics teaching, and they lead to further reappraisals. The changes arise in part from the tendency of mathematics to grow, reorganise and renew itself, and in part from the social development of the regions in which the changes occur. To discuss, plan and effect such changes, one needs knowledge and skills, and not only from mathematics; the resulting specialism has given rise to the new trade of mathematical educator† and to talk of the 'discipline' of 'mathematical education'. What, if anything, is meant by a 'mathematical educator' and what does this 'discipline' comprise? These are two questions which we hope to help answer in this book.

We shall see that the new courses are normally produced by a combination of various kinds of specialist – mathematicians, teachers and administrators. Our purpose is to explain the ways in which these specialists can cooperate *and the constraints under which they must work*. If they unite to produce a curriculum, but neglect any of the four factors – the nature of the society, the nature of its children, the nature of its teachers and the nature of mathematics – then they are likely to produce something ill-balanced, irrelevant, or impossible to implement.

Mathematical education, then, is far from being merely a subset of mathematics, because it must draw upon the knowledge and results of other disciplines; it is a practical activity and not merely a theoretical study. Yet it would be dangerous for anyone to try to practise it if he lacked an awareness of the factors involved. Some preliminary

---

† Until recently few people would have considered describing themselves as 'mathematical educators' – they were either teachers, teacher trainers or mathematicians interested in education.

1

study is essential, to cultivate this awareness. Clearly no one person is likely to possess the wide range of knowledge and experience that is needed to tackle the many problems which can be identified as relating to 'mathematical education'. Certainly the authors of this book would not claim to be omniscient. Always, we write as professional mathematicians who are teachers (even if of a very specialised type of student) and who are, and have been, involved in several kinds of administrative work; thus we have a sympathetic general acquaintance with these three kinds of specialist.

We assume that the reader has some mathematical knowledge and contemplates (or has) a career, perhaps as a teacher of mathematics at primary, secondary or tertiary level, perhaps as a teacher of mathematics teachers, or as a research mathematician (academic or industrial), or as an administrator in an educational system. It will help if he imagines himself as a potential participant in some mathematical development in his country or district, and considers how he might fulfil that role – for example to advise on the choice of a new syllabus and the consequent investment in books and examinations. Our book aims to supply him with some of the cautions, and some samples of the mathematics, that he may need.

2. ASPECTS OF A TOOL-KIT

For practical reasons we feel it essential that he should have a coherent view of mathematics, its place in the curriculum as a whole and the reasons – mathematical and social – for teaching it. The book is structured, therefore, so as to facilitate this. In Part 1 we consider why mathematics has been taught in schools and also what has been meant by mathematics within different systems and at different times. In Parts 2 and 3 we look at those factors that have led to changes in the curriculum and how they have reflected changes in society, pupils, teachers and mathematics.

In Part 4 we discuss how changes have been brought about, especially in recent years, and how the capacity to change might be built into an educational system. After this 'history' and 'theory', we show in Part 5 how actual mathematics can be embodied in curricula of various types, taking into account some of the constraints of which we have become aware. More detailed consideration of the planning of mathematical curricula is given in Part 6. In particular, we here pay especial attention to the problem of 'translating' a part of mathematics from a mathematician's mind to that of a child.

Finally, in Part 7 we look in detail at some problems posed by examinations.

3. THE NOTION OF A MODEL

Since the authors know the British (or, to be exact, the English) system of education best, it is inevitable that many of our examples are taken from it. We believe, however, that an understanding of a foreign system can illuminate one's own, so we hope that the book will also have value for readers outside Britain. Therefore, we have discussed other systems and drawn on our knowledge of education in other countries whenever possible. In any case, many systems are based for historical reasons on the English or Scottish systems, even though they are now being changed to make them more appropriate to the needs of the country concerned.

Now, such reasons and systems are complicated to describe, and we use the method of 'models'. Let us explain here our usage of this fashionable word. Some people speak of a 'model' school or system to mean one that is worthy of emulation. We, however, use the term to mean an *approximation* to a real situation, usually with some details suppressed for clarity. Except with music, sculpture and painting, it is usual to employ *words* to convey thoughts (or concepts) from one individual to another. Such words rarely describe *exactly* what we have in mind, for how can one person be sure that another has understood him *perfectly*? To achieve greater exactness we can either supply more detail or refine the language. The latter method is common in mathematics (see Chapter 19). Since mathematical education is concerned with *communicating mathematics* (subject to certain constraints), we must always be aware that what we communicate is frequently a deliberately chosen model, with details suppressed as appropriate. If too much detail is supplied at once, the message may become unintelligible. Frequently, a system is so complicated that it can be understood only by means of a model that suppresses detail, and we shall often need such approximate models, both of educational systems and of mathematics. By studying a simple model and gradually adding more complicated refinements, one can often be led to understand a complex system.

Thus we shall, for example, use the terms 'English' and 'Scottish' to describe two different models of a possible educational system. The 'English' model is one in which control of the material taught resides with the teachers rather than with the central administration; its opposite might be called the 'Prussian' model. The 'Scottish' model is one based on the idea of aiming at a high *average* level of attainment, but not aiming at a very high level for the few at the expense of a low one for the many. Anyone with a direct knowledge of English or Scottish education will recognise that these brief statements do not describe accurately the actual systems in those countries. Nevertheless, they approximate to the systems as they exist today and even more to the systems as they were in the past. Furthermore, such brief descriptions – like the 'Beeby' model given in Chapter 5 – are themselves compact concepts that can be manipulated usefully in a discussion, in contrast to more detailed ones. Similarly we use 'Greek mathematics' and 'Egyptian mathematics' as compact concepts to discuss two basically different attitudes to mathematics, even though historians are constantly adding details to the picture which show the actual historical situation to be less simple. Our usage is not to be taken as a licence for lazy inaccuracy; it is a technique for highlighting the forest at the expense of the trees. The reader, then, is warned of the dangers of the use of such models and will be expected to recognise and appreciate their use in this book.

One kind of relevant detail that we have excluded, although it is always nearby, is the description of the sheer effort required to induce desirable change. Thus a statement like 'there grew up a system' may well cover years of struggle against vested interests of privilege, inertia and ignorance. Such battles are too fascinating and important to be treated sketchily and the reader is referred to the history books for this essential material.

To some teachers the notions of 'model', 'syllabus' and 'curriculum' are anathema, for the words have a rigid inhuman connotation that goes against such teachers' own highly individualistic bent. They would hope to see mathematics taught – or rather learned – in a free, 'unstructured' way. This approach can be highly sophisticated, and is difficult to write about because it deliberately uses the complexities of human beings to develop their potential. We shall have something to say of the approach of such teachers later, since it is highly important; but theirs is a goal that will take long to attain, in the face of its complexity and the numbers of pupils and teachers involved.

4. AWARENESS

Our general aim in writing this book can be summed up as an attempt to make the reader more 'aware'. It is not our primary aim to

4

improve his technique as a mathematician, nor as a curriculum developer; but we hope that our book will motivate him to improve his techniques and knowledge, to want to supply hard arguments rather than platitudinous generalities. Thus we have supplied exercises that we hope will encourage him to reconsider his traditional beliefs, to investigate new approaches, to try his skill at designing 'teaching units' or constructing exercises, and to read other, more specialised works (throughout the book we draw the reader's attention to those articles and books that we ourselves have found of value). Above all, we want the reader to adopt a critical – not cynical or utopian – attitude to what is said and written about mathematical education. Some of our questions may appear naïve, but it is a sad fact of life that it is frequently the questions which can be expressed most simply that are most difficult to answer. Certainly there exists no companion book of 'Hints and Answers', for many of the questions are not amenable to simple solution. These, however, reflect our belief that it is often far better to ask a sensible question which cannot be fully answered, than to pose an irrelevant one whose sole virtue is that it possesses an easily obtained solution – a belief, it would appear, shared by few examiners and textbook authors. We have made no fixed assumptions concerning the level of the reader's mathematics, and the exercises with a mathematical content demand widely different mathematical backgrounds. We hope that, as a result, all readers will find some of the questions stimulating and will not dismiss them all as being 'trivial' or, on the other hand, 'impossibly difficult'. The literary, open-ended nature of some of them – so natural to historians and others – may also help the reader to appreciate and overcome the expository difficulties inherent in the subject, if he practises the appropriate languages. In any case, we stress here that a balanced reading of the book depends on (at least) reading the exercises.

It is possible that the reader will feel that, from time to time, we are suggesting too much in the matter of content and that, for example, there is 'not the time' to cover all the applications of mathematics we list in Chapter 19. Here it must be stressed that we do not consider the purpose of mathematical education to be the presentation of a fixed body of *content*. For, to quote from Ahlfors [1], †

† Numbers in square brackets after an author's name refer to the Bibliography. Even when only one work by a particular author is listed, we refer to it in the text by [1], to indicate that the reference is specific. Where only one author's name is given in the text, that does not necessarily mean that he was the sole author of the reference.

we believe that 'in mathematics, knowledge of any value is never possession of information, but "know-how". To know mathematics means to be able to do mathematics; to use mathematical language with some fluency, to do problems, to criticise arguments, to find proofs, and, what may be the most important activity, to recognise a mathematical concept in, or to extract it from, a given concrete situation.' It is the task of the teacher, therefore, to select suitable content and appropriate methods for the achievement of these ends. The authors hope that this book will help him in some way towards the accomplishment of that task.

2
Why Teach Mathematics?

This question is one which has often been asked. The answers have generally depended on the type of educational institution concerned and on the future envisaged for its pupils. We shall see how they have gradually developed to yield the kinds of reason which teachers might give today. We begin, however, by briefly surveying the growth of mathematics up to the end of the seventeenth century, for it was that body of mathematics which until recently most concerned mathematical education. The later developments of the subject are covered in Part 3.

1. SOME EARLY HISTORY

From the Egyptians† onwards, mathematics has had two aspects. One of these is practical, as a help in commerce, farming, building and control of the environment; the other is aesthetic, as men have enjoyed the contemplation of numbers and geometrical forms, the discipline of controlled imaginative thought and the thrill of discovering new mathematical relationships. The Egyptian priests combined both aspects, the mystical and the practical,‡ and their society honoured them highly for both. For that society, mathematics was taught in order to achieve the desirable state of being a priest.

Although other cultures – Babylonian and Assyrian, for example – generated some ideas, the next great stage of mathematical development was due to the Greeks. Through their philosophers and the Pythagorean cult, it was the aesthetic aspect of mathematics that was stressed. In particular, the Greeks formalised the notion of rational discussion and how to conduct it; the resulting style of argument, using 'postulates', 'definitions' and 'proofs' to establish

---

† With 'the Egyptians' we also associate Mesopotamians, Babylonians, etc. (see, for example, Bowen [1]).
‡ For details of their work on mensuration, the height of the Nile floods etc., see Gillings [1]. For the work of the Chinese, see Needham [1].

agreed 'propositions' ('theorems') enabled them to organise and generalise their existing geometrical knowledge. Euclid's *Elements*, exemplifying this approach, was regarded until a century ago as a standard of perfection that all mathematicians should seek to emulate.

The greatest Greek contributions to the practical aspect of mathematics were those of Archimedes, but they had relatively little influence and were hampered by the Greek 'failure' in arithmetic. Thinking that they could make a model of geometry using the only numbers they knew – the positive fractions and integers – the Greeks were frustrated by the discovery that the number 2 had no rational square root. For, in their model, such a root was needed to describe the hypotenuse of an isosceles right-angled triangle. The inadequacy of their model, therefore, led the Greeks to neglect arithmetic in favour of a 'literary' description of geometry. Lacking an adequate language of numbers, Archimedes had to invent a system of his own, described in his work *The Sand-Reckoner*, in order to prove to King Gelon that the number of grains of sand in the world could not be infinite, as was popularly supposed. His account (see Newman [1] for details) is a beautiful example of the conscious creation of a language which would allow him to make precise statements about apparently vague concepts.

Following the Dark Ages, mathematical activity was most vigorous in the East, where, around AD 1000, Arab, Indian and Chinese mathematicians developed arithmetic and algebra. They invented 'zero' and our 'place' system of decimal notation, and developed techniques which had ready applications when trade began to increase in the Middle Ages. At this time the profession of 'Reckoning Master' flourished, whilst geometry came to be applied by navigators and the builders of cathedrals.

The Renaissance brought back the Greek ideas, and saw the development of algebra and the theory of equations (with the associated need for complex numbers). This was followed by the work of astronomers, such as Kepler and Galileo, which led to the invention of the differential calculus. Here was the tool for the Industrial Revolution which was to begin in the eighteenth century (see, for example, the advertisement reproduced on p. 11) and for the rapid expansion of scientific knowledge.

The mathematics we have so far mentioned is that which, until recently, most commonly concerned mathematical education. Just as the mathematics had its two aspects – the practical and the aesthetic

8

– so we shall see that at different times and in different places both aspects were used to justify the teaching of the subject.

Exercises

1 Observe how Archimedes had to work out his own curriculum 'unit' to teach the king. Break down the steps in his argument and illustrate it with a flow diagram.

2 Investigate the reasons why great periods of growth in mathematics have been followed by times of stagnation. Why did the mathematics of the Egyptians and the Greeks stop where it did?

3 From the 'Moscow' papyrus (written about 1850 BC) it appears that the Egyptians were aware of the general formula for the volume of a truncated pyramid with square bases, having sides of length a and b, and height h:

$$V = \frac{h(a^2 + ab + b^2)}{3}.$$

What mathematics is needed to establish the validity of this formula?

2. MATHEMATICS IN EARLY SCHOOLS

Whereas the Egyptian priests studied mathematics especially because of its usefulness, the Greeks were attracted by the aesthetic aspect and by the power of mathematical argument. The educated Greek, moreover, was a wealthy man (rarely woman) who could leave practical things to his slaves; thus the lowly did arithmetic and mensuration, the highborn did the geometry of Euclid. Such geometry, therefore, became by association superior to the other, and the mark of a gentleman. This attitude, seasoned by intellectual curiosity, was absorbed into the Western universities of the Middle Ages in their role as producers of clerics, lawyers and medical men.

In sixteenth-century Europe, rich merchants and men of affairs were founding schools for boys, sometimes to prepare them for entry to universities, often to make them good Protestants, free of Popery. 'If I had children . . . I would have them learn not only the languages and history, but singing, instrumental music, and a full course of mathematics' wrote Martin Luther in his *Discourse on the Duty of Sending Children to School* (1524), in which he argued for schools and a curriculum which would be free from the authority and power of the Church. Sometimes the founders of schools wanted mathematics to be taught as a useful art, to provide financiers, surveyors and architects. For example, the statutes of Blackburn Grammar School, established in 1597, state that 'The principles of Arithmetic, Geometry and Cosmography with some introduction into the sphere are profitable' (quoted in HMSO [5], p. 8). In most of the grammar schools, however, only Latin and Greek were taught. By 1600, the provision of grammar schools in England was such that in most parts of the country any boy could, in principle, obtain a free grammar school education within a few miles of his home (Armytage [1], p. 6). However, few children availed themselves of this opportunity and, of those who did, only a small minority studied any mathematics other than the most elementary arithmetic.

The seventeenth century witnessed the beginning of modern science, the work of such men as Newton, Pascal, Boyle, Huygens, Hooke, Flamsteed and Halley, and the founding of such scientific societies as the Royal Society in England and the Lincei in Italy. The seeds were laid for more vociferous demands for practical subjects and, indeed, Newton, Flamsteed and Halley all showed interest in the establishment in 1673 of a special 'Mathematical School' for forty boys at Christ's Hospital. England was rapidly developing its overseas trade, and a knowledge of navigation was proving of more value than an acquaintance with the classics. Equally, the foundation of Christ's Hospital Mathematical School was welcomed since the institution of similar schools in France had been noted and the spirit of competition was as great then as it is today.

The example of Christ's Hospital was followed by the establishment of other schools such as Sir Joseph Williamson's Mathematical School at Rochester (1701) and Neale's Mathematical School in London (1705), and by the introduction of mathematics into the curriculum of old-established grammar schools on or near the sea coast, such as that at Dartmouth.

An important influence in society has been the non-conformist. The religious struggles of the seventeenth and early eighteenth centuries had marked out the Dissenters as vigorous citizens debarred from full civil rights because of their beliefs. Thus, they could not be accepted as true gentlemen. Now, where it was necessary to be educated as a gentleman, practical mathematics was to be eschewed – just as Jane Austen's ladies knew few household skills and thus proved to a socially calculating world that their men could afford servants. But not all the sons of gentlemen could inherit; and many non-conformists, debarred from the professions, had to earn a living as practical men. To meet their needs a number of private academies arose which laid a distinct emphasis on utilitarian studies for those who wished to train for a career in business or trade, in surveying or industry or at sea. The curriculum of one such school is described in an advertisement appearing in the *Leeds Mercury* in March 1743:

At PONTEFRACT

Writing; Arithmetic, Vulgar, Decimal and Instrumental; Logarithms, applied to Arithmetic Calculations, and their Use in higher branches of the Mathematics, also their method of Construction; MERCHANTS ACCOMPTS; ALGEBRA, numerical and in Species, with its Application to various Branches of the Mathematics; GEOMETRY in Theory and Practice, with its Application to Surveying of Land, Gauging, Mensuration of Artificer's Work, &c. TRIGONOMETRY, both plain and spherical, and their Use in Navigation, Astronomy, Dialling, &c. FLUXIONS, or the new Method of Investigation; and the USE of the GLOBES in Navigation, Geography, Astronomy &c. as also of all MATHEMATICAL INSTRUMENTS.

Taught by R. WRIGHT, after an easy and plain Method.

N.B. All sorts of Artificer's Work may be survey'd, for either Master or Workman, at reasonable Rates; and the most irregular Parcel of Land may be Survey'd, or Gentleman taught to Survey it by a new and exact Method founded on the Doctrine of Fluxions.

Also with the said *R. W.* Youth may be boarded.

Leeds Mercury (22 March 1743)

The grammar and public† schools were, however, loth to change their classical curriculum and, in fact, were often prevented from doing so by their statutes. Indeed, under a court ruling of 1805 (see, for example, Barnard [1], p. 19) grammar schools were specifically ordered to respect their founders' wishes with respect to the curriculum. The vacuum thus engendered by the conservatism and impotence of the established schools was soon filled by numerous private schools, many of which were well supported financially, and have lasted to this day.

In Scotland, things were different. The Calvinist fear of Jesuit indoctrination led to an Act of 1696 which decreed that it was the duty of every parish to provide a 'commodious house for a school' and salary for a teacher. (An earlier Act of 1496 had attempted, unsuccessfully, to require the sons of barons and freeholders to be sent to school from the age of 8 or 9 until they had perfect command of Latin.) By 1583, four Scottish universities were in existence, developed from French prototypes. They existed to provide any suitable 'lad of pairts' with vocational training as dominie (teacher), parson, lawyer or doctor. Fuelled by a Calvinist attitude to work, they were different places from the two English universities. True, they never produced a Newton, but by the eighteenth century

† Until the late nineteenth century 'Public School' is to be taken as referring to the nine schools: Charterhouse, Eton, Harrow, Merchant Taylors, Rugby, St. Paul's, Shrewsbury, Westminster and Winchester. After that time many of the old grammar schools and some of the newly created private schools were accorded 'public school status'.

Scottish intellectual life was flowering, and in a closely-knit community the 'Scottish model' of education, as mentioned on p. 3, was emerging.

The French Revolution at the end of the eighteenth century had a most important influence on mathematics. Napoleon himself was interested in the subject and saw its practical potential – not only for artillery officers like himself. He was also well acquainted with mathematicians such as Fourier (who accompanied him to Egypt) and Laplace, both of whom received honours from him. In particular, the French noted the great usefulness of mathematics in military engineering, the need for training in which led, in 1794, to the establishment of the Ecole Polytechnique of Paris. The influence of the Ecole was enormous; it served as a model for other schools in France and elsewhere, for example the US Military Academy at West Point, and its staff included such mathematicians as Lagrange, Laplace, Monge, Poisson, Fourier, Cauchy, Chasles and Hermite. It even had the distinction of twice refusing to accept Galois as a pupil, because he failed its entrance examination in Mathematics.

Exercises

1 Describe briefly the advances in mathematics due to those mathematicians mentioned above as being associated with the Ecole Polytechnique.

2 Prince Albert wrote (see Barnett [1]):
'What is to be gained by making the officers of the army, and the staff in particular, abstract *mathematicians* instead of soldiers? ... it is well ascertained fact ... that mathematicians, from their peculiar bent of mind, do of all men show the least judgement for the practical purposes of life, and are the most helpless and awkward in common life ...'
'Mathematics is the study which forms the foundation of the course (West Point Military Academy). This is necessary, both to impart to the mind that combined strength and versatility, the peculiar vigor and rapidity of comparison necessary for military action, and to pave the way for progress in the higher military sciences.'

(Congressional Committee on Military Affairs, 1834)

Discuss.

3. THE NINETEENTH CENTURY

The growth and organisation of education in England in the nineteenth century is of great interest, and the way in which it progressed has profoundly influenced *mathematical* education as we know it.

England had a more clearly defined class structure than it has now and, naturally, the educational system catered for that structure.

In particular, the reasons for teaching mathematics varied according to the type of pupil for which the system catered.

The educational route for the landed gentry and many of the professional classes was the public school – or a private tutor, for until the 1830s the public schools were hardly morally acceptable – and then Oxford or Cambridge, which at the beginning of the century remained the only two English universities.

The Industrial Revolution was generating wealth, with a class of newly rich. For these, the schools varied from cheap, privately owned 'writing schools', on the one hand, to the grammar schools and the private schools, on the other. Many private schools were maintained by joint-stock companies and some were soon to be accepted as public schools. An important part of their function was to provide an education that would mark out its possessor from those of lower class, and if possible to have him accepted in a higher social class. The Gentleman, and Respectability, were goals for aspiration.

For the working class, which contained not only the poor and ignorant, but craftsmen with a scientific curiosity and skill, there were elementary schools. These were provided at the beginning of the century by voluntary bodies (usually religious, to teach Bible study), but later the Government made an increasing contribution to their upkeep. At first, little of what we now regard as secondary education was available to members of this class, but after 1870 new types of school were created in an attempt to meet this need; it grew partly from the need to supply elementary teachers and partly because of the phenomenon of rising expectations.

One of the most significant features of this stratified system was that, by and large, the direction of all varieties – except for some of the smaller private schools – was in the hands of those who belonged to the first of these three classes and who incorporated their own prejudices and outlook into the systems they designed.

The mathematics taught in the different types of schools reflected the careers their pupils were expected to follow. Under the influence of Thomas Arnold, who was Headmaster of Rugby from 1828 to 1842, and others, the public schools had re-emerged as a potent force in English education, but they had done little to respond to the practical needs of users of mathematics. Of the boys leaving the sixth form at Harrow in 1861 'about half had done six books of Euclid, about one-third had done some trigonometry, and two-thirds had reached quadratic equations in algebra; a very few had advanced to geometrical conic sections and mechanics, fewer still to analytical geometry, while some reached the calculus by private tuition'. (Clarendon Commission 1864, quoted in HMSO [16].) Certainly the utilitarian aspect of mathematics did not unduly influence the planner of that curriculum. The public schools, moreover, did not hide their dislike for such motives. Dr B. H. Kennedy (author of the famed *Public School Latin Primer* and Headmaster of Shrewsbury School), when giving evidence to the Clarendon Commission, endorsed the unwillingness of the public schools to 'fritter away their power' on providing a utilitarian curriculum for the children of local inhabitants. (Kennedy had long been in dispute with the burgesses of Shrewsbury as to whether *libera schola grammaticalis*, as used in the school's original charter, was to mean literally a grammar school available without fee to local children, or a school 'free of all superiority, save that of the Crown'.) He dismissed the teaching of natural sciences as 'not furnishing a basis for education'. They lacked the 'mental discipline' of the classics (presumably shared by Euclid), and their utility could not compensate for this.

The grammar schools were slower than the public schools to emerge from the depths to which they had fallen, and so were unable to furnish any lead. Some of the newer private schools, however – for example, Mill Hill School (founded by the Congregationalists in 1807) and Cheltenham College (1841) – were quick to respond to the growing needs of the universities and the Services for entrants trained in mathematics. Cheltenham College, for instance, had from its inception a 'Modern Department' in which mathematics had a dominant place in the curriculum, and which was intended to train boys for the military academies at Woolwich and Sandhurst, for engineering or for commercial life.

Education for girls – other than by private governesses – was now thought respectable. The establishment in 1850 of the North London Collegiate School and, three years later, of the Cheltenham College for Young Ladies paved the way for the foundation of a number of High Schools for Girls. In these schools the curriculum – and in particular the mathematics course – was largely modelled on the contemporary curriculum for boys, the only difference being that less stress was laid on the classics and time was devoted to music, needlework and dancing. Just as in Cheltenham the College for Young Ladies complemented the College (for boys), so, at a higher level of education and in London, Queens College (1848) and Bedford College (1849) became the equivalents for young women of University College (1828) and King's College (1831). It was not long before educational institutions for women were opened at Oxford and Cambridge, and by 1890 mathematical education for women had progressed to such an extent that, in that year, Philippa Fawcett beat the Senior Wrangler at Cambridge. Her sex, however, still debarred her from receiving a degree (see Strachey [1]). Four years later, in a more enlightened Prussia,† Grace Chisholm made history as the first woman to receive a PhD in mathematics there, her supervisor being Felix Klein.

From the 1830s elementary education had increased, owing to partial financial support by the Government. Commerce and industry needed a large population versed in the three Rs, so the support grew until, following the Forster Education Act of 1870, compulsory elementary education was introduced for all children in Britain. However, with the exception of a few outstanding schools – for example, Kings Somborne National School (see Stewart [1], pp. 124–35) – one did not speak of 'mathematics' in elementary schools, but of 'arithmetic'. Mathematics other than arithmetic became a distinguishing mark of secondary education, for which the Forster Act indirectly produced a rapidly rising demand: a demand met by the growth of a complicated system of higher grade schools, science schools and technical schools which all provided what we now refer to as 'secondary education'. (Many of these schools were reorganised into the system of grammar/secondary schools following the 1902 Act.)

In 1870, however, mathematical education at the secondary level meant mathematics as it was to be found in public and grammar schools and in their newly established private counterparts. As we have mentioned earlier, in these circles it was the aesthetic aspect of mathematics or rather 'mathematics as a mental discipline' which was emphasised. This view manifested itself not only in the subject matter – for example, the use of Euclid's *Elements*, a book never used in the schools of France and Germany – but also in the way in which the subject was taught with a strong emphasis on rote learning. Opposition to Euclid and rote learning soon mounted, and in 1871 crystallised in the formation of the Association for the Improvement of Geometrical Teaching – the forerunner of the Mathematical Association. The AIGT failed to have any immediate success – mainly because of the conservatism of the universities – but it must be admitted that in its early days its aims were hardly world-shattering. The objectives were worthy: the recognition of 'hypothetical con- structions, the arithmetical definition of proportion, superposition

† There still had to be a bending of rules: see the account in Cartwright [1].

and the conception of a moving point and of a revolving line', coupled with the removal of limitations on 'the restriction of the number of axioms to those only which admit of no proof and the restriction which excludes all angles not less than two right angles' (see *Math. Gazette* **55**, p. 124). However, viewed as part of the general development of mathematical education, these objectives were mainly trivial ones. Nevertheless, they represented a step forward, even though they alone could not meet the criticisms of those who were demanding a more scientific education; a demand demonstrated and increased by the Great Exhibition of 1851, the founding of the Royal College of Science, the Royal School of Mines, and similar institutions. The superiority of foreign competition at the Paris Exhibition of 1867, and the Prussian victory over France in 1871, served as a spur for improving the quality of British manufacturing and engineering products, so as to compete with the Germans, Americans and French. The effect on education was not unlike that produced by the Sputnik some eighty years later (see Chapter 12). Thus, by the early 1900s there had grown the complicated system of secondary schools, all wanting to teach mathematics at a more advanced level, and from the point of view of scientific application. They were encouraged by such men as Professor H. E. Armstrong (see Quereshi and Richmond [1]), a chemist from the Royal College of Science and a leading exponent of 'free', pupil-centred methods of teaching science, who criticised the leisure-class concept of education as 'suitable only for men who would spend £1000 a year, and not for men who would earn £1000 a year'. There was talk, too, of training a race of inventors; and argument about the value of the cultivated amateur compared with the narrow specialist. In mathematics these arguments took the form of comparing its aesthetic and practical aspects.

Exercises

1 Investigate the differences in the mathematical curricula of schools in England, France, Germany and the United States during the nineteenth century.
2 What are 'hypothetical constructions'? Discuss the objectives of the AIGT as described above.

4. THE EARLY TWENTIETH CENTURY

As far as mathematical education was concerned the twentieth century opened with a bang. At the meeting of the British Association for the Advancement of Science held at Glasgow in 1901 a great deal of time was allocated to a discussion of the aims of teaching mathematics. In a report, we are told that several speakers urged that mathematics should be introduced through experiment and intuition, and that if this were done, the more logical aspects of the course would gain rather than suffer.

One speaker, however, spoke out more loudly than the rest. He was an engineer, Professor John Perry; a former physics teacher at Clifton College, one of the newly established private schools. Perry did not mince words and his remedies were generally of the simplest. Thus we find him in 1909 advancing as a panacea for most educational ills 'A doubling of salaries (of teachers), halving of classes, the ousting of mere specialists, and completely getting rid of the outside examiner'† (*Math. Gazette* 5, p. 1). In 1901, however, his thesis was the need to emphasise the practical aspect of mathematics, for, as he said, 'The study of Mathematics began because it was useful, continues because it is useful and is valuable to the world because of the usefulness of its results, while the mathematicians, who determine what the teacher shall do, hold that the subject should be studied for its own sake.' Professor Perry noted eight 'obvious forms of usefulness':

(1) In producing the higher emotions and giving mental pleasure. Hitherto neglected in teaching almost all boys.

(2) (*a*) In brain development. (*b*) In producing logical ways of thinking. Hitherto neglected in teaching most boys.

(3) In the aid given by mathematical weapons in the study of physical science. Hitherto neglected in teaching almost all boys.

(4) In passing examinations. The only form that has not been neglected. The only form recognised by teachers.

(5) In giving men mental tools as easy to use as their legs or arms; enabling them to go on with their education (developing of their souls and brains) throughout their lives, utilising for this purpose all their experience. This is exactly analogous with the power to educate one's self through the fondness for reading.

(6) Perhaps included in (5): in teaching a man the importance of thinking things out for himself and so delivering him from the present dreadful yoke of authority and convincing him that, whether he obeys or commands, he is one of the highest of beings. This is usually left to other than mathematical studies.

(7) In making men in any profession of applied science feel that they know

† The role of the 'outside examiner' is explained in Chapter 3.

the principles on which it is founded and according to which it is being developed.

(8) In giving to acute philosophical minds a logical counsel of perfection altogether charming and satisfying, and so preventing their attempting to develop any philosophical subject from the purely abstract view, because the absurdity of such an attempt has become obvious.

Perry's attack on current mathematical education certainly heartened reformers in Britain, but his ideas were also influential across the Atlantic and, for example, affected those responsible for producing a revised curriculum for use in the schools of Ontario. Perhaps even more significantly, Perry's views were expounded and elaborated by Professor E. H. Moore of Chicago, one of the leading mathematicians of the United States. In 1902 Moore [1] gave an address, as retiring President of the American Mathematical Society, and devoted it to suggestions for the improvement of mathematics teaching in American schools. He deplored the 'compartmentalising' of mathematics into algebra and geometry and urged that stress should be laid on the interrelationship of mathematics with science. It is indicative of the slow rate at which mathematical education evolves that reformers in the United States (for example, Schuster [1]) are still protesting at the way in which 'geometry' is fitted into its Tenth Grade 'compartment'. Moore's second point needed to be reiterated by mathematicians some sixty years later, as we shall see on p. 140.

In Britain, however, Perry's plea to emphasise the utility of mathematics did not go unheeded. Applied mathematics has been a feature of the mathematical curriculum in the grammar and public schools throughout this century and, in more recent times, it has been recognised that this term can mean more than Newtonian mechanics. For complicated reasons the subject is, however, rarely taught so as to be useful. This is a problem currently occupying many mathematicians involved in mathematical education, and one to which we shall return in Chapter 19.

For Perry and others, though the question was 'What mathematics should be taught to an intellectual and social elite?', and they left in abeyance the question of what mathematics should be taught to the working class. The need to provide workmen with special training (short of an education which would lift them 'above their class') had been recognised in the late eighteenth century and had led to the formation of Mechanics Institutes – by 1850 there were over 600 of these in existence. The hopes of their founders were, however, not always met, for the students they attracted came more and more from the middle class.† Nevertheless, the Institutes did help to produce the polytechnics and the university extension movement.

The attitudes of many of the reformers of mathematical education at the beginning of this century can be judged from a set of reports assembled by the Board of Education in connection with the 1912 International Congress of Mathematicians which was held at Cambridge. Several contributors to the two volumes were (or became) authors of widely used texts, such as Godfrey, Siddons, Barnard, Durell, Gibson and Hardy. We cannot stress too strongly the evidence of an enormous amount of activity and excitement, because so many people were involved in planning courses to meet the new demands of the time. Many modern grammar schools were being created and technical education was increasing. Extracts from one of the papers – an account of the reconstruction of the mathematics curriculum at Dartmouth Royal Naval College – are reproduced in Chapter 13 of this book.

Exercises

1 Investigate the treatment of applied mathematics in the schools of France, Germany and the United States, and try to account for the differences in attitudes which you observe.

2 In his contribution to the 1901 British Association debate, Oliver Heaviside (a self-educated physicist and electrical engineer, and nephew of Wheatstone 'of the bridge') said:

'They (the boys) have also the power of learning to work processes, long before their brains have acquired the power of understanding (more or less) the scholastic logic of what they are doing . . . Now, the prevalent idea of mathematical works is that you must understand the reason why first, before you proceed to practise. This is fudge and fiddlesticks . . . I know mathematical processes, that I have used with success for a very long time, of which neither I nor any one else understands the scholastic logic. I have grown into them, and so understand them that way.'

(Quoted in HMSO [16])

Discuss Heaviside's argument, and its relevance for mathematical education.

† This same pattern has frequently been repeated in education, and an institution created for the working class has become a middle-class preserve. Thus, for example, the extension classes at Reading and Exeter developed into the universities to be found there; the colleges founded by the wealthy industrialists Owen, Mason and Hartley became, respectively, the Universities of Manchester, Birmingham and Southampton; the Colleges of Advanced Technology aspired to, and were granted, university status: at the present time the polytechnics are seeking to become indistinguishable from the universities, while the Open University is concerned about its students being predominantly middle-class.

3 'The instruction of children should aim gradually to combine knowing and doing. Among all sciences mathematics seems to be the only one of a kind to satisfy this aim most completely.'   (Kant, *Werke*, Bd 9, 1838)

Discuss. Do you think Kant would still hold this view if he were alive today?

5. SPENS, DAINTON AND OTHER REPORTS

By 1912 or so, mathematical education in Britain had jelled, and was hardly to be radically disturbed until the 1950s. The main types of mathematics were agreed upon, as were the levels of school for which each type was suited. Perhaps a pause for consolidation was necessary after the ferment of the past forty years, which had seen the evolution of what was virtually a new educational system and, in the field of mathematics, had seen the displacement of Euclid's *Elements* by Godfrey and Siddons, the takeover of school algebra by Hall and Knight, and the inclusion of calculus and mechanics in the school curriculum. The period between the two World Wars became, then, for mathematical educators, a time of consolidation. The educational system as a whole was fairly static – often for lack of finance to make desired changes. Even the Fisher Education Act of 1918 had little immediate impact on the educational system, for its most important provisions – to raise the school-leaving age, and to start day continuation schools – were in the event inoperative, owing to national economic difficulties.

Nevertheless, ideas for the advancement of education were still being floated. Various Government committees commented on the teaching of mathematics from a 'social' point of view: their views are not radical about mathematics itself but they are those of tax-payers worried about the role of the mathematically trained, and we now consider some of them. Far-reaching recommendations were contained in the 'Hadow' report† on *The Education of the Adolescent* (1926) (HMSO [4]) which saw secondary education, not as the privilege of some 10 per cent of the school population, but as a right of all (see pp. 52–6). Unfortunately, the economic depression and the Second World War meant the postponement of national attempts to put the major recommendations of the Hadow Committee into practice.

A second committee to report in the inter-war years (1938) was that chaired by Sir Will Spens. Like the Hadow Report, that of the

† It is customary to refer to such reports by the name of the Chairman of the Committee established by the Government.

Spens Committee illustrated the high-minded liberal tone of leisure-class educational thought at its best. Both reports are recommended to the reader. Here we must be content with a brief extract, observing that the climate of thought is pre- 'New Math' and is largely concerned with secondary schools.

We have said ... that we believe that Mathematics should be taught as Art and Music and Physical Science are taught because it is one of the main lines which the creative spirit of man has followed in its development, and that if Mathematics is taught in this way it will no longer be necessary to give the number of hours to the subject that are generally assumed to be necessary.
(HMSO [5], p. 235)

The report went on to argue how, with different emphasis on the objectives of mathematics teaching and with the use of new teaching methods, the mathematical education of secondary school children could be reduced in quantity but greatly increased in quality. It cannot be said that many schools took up the challenge. Nor was the recommendation that less time should be allotted to mathematics acted upon generally, although according to the 1958 Ministry of Education Report *Teaching Mathematics in Secondary Schools* this was done by some 'unsympathetic headmasters'. (The report does not tell us to whose views the headmasters were 'unsympathetic'.)

A justification for the criticism contained in the Spens Report is made clear in certain statistics published by the Organisation for Economic Cooperation and Development in 1965 (OECD [1]). In these we see that pupils in the science streams of secondary schools in the United Kingdom spent considerably more time studying mathematics than did their counterparts in, say, the Federal Republic of Germany (22 per cent of total time against 12.8 per cent). Indeed no secondary school pupil in the OECD countries devoted more time to mathematics than did the British pupil in the 'Science stream'. Yet if we consider the British pupil who opted for an arts-based course, the position changes, for such a pupil devoted on average only 10 per cent of his total time in secondary school to mathematics, a smaller percentage than pupils in similar streams in Belgium, Denmark, and several other countries. The ratio of the time spent studying mathematics by pupils in the science streams, to that of pupils in the arts streams, was, in fact, considerably higher in the United Kingdom than in any other OECD country.

This distinction did not go unnoticed by the Dainton Committee when they reported (1968) on *The Flow of Candidates in Science and Technology into Higher Education* (HMSO [11]). They argued

In our view normally all pupils should study mathematics until they leave school, and only in exceptional circumstances should it be held to be possible or desirable for a pupil to opt out. At present a high proportion of pupils with 'O' level passes in the subject abandon it in the sixth form, and some pupils drop it even earlier. We believe that the overwhelming majority are capable of benefiting from the continued study of the subject.

Although at first sight the recommendations of the Dainton and Spens Reports may seem opposed, this is not really the case. What both committees asked educators to do was, to reconsider the amount of time traditionally allocated to mathematics and to note changes in the objectives of mathematical education, and new methods of teaching the subject. They then asked if these observations did not call for radical changes in the way in which mathematical education in the school was organised.

It is important, then, to realise that the amount of time pupils should spend studying mathematics at various stages of their school careers is a matter for discussion and debate rather than something which has been divinely or otherwise ordained.

Often educationalists will feel, as did the various committees, that a wrong allowance of time is being allocated to the study of mathematics by certain pupils in our schools. They will then be called upon to justify the changes they propose and to estimate the benefits which would follow. Indeed, at the present time, when the entire school curriculum is under close scrutiny, it would seem essential that all mathematics teachers should be able to advance the claims of mathematical education with clarity and conviction. They should know of claims and counter-claims by bygone visionaries, reformers and those opponents who have warned against overbearing optimism. This is not so that they may be successful in hogging time for mathematics; to increase quality, it may sometimes be essential to decrease quantity!

As representatives of the taxpayers, the Dainton Committee was naturally concerned about the ways in which mathematics should help to prepare pupils for their future employment. Although the committee's recommendations were intended to meet the needs of a particular class of student, their list of reasons for the study of mathematics would be acceptable to most mathematical educators today:

1 Mathematics as a means of communicating quantifiable ideas.
2 Mathematics as a training for discipline of thought and for logical reasoning.
3 Mathematics as a tool in activities arising from the developing needs of engineering, technology, science, organisation, economics, sociology, etc.

4 Mathematics as a study in itself, where development of new techniques and concepts can have economic consequences akin to those flowing from scientific research and development.

Notice that these reasons are intended to apply to all pupils and there is now no underlying attitude that 'our' reasons for studying mathematics differ from 'theirs'.

6. MATHEMATICS IN A WIDER SETTING

To summarise, then, we have seen how mathematics has come to be regarded as a language – a means of communication and description increasingly used by economists, geographers, businessmen and others; as a training ground in which extra-mathematical educational objectives can be attained; as a tool whose value increases rapidly over the years; and as a subject worthy of study for its own sake, capable of giving pleasure and creating interest. When designing a curriculum it will be necessary to decide what weight we wish to give to each of these objectives at any one time and, indeed, how we are to interpret the words 'language', 'training ground', 'tool' and 'subject'. As is so often the case, general principles will be accepted readily – it is in the working out of those principles that differences, often irreconcilable, will arise.

In conclusion, it is worth drawing attention to the essential difference between the 'training ground' aim and the others. It alone is not aimed at producing behaviour or responses which can be labelled specifically 'mathematical'. Rather is mathematics being used as one of the most suitable means for the inculcation of ideas and ways of thought considered educationally desirable.

Far too often when considering why we should teach mathematics, we forget that education has wider aims than are immediately apparent from the school timetable. For example, the supporters of the method of teaching children in small groups rather than as a class frequently stress the subsequent improvement in mathematical learning, but rather less often do they argue that it is teaching children to work cooperatively in small groups which is probably the more important educational aim. This is not meant to imply that we should all follow the Russian, A. M. Khinchin [1], in asking 'How can we teach mathematics in order to achieve the educational aim of developing the spirit of patriotism?' For, as was demonstrated in the British universities of the eighteenth and nineteenth centuries, patriotism can soon turn to chauvinism, the rejection of ideas from abroad, and stagnation. There are clearly some desirable (and possibly undesirable) aims of education which cannot be effected through the teaching of mathematics. Yet, when considering the role of mathematics in the curriculum, it is essential to remember that our primary concern should be the improvement of the overall education of the pupil – that the teacher's job is to educate and not merely to instruct.† This aim is sometimes (paradoxically) best approached by specialisation; it is only discredited by mere attempts at 'preserving the subject' or cushioning the employment of those qualified as mathematics teachers.

Exercises

1 Smith and Tyler [1] suggest as objectives of education:
  ' 1. The development of effective ways of thinking.
  2. The acquisition of important information, ideas and principles.
  3. The development of effective work habits and skills.
  4. The development of increased sensitivity to social problems and aesthetic experiences.
  5. The inculcation of social rather than selfish attitudes.
  6. The development of appreciation of literature, art and music.
  7. The development of an increasing range of worthwhile and mature interests.
  8. Increased personal–social adjustment.
  9. Improved physical health.
  10. The formulation and clarification of a philosophy of life.'

  Does this list have any obvious omissions or items whose presence would not appear to be justified?
  In what ways can the teaching of mathematics contribute to the achievement of these objectives?

2 'Education is the acquisition of the art of the utilization of knowledge.'
  (Whitehead [1])

  'Education is the establishing of behaviour which will be of advantage to the individual and others at some future time.' (Skinner [1])

  Consider these definitions and their significance for the teaching of mathematics.

3 'I shall take it as self-evident that each generation must define afresh the nature, direction and aims of education to assure such freedom and rationality as can be attained for a future generation.' (Bruner [1])
  In what ways have the nature, direction and aims of education changed in your country during the last twenty years?

4 'One principle of education which those men especially who form educational schemes should keep before their eyes is this – children ought to be educated, not for the present, but for a possibly improved condition of man

---

† This point of view (which the authors hold strongly) would be challenged by many teachers in systems such as the French.

in the future; that is, in a manner which is adapted to the idea of humanity and the whole destiny of man.' (Kant [1])

'The test of a successful education is not the amount of knowledge a pupil takes away from school, but his appetite to know and his capacity to learn.' (Livingstone [1])

Discuss these extracts and their possible consequences for the teaching of mathematics. Must the 'test' stated by Livingstone *inevitably* lead to the 'cultivated amateur'?

5 'Throughout nearly the entire history of education mathematics has held its place in the schools chiefly because it has been considered indispensable in the formation of the educated man – that is to say, of the knowledgeable man trained to approach the affairs of his daily life with some sense of detachment and objectivity and to reason about them soberly and correctly. No-one capable of seeing below the surface of things would think of denying that mathematics has indeed played an important role in precisely this sense, and now has greater, rather than lesser, potentialities for contributing to the general education of our citizenry.' (Stone [1])

Discuss the role of mathematics in the sense described above. What are the increased potentialities of which Stone speaks?

6 'We cannot go to the stake for mathematics as "the arithmetic of everyday life"; for the amount of mathematics that is necessary, or even marginally useful, in everyday life is progressively diminishing – and it can be certainly taught to most children before the age of eight or nine.' (Wheeler [1])

What is this 'everyday' mathematics that is needed by everyone? Is it decreasing in quantity? Can it all be taught in the primary school?

7 The time allocated to the study of school mathematics varies from country to country; science streams and arts streams spend different amounts of time on the subject. Do you think that the schools of your country devote too little or too much time to the subject? Is the time spent efficiently? Would you like to see a re-allocation of time within the school years as a whole, whilst keeping the overall time spent constant?

8 'We therefore recommend that normally all pupils should study mathematics until they leave school.'

Discuss the arguments for and against this proposal by the Dainton Committee. (Do not forget the problems of implementation.)

9 Moore (p. 18) urged that stress should be laid upon the inter-relationship of mathematics and science. At the primary school level the teaching of these two subjects is often indistinguishable. Thus, say, practical work on volumes and mass, carried out for its mathematical value, leads naturally to an understanding of such physical concepts as density. To what extent can, and should, the teaching of science and mathematics be coordinated at the secondary school level? What place has mathematics in inter-disciplinary projects and teaching in general?

10 'The arresting thought that we as mathematicians have done next to nothing to inform and convince the sweating men and sweated women, whose hard labour makes possible our own leisurely pursuit of the "scientific divine", that mathematics does mean something in their lives and might mean much more, may well make us apprehensive of the future . . .' (Bell [1])

E. T. Bell wrote the above at the time of the American depression of the 1930s when the well-being of American mathematics was in jeopardy. Do you think that the average working man today is aware that mathematics means something in his life? To what extent, in fact, is Bell's statement true? Does Stone's argument (p. 120) add to, or detract from, Bell's?

11 'If a questioning attitude is fostered in the teaching of mathematics or science, this may lead to a questioning of the authority of the State.' This argument has often been heard: what reply might be given to its proponents?

SOME FURTHER READING

History: Curtis, S. J. [1], Heath [1], HMSO [1–6, 14, 16], Howson [1, 7], Kline [5], Knott [1], NCTM [2], Needham [1], NSSE [1], Prost [1], Siddons [1], Stamper [1], J. M. Wilson [1]

General Principles: ATCDE [1], Benacerraf and Putnam [1], R. B. Davis [1], Dubisch [1], Gattegno [1], Halls and Humphreys [1], Hardy [1], HMSO [19, 20], Newsom [1], Pólya [1], Wheeler [1]

3

Examinations and Objectives

Apart from any general answers that *society* may give to the question
'Why teach mathematics?', the immediate practical reason that has
been important to *schools* for many years has been that the examination system has required that mathematics be taught. In this
chapter, we shall look at the examination system and its history
in a general way, to show something of its effects – good and bad – on
society and on the teaching of mathematics. The more detailed
aspects of examinations and their relationship with new curricula
will be postponed until Chapter 20.

1. SELECTION AND RECRUITMENT

For thousands of years, the priesthood, stonemasons and other
skilled trades have needed to recruit new members, to train them
and to judge whether or not to admit them to full membership. The
judgement has had an aesthetic aspect – 'Can the candidate perform
the craft so well as eventually to improve on known techniques?' –
and a practical aspect – 'Is the candidate sufficiently competent to
uphold the credit of the craft, so that he does not bring it into dis-
repute by producing shoddy work or charlatanry?' Different crafts
lay different stress on each aspect, depending on their stage of
development; rapidly developing crafts like that of the research
scientist will stress the first, well-established ones like those of the
seaman or lawyer, the second. Some crafts, like medicine or engineering, must stress both aspects, although they must stress the second
aspect at the expense of the first if there is a clash. Clearly, if the
first aspect is neglected, then the craft may well decay or be super-
seded, while neglect of the second ignores the accumulated wisdom
of previous generations and may lead to work with unsound foundations. We see here the tension, which properly channelled can be of
immense value, between the heretic and the orthodox, the radical

27

and the conservative, the liberal and the authoritarian, the idealist and the realist.

Some crafts have visible end-products, like jewellery, tools, buildings, and other artifacts. Here the basic standards are obvious to both teacher and pupil, and the course of instruction has a clear objective, to produce work as free from defects as similar work produced by earlier craftsmen. But other crafts, like law and administration, require skills that are not so clearly defined, as well as such basic techniques as literacy and knowledge of well-established procedures such as accounting and conveyancing. Lawyers and administrators deal with human beings, and consequently meet unusual situations needing departures from known rules; the same holds nowadays, of course, in many other crafts, especially the science-based ones. The adviser to the State or Ruler was the earliest who had to take *responsibility* for coping with the unforeseen – war, famine, and the other problems that must be overcome by a developing society. How then are such advisers to devise standards for recruits to their craft? The ancient solution to this problem led to the examination as we now know it, which nowadays reaches far beyond the administrative crafts.

In China the mandarins devised the technique now known as 'competitive examination for public office', which flourished, as did the mandarins, and in a fairly unchanging manner, from at least 200 BC† until AD 1905. It was copied in modern times by the Jesuits and later spread (as we shall see) into the universities of the West. The mandarin pattern was this. Emperors might come and go, but Government was effectively in the hands of the mandarins, and entry to their class could be gained only by educational prowess – presumably on the ground that this guaranteed basic skills and indicated the less-clearly defined abilities needed in administration. In this way an examination system (involving written papers) was created which was to act as

(a) a screening service (to select for a future role),
(b) a provider of incentive,
(c) a stabiliser of the membership of the power hierarchy,
(d) a maintainer of the tradition of scholarship,

– four features still to be found in examination systems today. As now, success conferred privileges, thus giving society at large an interest. Conversely, the students' interests in society were also

† Perhaps earlier. By then Chinese Culture was at least 2000 years old.

sometimes the cause, even many centuries ago, of periodic rioting (see Lien Pu [1]), just as they can be today!

Where a craft is not too specialised, it may dispense with asking questions about standards, and use notions of 'merit' such as birth or relationship. Thus one finds it difficult to enter certain professions if one's relatives are not already members, as with dockers in Britain or aspirants to the ruling class of certain societies. Even here, however, and especially if the external competition is severe, good advisers are essential for long-term survival; this general rule may account for the Chinese preference for ability rather than birth. Such an explicit preference at the top by the 'mandarinate', i.e. the men responsible for making society 'work',† has an important effect on the structure of society, however. First, talented children can rise rapidly up the social ladder of the particular society, if they can somehow find a patron to supervise their education. Secondly, such recruits to the mandarinate tend to value learning, as with (d) above, and to arrange their working conditions so as to have the leisure to pursue learning. Thirdly, less-talented children whose families can get education for them, can often be trained to pass the examinations that will get them a privileged position in society. Fourthly, an 'educated class' develops, which may control the education system, and feeling reasonably satisfied with the quality of recruitment to the mandarinate, fail to provide adequate education for the less fortunate classes; access to the social ladder may then become impossible for the less fortunate. This last situation is most likely to occur in a stable condition of society, as with the Chinese before Western penetration, so that survival is possible for the educated class without replenishment of talent from a wider class. In an unstable situation, either an enemy class supersedes the educated class by superior leadership, or, say, a rapidly developing technology produces a demand for trained recruits in professions other than the mandarinate and a supply that may or may not be adequate, either in quantity or in quality. It is this last situation which has been the commonest (and most interesting) during the past century of world history. The inadequacy of the supply in numbers has led to the educational problem of supplying relatively advanced education to society at large, rather than to a small elite. The inadequacy in quality has led to the questioning of traditional curricula, and the design of new ones. In particular, the assumption of a stable society

† A precise definition is not necessary here. For a full discussion see Weber [1].

is now no longer valid, so that the examinations for entrants to almost all crafts need to be able to take account of future adaptability, because a man who copes well at 21 may live in a totally different world at 61. For example, Winston Churchill took part in a cavalry charge as a young man, and yet in old age had to make decisions about the Hydrogen bomb, whereas the problems facing Julius Caesar at all stages of his life occurred in a society of roughly unvarying complexity.

Now, the very prestige of the mandarinate will tend to produce schools catering for those who wish to become mandarins. The teachers in such schools may well be regarded by others or by themselves as part of the mandarinate; and the curriculum in such schools acquires the general 'functional' aim of providing education suitable for intending mandarins. Such schools themselves acquire prestige and others may emulate them (for bad reasons).

Where the mandarinate selects recruits by ability rather than by patronage, it must use assessors to examine the candidates, and it is easiest to find agreement that they will be impartial if they do not know the candidates, and if they (rather than the teachers) ask the questions. Such assessors are then called *external examiners*, and they themselves may be supervised by a *moderator* who tries to ensure that all the examiners apply similar criteria: the moderator's job is to ensure uniformity of *standards*.

The examiners may well decide to use written examinations to test the candidates, for several reasons. But when they do, their reasons may well be forgotten by the schools preparing the candidates. The result is that the schools teach their pupils how to pass the examination, with possible neglect of other aspects of education. Here we have the 'freezing' effect of examinations, in part due to the permanence of print, which can have disastrous effect on the adaptability of the pupils when, as mandarins, they may be required to be adaptable in an unstable situation. Other drawbacks of the freezing effect will be discussed in later chapters; some were already noted in Chapter 2 and we recall that, for example, Professor Perry sought to get the mandarinate of his time to accept that they should include technologists among the men who make society 'work'.

Note, however, that if the educational system expands and the profession of teacher has prestige from its association with the mandarinate, the teaching profession itself is an important social ladder, both to allow children to enter it and for their own children to derive benefits.

The mandarin approach, then, sees examinations as a means of selecting talent for a particular profession, and as a counter to corrupt systems of selection. It depends on having external examiners who can be assumed to be impartial towards the candidates. Such impartiality is often confused with *objectivity*, but that is a much more subtle matter, as we shall see in Chapter 20. Against this most valuable impartiality, and other good effects such as an initial rise in academic standards, has to be set the potential freezing influences of the examinations on the educational system and on the adaptability of the candidates. As was said by an HMI† early this century, the system stops direct corruption, only to cultivate in candidates a wish to cheat the examiner by other means, such as question-spotting. Also it gives a strong advantage to those who come from an environment which is already strongly oriented towards the 'middle-class' values: one might even *define* middle-class in this context as ability to profit from the educational system!

Exercises

1 Trace the path of competitive examinations into the West from China, via the Portuguese trading posts, the Jesuits and the East India Company (for England), Voltaire (for France). See Têng [1].

2 Is a training in mathematics likely to induce attitudes of incorruptibility or morality?

3 If the population in a society at time t is P, and if the number of mandarins is M, discuss the appropriateness of assuming that (a) M is proportional to P, or (b) $\dot{M} = aP + bM$, $\dot{P} = uP + vM$, where a, b, u, v are constants. Consider the effects of a mandarinate that increases total wealth as against one that decreases it. (See also Chapter 19, Exercise 5, p. 317.)

2. EXAMINATIONS AND THE UNIVERSITIES

The universities of the West began in the Middle Ages and they quickly adopted an examination system. Even at that time a certain amount of technical expertise was needed by those responsible for the smooth running of society – there was a need for administrators in the Church and State, and for doctors and lawyers – and examinations provided a necessary safeguard against charlatans. The various degrees offered by the universities also served to establish a well-defined hierarchy in the medieval academic society. For example, a man with the degree of Master of Arts was marked out as a

---

† HMI is the standard abbreviation for 'Her Majesty's Inspector'.

mature scholar who was fitted to teach and to guide the work of his Bachelor assistants. Naturally, methods had to be devised for judging whether or not a man possessed the necessary qualifications for a degree, and these varied. Originally, paper work was of little importance and emphasis was laid on the student's skill and agility in the art of disputation. This method of examination later evolved into the oral examination, which nowadays tends to be associated in Britain with higher, research degrees.

The Northern Renaissance was accompanied by a decline in interest in examinations, and, indeed, by a loss of initiative on the part of universities. Examination by disputation became a test of 'a young man's ability to learn well-worn arguments by rote' (Curtis, M. [1]). In England the universities were, in the century following the English Civil War, to sink to their lowest ebb, becoming servants of the State and the Church of England.

The emergence of Oxford and Cambridge from mediocrity coincided with, and is related to, widespread changes in their examination systems (see, for example, Trevelyan [1], p. 366). In an attempt to meet the needs of the serious-minded students and to raise and standardise the performance of candidates, the University of Cambridge established in the mid-eighteenth century a new 'tripos' examination,† first in mathematics and then in classics, history and other subjects: indeed, the examinations enforced the notion of a 'subject' as a special skill which might excuse its possessor from knowing much else. These methods were soon adopted in Oxford as well.

Not all students were expected to take the new examinations and so it was possible to introduce standards much higher than had been usual before. An *honours* system began to evolve, whereby candidates were graded by degrees of excellence. For those who could not aspire to an honours degree, there was a second type of degree – an ordinary, poll, or pass degree – but the two types were sharply distinguished.

The control of these new examinations became a 'university' and not a 'college' matter. As a result they became 'external' in the sense that those who set the questions – specially appointed examiners and moderators – did not usually teach the courses. Whilst this ensured impartiality, it had far-reaching, and not always beneficial, effects on the teaching of mathematics. Another important innova-

† For an explanation of the term 'tripos' – involving three-legged stools and Christian symbolism connected with the number 3 – see Ball [1].

tion was the introduction of written questions, following the precedent set by the Chinese 1000 years earlier.

The major effect of these changes was that the examination ceased to be a test of competence and became a means of classification, with an associated element of competition and even the excitement of a horse race. In mathematics until the twentieth century the 'Senior Wrangler' (i.e. the top man in Part I of the Tripos) was accorded immense prestige, and might even be awarded a college fellowship with life-long tenure on the strength of it. These examinations also had a major effect on the British Civil Service, and in the following section we shall be discussing the *English* system, because of its importance within the nineteenth-century British Empire and its effects on the succeeding Commonwealth.

Exercises

1 Why would you expect the Renaissance to be 'accompanied by a decline in interest in examinations' (p. 32)? Relate this to the new skills that were emerging at that time.

2 Discuss the effect of written examinations on mathematical notation and terminology, especially on that of calculus. (See Ball [1] for the introduction of the Leibniz 'dy/dx' notation as a competitor to the fluxional, '$\dot{x}$' notation of Newton.)

3 Find reasons for 'the emergence of Oxford and Cambridge from mediocrity'. Consider the effects of Edinburgh (the 'Athens of the North'), and ideas from the mainland of Europe; see also Ball [1] for the influence of the mathematicians Babbage, Herschel and Peacock.

3. THE ENGLISH MANDARINS

Certainly, the period following the introduction of the modern type of examination system was marked by an improvement in the quality of work at the universities concerned, and much of the credit for this was given to the new system. The Oxford University Commissioners of 1850 could claim, therefore, that 'The (University) examinations have become the chief instruments, not only for testing the proficiency of students, but also for stimulating and directing the studies of the place' (quoted in Montgomery [1]). This stimulating effect of examinations was preached by many notable figures: Adam Smith spoke of the way in which 'rivalship and emulation render excellency, even in mean professions, an object of ambition, and frequently occasion the very greatest exertion' (quoted in Armytage [1]), and John Stuart Mill accordingly advocated the compulsory examination of children every year.

Examinations, then, could classify and stimulate.

From the early nineteenth century onwards, more and more graduates of the new kind began to enter English life. Pride in their own skills was often accompanied by a strong moral sense and an incorruptibility like that of Arnold of Rugby (see p. 127). They began to change a mandarinate previously recruited through social connections, and at its least efficient when tested by the Crimean War. Already such graduates formed the Inspectorate of Schools (the HMIs) which was needed once the Government began to finance education from the 1830s. One leading HMI, whose imaginative ideas have influenced succeeding generations and set a high-minded tone which the Inspectorate has maintained, was Matthew Arnold, son of the Rugby Headmaster. Other proponents of this moral attitude were Jeremy Bentham, who in 1827 formulated schemes for screening candidates for public employment, and the radical Chartist, William Lovett, who proposed that Government ministers should not be allowed to take office until they had passed suitable written tests. Eventually the enthusiasts made the point that since only 500 British officials administered India, it was essential that they be of high calibre in more than just the social graces: here we have the situation mentioned in Section 1, where a mandarinate must look to its survival. In 1853, therefore, the India Bill was passed, decreeing that men entering the India Service as civilians or soldiers were to be screened by examination. Such competitive examinations gradually spread to the Civil Service generally, and a type of professional Civil Servant emerged, incorruptible and intelligent.

Our ethic is simply stated. We stand committed to neutrality of process. We profess that public power is not to be used to further the private purposes of those to whom it is entrusted. It is to be used solely for the furtherance of public purposes, as defined by constitutional process . . . to evolve a set of procedural rules such that those who are not in themselves value-neutral – because they are human beings with needs and aspirations – can, by accepting those rules, contribute to a process which has public rather than private outcomes.

This was the credo of D. H. Morrell,† a Civil Servant in the 1960s, and it ably describes the beliefs and ethos of his profession. In any discussion about examinations, that ethic stated by Morrell must not be forgotten.

During the century 1815–1915, between Napoleon and the last

---

† Morrell was appointed one of the Secretaries of the Schools Council (see p. 144) on its establishment in 1964, and until his early death in 1969 he played a leading role in curriculum development.

Kaiser, Britain steadily developed and prospered. An educated class emerged, as predicted by our general principles in Section 1. The newly-instituted competitive examinations to the Civil Service ensured that entrants were still drawn from those who had benefited from a polished liberal education, because the examinations reflected the interests and prejudices of the examiners. The senior mandarins still came from the 'gentlemanly' class of society, but the system ensured that only the more intelligent were admitted. For example, an ex-Ambassador told how, even within this century, the British Foreign Office could demand that its recruits had First Class Honours from Oxford or Cambridge, a private income of £400 per year, and letters of introduction that in effect required some relative to be known personally to the Foreign Secretary.

The 'gentlemanly' class itself, however, was expanding as nineteenth-century industry produced more wealth, and its children could acquire gentlemanly attributes at the 'mandarin' schools, provided they could pay the high fees.

Nevertheless, perhaps because of the elements of Christian morality, liberalism and notions of the equality of mankind, this impartiality of the mandarins and their benevolent social outlook of the Matthew Arnold kind, led to an interest in the education of the poor. The resulting expansion of the educational system has been a remarkable and potent force in recent British history.

Another effect of examinations was to break the hold on schools of the classical curriculum, for the armed services demanded entrants with a knowledge of mathematics and science; cf. p. 12 and Major-General Stanley's song in *The Pirates of Penzance*. As an example, the curriculum of the Royal Naval College, Dartmouth, which we quote in Chapter 13 is typical of that of many grammar schools in the first half of this century.

Naturally, it took some time to eliminate patronage: when Lord Raglan (of the Crimean War and the 'Raglan' sleeve) was Master-General of the Ordnance, the future Lord Cromer was rejected for the Military Academy at Woolwich; whereupon his mother ordered her carriage, called personally on Raglan, and got the order rescinded! (HMSO [7].) Eventually, however, the Lord Raglans themselves were replaced by those who were successful in the examination system, against whom a later Lady Cromer would not have prevailed.

Even as this vast nineteenth-century examination system was getting under way, dissenting voices were to be heard. For example, Sir Frederick Pollock wrote to Augustus de Morgan (see Ball [1]),

> My experience has led me to doubt the value of competitive examination. I believe the most valuable qualities for practical life cannot be got at by any examination – such as steadiness and perseverance. It may be well to make an examination part of the mode of judging of a man's fitness; but to put him into an office with public duties to perform merely on his passing a good examination is, I think, a bad mode of preventing mere patronage. My brother is one of the best generals that ever commanded an army, but the qualities that make him so are quite beyond the reach of examination.

and the Master of Marlborough College spoke of the Indian Civil Service and Woolwich Military Academy examinations as 'sitting like a blight on education' and forcing the teachers to neglect what was educationally in the best interests of the pupils, in favour of work directed specifically to the examination.

Thus two aspects of the examination system came in for early criticism:

(a) many qualities and educational objectives were not readily assessable by examination,

(b) examinations tended to dominate syllabuses rather than to reflect them.

These particular criticisms are, of course, still being made; more significantly they are now being made by 'subject specialists', and not only by those wanting a more general education.

In spite of such criticisms, the Oxbridge type of written examination was adopted by the newer universities† which were founded during the nineteenth century (see p. 14). Partly to show that their courses were at least as hard as those at Oxbridge, they used the Civil Service pattern of having an external examiner – usually a professor in another university – for each course. At first the external examiner could set and mark the questions, but this became difficult, especially in science subjects, where also the written format was modified to allow for experimental work. Nowadays the entire examination is set and marked by those who teach the courses, since otherwise most candidates would either do only bookwork or – like H. E. Armstrong's (see p. 16) – fail. Thus the external examiner acts more as an auditor, who sees that justice is done, advises on the difficulty of the questions set, and arbitrates borderline cases. Such

---

† University College London was instituted to provide education for Dissenters, Jews and others who could not subscribe to the Thirty-Nine Articles of the Church of England. The religious tests at Oxford and Cambridge were not finally abolished until 1871. Before then, Dissenters, such as de Morgan and J. J. Sylvester (both of whom studied at Cambridge), could not be appointed to fellowships at these universities.

borderlines are inevitable with the 'Honours' Degree, now no longer a degree with honours for especially good work, but still classified into the divisions First, Second (Upper and Lower), Third, Pass and Fail. Since, for example, a teacher's retirement pension is affected by the class of degree he obtained years before, their award has to be carefully watched even though many academic staffs are dissatisfied with the whole procedure. The differentials introduce non-academic criteria, and the proportion of Firsts, for example, is largely governed by different traditions in different subjects (it is highest in mathematics). The external examiners certainly ensure greater uniformity between universities within each subject, than in the same university in different subjects! Greater sophistication about examining is growing in universities, as elsewhere, and this particular form of examination is almost certain to be modified before long. Curiously, the ancient Scottish universities held out for a century against the 'Honours' Degree, but on the ground of its inherent specialisation, not because of the method of examination.

The narrowness and depth of its first degree – which is covered in three years, usually with no second attempt allowed – still distinguishes an English university from one in America or the rest of Europe. It has also had its effect upon the English sixth-form curriculum, as is indicated by the data about 'hours devoted to mathematics' on p. 21. This effect is less marked in countries such as Scotland, where the course for a first university degree lasts four years.

4. EXAMINATIONS IN SECONDARY SCHOOLS

Although Prussia had established an examination – the *Abitur* – which served as a school-leaving and university entrance qualification as early as 1788, there was for a long time no similar examination in Britain. Indeed, until the late nineteenth century it was still the case that few colleges at Cambridge demanded more from prospective students than a certificate of recommendation 'from any MA of Cambridge or Oxford' (Stamper [1], p. 123).

The coming of the new nineteenth-century universities helped to change matters, and in 1838 'London matriculation' began as an examination for entrance to London University. It had no specific relation to the work of the schools – many of the candidates were private students – but it came to be used as a leaving examination for pupils who did not intend to proceed to a university. Such developments were appropriate, for the contemporary development of the railway system enabled clerks and others to seek employment in different parts of the country; thus employers needed some kind of uniform system to compare applicants for jobs.

The first public examination designed as a test of competence for secondary school pupils came some twenty years later (1857), and was for schoolboys in the West of England. It was a four-day examination, held at Exeter, and cash prizes were awarded to the most successful boys. The interest aroused by this examination led Oxford University to agree to promote a similar scheme – without the cash incentives – in ensuing years. Successful candidates would have the title of 'Associate in Arts'. Six months after the first Oxford examination, Cambridge held its first 'local' examinations, which closely resembled those of Oxford. The same year, 1858, also saw the first Durham 'locals'.

Girls were admitted to the Cambridge 'locals' in 1865 and to those of Oxford in 1870. Recall (p. 14) that girls' secondary schools were then beginning to flourish, with a consequent demand for university education.

By 1870, however, the Taunton Commission had reported on the endowed secondary schools and had recommended the establishment of a state examining board. The Endowed Schools Bill 1869 provided for the compulsory examination of all such schools by an external authority which would also be responsible for certifying secondary schoolmasters as efficient. These proposals did not become law, but the climate of opinion was such as to make the headmasters of many of the more influential schools apprehensive concerning their future. In an attempt to withstand 'the bigotry of the liberals', headmasters of some of these leading schools met together in December 1869 and thus brought into being what later came to be known as the Headmasters' Conference – the 'House of Lords' of British secondary education.

Since the 'local' examinations referred to above were created primarily to meet the needs of pupils in the 'middle' or 'middle-class' schools rather than those in the public schools, the Headmasters' Conference was particularly concerned to make available examinations suited to its own needs. The outcome was the Oxford and Cambridge Schools Examination Board – a joint board distinct from the two 'locals' – which was created in 1873 and held its first examinations in 1874: the Higher Certificate for boys of 18 and over. Ten years later the board instituted a Lower Certificate for boys of 16.

To meet the mushrooming needs of secondary education, further examination boards were established by the 'modern' universities in the first decade of the twentieth century. By that time, the number of examinations available to schools had grown apace. In an attempt to bring order out of chaos, a Government committee reported on *Examinations in Secondary Schools*, with a report (HMSO [13]) published in 1911 that contained many far-sighted recommendations. In particular, the committee wished teachers to have a greater influence in the running of examinations and thought that

teachers should have the option of supplying the examining authority with full syllabuses of any subjects in the teaching of which the school was making any special experiments, and should be able to claim that the examination papers should be drawn up so as to meet the special requirements of the case . . . When such a claim is admitted, we think that special examination papers should be prepared without extra cost to the school.

Yet even such an enlightened report still thought of 'examinations' as 'timed examinations'. Apart from the examining of practical science, there is no mention of continuous assessment or any other means of examining. The main outcome of this report and of the resulting 1914 circular issued by the Board of Education, was the establishment of the Secondary Schools Examination Council (SSEC) in 1917 to act as a coordinating body. Existing boards agreed at that time to modify their examinations and recast them into the First (or Lower) School Certificate Examination and the Second (or Higher) School Certificate. So as to enable pupils of limited means to climb even further up the educational – and social – ladder, a system of state scholarships to the universities was instituted in 1920. These scholarships were awarded on the results obtained in the Higher School Certificate.

5. SOME CRITICISMS: OFFICIAL

Despite the hopes of the 1911 Committee that the examination should follow the curriculum and not determine it, the Spens Committee of 1938 (see p. 20) had to admit that the School Certificate Examination 'now largely determines the curriculum for pupils under the age of 16' and resulted in 'overstrain and pressure on individual pupils'. (The air of impotence and awe with which teachers so often approach examiners and examinations is frequently exemplified in educational writings. Thus, to take but two mathematical examples, we are told

in *Math. Gazette* **10**, 1921, of an occasion when a resolution stating
what examiners should do had to be modified on the grounds that 'it
would mean dictating to examiners', and when H. F. Baker (*Math.
Gazette* **12**, 1924) reviewed the Association's *Report on the Teaching
of Geometry in Schools* (Math. Assn [1]), he remarked on the way that
'right through this Report we see the writers glancing up at the
Examiner standing over them'. In defence, teachers need confidence
and training before they are prepared to challenge established
practices.)

The Higher School Certificate, like its later variant, the A-level
examination, had dual functions – as a record of school work and as
an entrance qualification for university. These functions were
recognised by the Spens Committee as being incompatible, and the
latter was seen as taking undesirable precedence over the former. As
antidotes the committee recommended that there should be a reduction in the content of examination syllabuses – never, in fact,
realised – and that the pupil's work at school should count along
with his examination marks when the certificates were awarded – a
suggestion that took thirty years to be adopted, even to a limited
extent. Such is the inertia of the system that the problem of the
incompatibility of the dual functions has still not been resolved, in
spite of much contemporary discussion.

The part teachers should play in the award of certificates was
given great emphasis by the Norwood Committee (HMSO [6]) on
Curriculum and Examinations in Secondary Schools (1941). It recommended that the (Lower) School Certificate should gradually become
an entirely internal examination, 'that is to say, conducted by the
teachers at the school on syllabuses and papers framed by themselves'. Again, the recommendation was to prove ahead of its time.
Significantly, this recommendation did not receive the support of
the one headmaster serving on the committee.

6. THE GCE AND CSE

In 1948 the SSEC (p. 39) produced new proposals for the examination system. The two School Certificate Examinations were to be
abolished and replaced by a new General Certificate of Education, to
be taken at Ordinary-level by pupils of 16 and at Advanced-level by
those of 18.

In order to prevent 'premature specialisation' – the 1911 report
had found that 'the presentation of young and immature pupils for

external examinations was mischievous' – the age level for the Ordinary-level had been fixed at 16. Alas, the fight against this mischief was soon abandoned, for, within two years of the GCE's being launched in 1951, pressure from grammar schools resulted in the relaxation of the age limit.

Unlike the School Certificate, which had to be passed as a whole, it was possible to obtain a GCE O-level pass in any number, however small, of self-chosen subjects (a change recommended by the Norwood Committee). This allowed pupils in secondary-modern schools (see p. 54) to obtain GCE qualifications. By 1959 candidates were being entered from a quarter of such schools, and by 1960 from a third. There was also an increasing tendency for pupils in those schools and in the lower streams of grammar schools (p. 54) to remain at school beyond the age of compulsory attendance (then 15). Thus further consideration of the role of the external examinations became vital. In 1960 a committee under the chairmanship of Robert Beloe reported on *Secondary School Examinations other than the GCE* (HMSO [7]). The report presented the case for and against external examinations† and recommended that a new examination should be established which would cater for pupils of 'average ability' completing a five-year course of secondary education. The examination would lead to a Certificate of Secondary Education (CSE) and would provide teachers with an opportunity to control their own syllabuses and examinations along the lines envisaged by the Norwood Committee.

The examination, which began to operate in 1965, is administered by regional boards. Schools can opt to be examined in one of three ways:‡

Mode 1: external examination on the board's syllabus,
Mode 2: external examination on the school's syllabus,
Mode 3: internal examination on the school's syllabus with external moderation of the syllabus and the assessment procedure.

Most importantly, assessment procedures are no longer restricted to the 'timed examination'.

---

† An appendix also contains – as does Tibble [1] – a reprint of similar pros and cons advanced in the 1911 report.

‡ During the past few years these divisions have become somewhat blurred. Thus, for example, in some subjects, and with certain boards, there is now considerable teacher participation in the assessment of Mode 1 examinations.

Thus, after many years of struggle, full teacher participation in external examinations became possible. It must be stressed, however, that many teachers are not yet willing to undertake Mode 3 work and prefer the traditional Mode 1 pattern of examination. For, even if teachers feel well-disposed towards internal assessment, they may still worry about the difficulty of maintaining comparability of standards. Also they may be wary of undertaking the responsibility of grading their pupils and, of necessity, labelling some of them as 'inferior'. Moreover, Mode 3 examinations are more expensive to administer and do involve the teacher in a considerable amount of extra work. It is gratifying, then, that in face of such difficulties, the number of schools opting for Mode 3 work has grown steadily. If it is to be successful, Mode 3 work does, however, demand teachers who in the Beeby model (see p. 62) have reached the Stage 4 level.

The establishment of the CSE proved, in the event, to be one of the last acts of SSEC, for its duties were taken over in 1964 by the newly created Schools Council for the Curriculum and Examinations (see Chapter 12). With the creation of the Schools Council, it was acknowledged that the curriculum and the examination system must be considered together and not separately – and that 'curriculum' must come first! The Council since its formation has encouraged experimentation within the examination system and has not restricted itself to the consideration of big issues only, but has been willing to assist individual schools to solve problems that have arisen in the wake of such experiments.

Exercises

1 Criticise the argument that premature specialisation can be avoided by delaying O-level to 16. Why is it desirable not to specialise too early? Does an examination in Literature lead to a broadening of literary knowledge?

2 Conventional timed examinations in mathematics are being criticised because they teach pupils only to answer questions formulated by other people (who already know the answers). Consider situations where skill in formulating questions is necessary.

3 Discuss in detail possible changes in pupil–teacher relations that might arise through the Mode 3 work mentioned above.

4 How might collaborative work between pupils be assessed? How might team-work on the industrial pattern be fostered?

5 GCE Boards have allowed 'coursework' to be included in the assessment of candidates. Why should this facility have been largely ignored in the case of mathematics examinations?

7. EXAMINATIONS IN OTHER SECONDARY SYSTEMS

The growth of the English examination system has been described here in some detail, but the various trends that have governed its evolution have also affected other systems. The Scottish system, with its Scottish Leaving Certificate (1888), now known as the Scottish Certificate of Education, has two grades – Ordinary and Higher – and it has much in common with the English system. Again, the French *Baccalauréat* has much in common with the German *Abitur*. The need to cater for all aspects of education has, however, in many countries led to a multiplicity of external examinations, and it is doubtful, for example, if anyone outside France could understand – let alone try to explain – that country's examination system, which includes, in addition to the *Baccalauréat*, a bewildering collection of *brevets*.

Our remarks in earlier sections about the origin and ethos of examination systems suggest that they would have had few attractions in post-revolutionary Russia, and it is not surprising to read (Strezikozin [1]) that

When the new system of education was in the making, methods and forms were sought to overcome, in a short time, the formalistic and scholastic tendencies inherent in the pre-revolutionary Russian general education school. At the time it was believed that examinations foster formalism in teaching and get the pupil into the habit of mechanical memorization of material, rather than promote a creative approach to study.

Attempts were therefore made to abolish examinations.

Alas for such Brave New Worlds, for we later read in the same article that the 're-introduction of examinations in Soviet schools was intended to raise the educational level of young people' and that those fears expressed earlier about the stultifying effects of examinations 'were unfounded'. The resulting Russian examination system is an extremely centralised and all-embracing one, as reference to Strezikozin's paper will make clear.

To the European, at least, the United States has always seemed blessedly free from examination constraints. True, in an attempt to bring order out of the chaotic state of college entrance requirements, the College Entrance Examinations Board was established in 1900, by the Association of Colleges and Secondary Schools of the Middle States and Maryland. Initially, the CEEB was concerned only with twelve colleges and for many years it had only limited influence outside the eastern states. Nowadays, it has outgrown these limitations and also its original aim; for it is now no longer concerned exclusively with examinations, while the examinations it offers are concerned with requirements in addition to college entrance.†

Finally, one might add that the growth of 'international schools' in various parts of the world has created the need for an internationally accepted university entrance qualification. To meet this need, work began in the late 1960s on the establishment of an 'International Baccalaureate'. This examination is now organised by an international council under the control of the Swiss Federal Government. In 1971 it was taken by schools in eleven countries (including the UK and the USA) and had been recognised as a university entrance examination by universities in nineteen countries. It is hoped that after an initial experimental period ending in 1976, the examination will be established under inter-governmental control (see, for example, Peterson [1] or Morgan [1]).

8. EXAMINATIONS IN PRIMARY SCHOOLS

Once the State system of secondary schools had been established in Britain by the Act of 1902, and the 'free place scheme' had been initiated in 1907, the number of scholarships (bursaries) to secondary schools grew rapidly. Pupils were selected for these scholarships on the basis of examinations held around the age of 11. Those who were selected were regarded as of 'higher ability' and got secondary education free. The rest might get secondary education if they could afford it, but usually continued their elementary education until they could leave at the legal minimum age of 14 (or 12 if they could pass a 'Labour examination' – abolished in 1918). Some of the early leavers were very able, but either because of parental ignorance or parental poverty were not entered for the secondary school selection examination. The method of selection varied from authority to authority, and, for example, in an early attempt to combat social disadvantages, the education authority in Brighton used to allot a number of scholarships to each elementary school so as to counter the advantage which schools in the 'better' districts would have (Montgomery [1]). More secondary places were gradually provided, until the Butler‡ Act of 1944 guaranteed 'secondary education' for all, with an '11+' examination for all pupils to take to decide in

---

† A brief survey of the Board's current activities can be found in Valentine [1].
‡ R. A. Butler (later Lord Butler); his autobiography [1] gives insight into the benevolent mandarin mind.

44

which of the secondary schools – grammar, technical, or modern – the pupil should continue his education. The decline of this 'tripartite' system of secondary education and the gradual abandonment of the 11 + examination are described in Chapter 4.

9. EVALUATION

Let us now consider briefly, and in broad outline, some moves away from the mandarin concept of examinations: a more technical discussion will be given in Chapter 20. The 1911 report (HMSO [13]), besides dealing with external examinations, had something to say about internal examinations during school life – 'a recognised and desirable part of the machinery of a good school'. The committee recognised that such examinations could not only help the student but could also 'help the teacher by showing him where his teaching had failed and where it had succeeded'. That is, examinations have a role to play besides that of grading and classifying pupils.

Within mathematics, such views had vigorous support from G. H. Hardy, a leading Cambridge mathematician, who was the first to challenge the Tripos system from a 'professional' point of view. He declared that the Senior Wranglers could not even give a proper proof that if $|x| < 1$, then $x^n \to 0$ as $n \to \infty$; here we have a 'demarcation dispute' between the professional attitude of a mathematician and that of the mandarin, since the Tripos examination was mandarin-oriented and static, while Hardy wanted it to be mathematics-oriented and dynamic. Hardy finally won, but not as completely as he had hoped (see *Math. Gazette* **32**, 1948, p. 34).

Exercise

Summarise Hardy's views on examinations and relate them to the earlier criticisms of Tait (see Knott [1]) and Whitehead [1].

Much thought has recently been devoted to these other aspects of examining or what, to avoid misunderstanding, is often nowadays referred to as 'evaluation'. By this term, then, we are to mean much more than the old-fashioned examination. We are to think of evaluation as an aid in the construction of a curriculum.

Thus in a book on evaluation, by Benjamin Bloom and others (Bloom [1], 1971), which sets out to describe evaluation as the authors envisage it, and to provide a guide to the 'state of the art' of evaluating pupil learning, we are asked to consider:

1 Evaluation as a method of acquiring and processing the evidence needed to improve the student's learning and the teaching.

2 Evaluation as including a great variety of evidence beyond the usual final paper-and-pencil examination.

3 Evaluation as an aid in clarifying the significant goals and objectives of education and as a process for determining the extent to which students are developing in these desired ways.

4 Evaluation as a system of quality control† in which it may be determined at each step in the teacher–learning process whether the process is effective or not, and if not, what changes must be made to ensure its effectiveness before it is too late.

5 Finally, evaluation as a tool† in educational practice for ascertaining whether alternative procedures are equally effective or not in achieving a set of educational ends.

In particular, then, one can conceive of two types of evaluation: that which takes place at the end of a course and has as its primary purpose the certifying or grading of students – what is known as *summative evaluation* – and that which is intended to help the teacher to remedy faults in his pupils' understanding and to present a better course – what is known as *formative evaluation*.

The examinations we discussed in earlier sections of this chapter were, of course, examples of summative evaluation, and we notice that although this type of evaluation can have valuable feedback for the teacher, this comes too late to help the particular group of pupils taking the examination.

All those who have passed through the conventional school are well acquainted with examples of 'formative evaluation'. Indeed, every teacher who has asked a question of his class and let the replies guide his next step, has made use of it. Just as M. Jourdain was delighted to learn he had been speaking 'prose' all his life, so can teachers rejoice in being 'formative evaluators'. It would be a pity, though, to shrug off these ideas as yet more new clothes for the Emperor. As we shall see in several other instances, the mere existence of a vocabulary and of a means of classification can be a help in ensuring that one uses the techniques at one's disposal and does not overlook vital educational objectives. Again, it must be stressed that such formative evaluation as there has been in the classroom might have led to changes of method, but has rarely been allowed to influence the content of traditional syllabuses.

---

† Observe how the thinking and terminology of production engineers is being borrowed, in contrast to the nineteenth-century Victorian philosophies.

10. EDUCATIONAL OBJECTIVES

In the preceding chapter we discussed at length what we might hope to achieve by teaching mathematics. Evaluative processes will help us check whether or not our objectives have been attained. Clearly, however, there would be considerable differences of opinion on how to check whether or not one had attained the objective of 'showing how mathematics is used as a means of communicating quantifiable ideas' (see p. 150). Such an aim can serve as a guide to policy when we are drawing up a curriculum, but is far less useful for evaluation purposes. For the latter, it is necessary to state our objectives in more detail. Indeed, many authors would not term such vague statements of purpose as are set out on p. 150 as 'objectives', but would describe them as 'aims' or 'goals'. They would reserve 'objective' for something more specific.

The need to clarify what is meant by an 'educational objective' has led to a considerable amount of research work, in particular in the United States of America.

In Chicago, Bloom and his colleagues have attempted to provide a comprehensive model of the levels of educational performance, so that teachers and curriculum planners can more clearly indicate the type of performance they expect from their pupils. Their work would, hopefully, (i) clarify and tighten the language of 'educational objectives', thus helping to remove one source of confusion amongst educators, (ii) provide a convenient scheme whereby test (examination) items could be classified, (iii) provide a means whereby different curricular programmes could be compared, and (iv) reveal some pattern that would be of value in the development of a theory of learning.

As a first step, Bloom distinguishes between three *domains* – the *cognitive*, the *affective* and the *psychomotor* – and attempts to classify educational objectives within these. His domains are defined as follows.

The *cognitive domain* comprises objectives which:

(1) emphasise remembering and reproducing,
(2) involve the solving of some intellectual task in which the pupil has to select and reorder materials, and combine them with ideas, methods, or procedures previously learned.

The *affective domain* comprises objectives which emphasise feelings, emotion, or a degree of acceptance or rejection. They are expressed as interests, attitudes, appreciations, values, and emotions or biases.

47

For example, if a pupil learns Pythagoras' Theorem, he has achieved an objective in the cognitive domain; if eventually he comes to enjoy logical proof and to appreciate what constitutes a 'beautiful' proof (see Hardy [1]), then he has achieved objectives in the affective domain.

The *psychomotor* domain comprises objectives which emphasise some muscular skill, manipulation of material and objects, or acts requiring neuromuscular coordinations.

Detailed classifications within the first two of these domains can be found in Bloom [1, 2, 3]. At the time of writing, no detailed classification of the psychomotor domain has been published.

Bloom's work cannot be lightly dismissed because of the 'static' connotation of the term 'taxonomy', nor because he says little about teachers or pupils. The value of his approach is that by looking at shades of meaning of phrases that are normally used without definition, one can hope to pin down certain aspects that we can then plan to teach and evaluate for feedback (not to grade pupils like eggs).

The significance of such classifications for mathematics teaching is clear. In our normal testing and examining we have traditionally placed great emphasis upon the objectives lying within the cognitive domain. Geometric drawing and, say, the use of Cuisenaire rods, the slide rule or the computer terminal, do provide us with some objectives in the psychomotor domain which we – perhaps unwittingly – test in a limited way. If, however, one studies the 'goals' of the various curriculum projects (as set out in, say, Maryland [1]), then we note that many of these lie within the affective domain – they are concerned with the pupils' attitudes towards mathematics. Yet how frequently are objectives related to these 'affective' goals set out and tested? In this particular respect, Bloom's original work on the affective domain is disappointing in that it includes no examples drawn from mathematics. The later handbook on evaluation is more helpful, and we shall refer to some of its examples in Chapter 20. Clearly, though, there is a need for a closer inspection of the way in which mathematical objectives in the affective domain can be tested. There is also need for further research into the relationship between the cognitive and affective domains. For example, one might argue that once the affective objectives are achieved, then many of the cognitive objectives will follow more rapidly, and that the process of attaining some limited cognitive objectives could produce a distaste for mathematics and so militate against the attainment of affective objectives. This distaste for mathematics is widespread in populations exposed to traditional ways of 'mandarin' teaching, i.e. where cognitive objectives have been pursued and affective ones neglected. But the argument is part of the age-old question: 'Creativity or Technique?'

In Chapter 20 we shall give a brief description of some attempts that have been made to translate part of the Bloom handbooks into the terms of mathematical education.

Exercises

1 Discuss the possibilities of using Cuisenaire rods to develop objectives within the cognitive and psychomotor domains.
2 'Beauty in mathematics is seeing the truth without effort' (Pólya in ICMI [1]). Discuss.

11. PROBLEMS THAT REMAIN

Such classifications and attempts to specify detailed objectives as were briefly described in the last section will not solve all the educational and social problems posed by examinations. Indeed, there is a danger that they might lead us to ignore those objectives that are not readily classified and measured.

Unfortunately, it is not the case that all are agreed on the purpose of an examination. On p. 40 we mentioned the traditional problem of the school-leaving examination – is it intended to certify or to classify? Even more troublesome, is it meant to predict future attainment? In what way can it be used for selection purposes?

This conflict of aims is not peculiar to the school-leaving examination. It occurs in other instances, for example, with examinations for University degrees. Society still needs mandarins, and the indications are that it will still select them through examinations of one kind or another. How are the social objectives of examinations to be satisfied? Are the qualities that made Sir Frederick Pollock's brother a good general (see p. 36) still 'beyond the reach of examinations'?

These and related questions will still demand consideration and research in the years ahead.

SOME FURTHER READING

Bruce [1], Hardy [2], HMSO [6, 7, 13, 14], Montgomery [1], NCTM [2], Roach [1], Schools Council [1, 9], Têng [1], Wilkinson [1], M. Young [1]

PART 2
DETERMINANTS OF CHANGE: EXTERNAL

In Chapter 2 we saw how, as a result of various influences, the reasons for teaching mathematics in schools gradually changed. In Britain at least they came to include both the utilitarian and the aesthetic aspects. Now we shall look in more detail at some of the factors that help to bring about changes in the mathematics curriculum. Many of these factors are not specifically mathematical in nature, but arise from changes in the educational system as a whole. These factors – which we term *external* – will also promote changes in other subject curricula. They are considered in this part of the book. Other determinants of change spring from the growth, increased use, and reorganisation of mathematics itself. We shall study these *internal factors* in Part 3.

A country's educational system clearly depends on its pupils, its teachers and the money it pays for the system. Any great changes in these three 'inputs' would have a marked effect on the system as a whole and, in particular, on the curricula within the system. A more refined analysis would have to allow for other inputs and, more importantly, for feedback from within the system – for example, the outcomes of educational research and the development of new teaching methods and educational aids. Before considering these secondary factors we first study the three principal inputs and how changes in these have influenced, and are influencing, the development of new curricula. What we describe relates to a period of unprecedented economic growth in the rich countries of the world, and the attempts of the developing countries to catch up. The underlying belief that world resources were infinite began to be questioned in the late 1960s; in the early 1970s these resources started to become scarce and expensive. Attitudes to education may therefore change. We cannot predict what effect these new attitudes will have upon educational growth and change, but the reader should compare what we describe, with new developments as they happen.

51

4
Pupils

1. THE SPREAD OF EDUCATIONAL OPPORTUNITY

In an old Music Hall song, one 'Mrs Moore' was enjoined not to have any more children, 'for the more you have, the more you want, they say', and whether or not this is true of children, it certainly appears to hold for education. What was once seen as a privilege of the moneyed classes, is now widely claimed as the birthright of all. Between 1904 and 1925, for example, the number of pupils attending secondary schools in England quadrupled; and although in 1914 only about 56 children per thousand received secondary education, the 1944 Education Act ensured that all children now do so. In the United States the proportion of the age group attending secondary school rose from 12 per cent in 1900 to over 90 per cent in 1967 (Coombs[1]).

The rate of expansion in the industrialised countries has been matched in the developing ones; for example, secondary school enrolments increased sixfold in West Africa between 1950 and 1963 (UNESCO [1]) – the only difference here being that the developing countries still have a long way to go before they even achieve the goal of universal primary, let alone secondary, education. For, despite the progress made in the years since 1945, about one-third of the world's children still do not attend even primary school (OECD [2]).

The number of children enrolling for secondary school education has not only risen, but so has the average time which pupils spend in the secondary school. In England, as elsewhere, the school-leaving age has crept inexorably upwards; and each increase has been accompanied by a corresponding rise in the number of pupils opting to stay on at school beyond the statutory leaving age. Education creates a wish for more education.†

---

† Some reaction against this growth in education, or, to be more precise, in schooling, is now to be observed, especially in North America.

Such expansion in the number of pupils receiving secondary education and the consequent changes in the type of pupil involved have necessitated reappraisals of the school curriculum. For example:

The secondary school is no longer simply an institution to apply gloss to the unruly son of a chiefly family, to cultivate a small coterie of classical scholars, to teach cricket – physically and morally – to civil service candidates, so that they could better help their rulers rule their brethren. Now they have to produce citizens to fill the multitudinous key roles in the modern state.

The last sentence describes the major role of the secondary school in a developing country as seen by Adam Curle [1]. But schools in such a role are still elitist to the extent which secondary schools were in England and the United States prior to the First World War. The problems of developing countries are no less real, but they are different in nature from those that now confront countries which have a long tradition of universal elementary education but which have only recently been able to offer secondary education to all. In the secondary schools of the latter countries a type of pupil appears who possesses a quite different attitude and ability from the conventional pupil of former years. These schools were previously used as a social ladder by ambitious children, from ambitious homes, who were often selected by tests wherein they demonstrated their ability at academic work (which they liked and were trained to like). But if *all* children are allowed into these schools, the schools will, for some time at least, have pupils without the motivation, ambition and ability they prize; and great tensions will arise. The social attitudes that led to universal secondary education may be different from those of the pupils.

Exercise

Many people think that education is essentially a means of changing one's status. Are they right? How would you reply to political opposition to educational change that was based on this belief?

The effect of the 1870 Education Act was to bring into primary schools children whose poverty left them with needs even more basic than the three R's. According to contemporary writers (see, for example, Wardle [1]), such children had to be taught to sit without crouching, expected a constant stream of blows, or were like the inmates of Fagin's kitchen. The problems following in the wake of universal secondary education are of a different kind, but the question of what is to be done for our contemporary newcomers to secondary schools – pupils of 'average or less than average ability' – remains.†

The significance of the effect of educational expansion on the curriculum has certainly not passed unnoticed in England, as is demonstrated by the series of great reports which have been dedicated in part to that problem. These are *The Education of the Adolescent* (HMSO [4]), the Hadow Report, in 1926; *Secondary Education With Special Reference to Grammar Schools and Technical High Schools* (HMSO [5]), the Spens Report, in 1938; and *Half our Future* (HMSO [8]), the Newsom Report, in 1963.

Such reports could not by themselves solve any problems. Indeed, they sometimes give the impression of being little more than safety valves provided by Governments to prevent forward-looking educationists from becoming over-heated. It was only with the establishment of machinery for handling curriculum development (see Part 4) – the Schools Council in England, the Projects financed by the National Science Foundation and by the Education Development Center in the United States – that effective and relatively rapid changes in the curriculum became possible.

The essence of the recommendations of the Hadow Committee was that all children should receive some form of post-primary education in what were to be known as 'secondary schools'. Moreover, the Committee envisaged several types of secondary schools 'in which the curricula will vary according to the age up to which the majority of pupils will remain at school, and the different interests and abilities of the children'. The report goes on to add that it would be necessary to discover the type of school most suited to a child's abilities and interests, and suggested that this could be done by means of a written examination taken at 11 +, supplemented, wherever possible, by an oral examination and, in special borderline cases, by a psychological test.

These particular recommendations of the Hadow Report were gradually acted upon; and by the time that the Spens Committee came to report, it was able to observe that there were in effect three types of secondary schools in England. The first and most prestigious were the 'grammar schools', a type copying and including many public schools; entry mainly depended on academic ability, scholarships being available to pay the fees. Least prestigious were the 'modern schools' or 'senior schools', which were non-selective in the

† In England, these children are often referred to as 'Newsom children' because they were the subject of the Newsom Report mentioned below.

sense that they received pupils who had not been selected for any other type of secondary education. Their curriculum was not very advanced, the teachers less well qualified than those in the grammar schools, and the premises often poor. The third kind were schools of a vocational or quasi-vocational type known as 'junior technical schools'. The committee considered the possibility of bringing the first two types of school together in what it termed a 'multilateral' school which, by means of separate streams, would provide all types of secondary education, other than that provided by the junior technical schools. (The last-named schools needed specialised equipment and so were better attached to technical colleges – part of the further education system.) The advantages of this arrangement would be to ensure closer association of pupils of widely varied academic ability and interests and to facilitate transfer between academic and less academic streams.

For us, looking back to 1938, it is interesting to note that, while recognising these advantages, the committee nevertheless rejected the introduction of multilateral schools on the grounds that it would lead to:

(a) Schools which were too large (it was suggested that pupils gain more from being in a school of less than 800 pupils).

(b) Schools in which the sixth form was too small in relation to the rest of the school and had an academic character at variance with the multilateral nature of the school as a whole.

(c) Small sixth forms which would not then be able to benefit from a variety of options.

(d) The 'modern' streams having the 'grammar school' curriculum forced upon them.

The idea of a tripartite secondary education system was given further credence in *Curriculum and Examinations in Secondary Schools* (HMSO [6]), the report of the Norwood Committee (1941). In this report we even find an attempt to characterise the three types of pupil:

Grammar school† ... interested in learning for its own sake, can grasp an argument or follow a piece of connected reasoning, ... is sensitive to language as expression of thought, to a proof as a precise demonstration ... He can take a long view and hold his mind in suspense. He may be good with his hands or he may not; he may or may not be a good 'mixer'.

† Most of the committee were of this type, because it is this type that is usually in charge of the educational system.

Technical school . . . interests and abilities lie markedly in the field of applied science or applied art . . . He often has an uncanny insight into the intricacies of mechanism whereas the subtleties of language construction are too delicate for him . . . He may have unusual or moderate intelligence.

Modern school . . . deals more easily with concrete things than with ideas . . . interested in things as they are, finds little attraction in the past . . . may see clearly along one line of study or interest . . . but he often fails to relate his knowledge or skill to other branches of activity.

It was then necessary, the committee thought, to differentiate the pupils for the kind of secondary education appropriate to them. Such differentiation would be made upon the basis of the judgement of the teachers from each child's primary school, supplemented if desired by 'intelligence', 'performance' and other tests. Consideration would be given to the choice of parent and pupil.

It was also proposed that a core curriculum common to all three types of school should be followed during the first two years of secondary education, to facilitate transfer at 13+.

Exercise

What do you see as the strong points and the weaknesses of those of the Norwood proposals summarised above?

2. 1944 AND SELECTION

As a result of the Education Act of 1944, secondary education up to the age of 15 was made available for all, and it was provided in the tripartite form recommended by the Norwood Committee. It was probably the cheapest change that could then be made in the British system, short of actually closing schools, since it mainly rationalised an existing pattern. One significant change, however, did involve finance; this was that fees were abolished at most grammar schools, and entry became entirely selective by means of an academic, not financial, examination.

Thus, the tripartite system demanded strict selection procedures. Nature, in fact, did not label children in such an obvious way as the Norwood Report would have us believe, and some apparatus of selection was still necessary. As the years went by, the methods of selection came to be attacked more and more, as did the assumption that different schools are needed to cater for children of different abilities. Apart from the educational principles, there were strong economic reasons for disquiet, because a child who was not selected for a grammar school was virtually debarred from taking up any profession. Since admission to grammar schools could no longer be purchased, and because fees at private schools were rising rapidly, many middle-class parents were now having to send their children to other types of secondary school. For the first time the State school system as a whole became the interest of a literate and loquacious body of parents, and this interest generated much new discussion. The optimistic belief of the psychologists, that IQ could be measured and could, together with a few other tests taken at 11+, be used confidently to predict a pupil's future role, was attacked. 'Selection at 11+ was immoral and had to go', was the emerging theme. Arguments began to be heard for 'comprehensive' schools to which all children would go and which would cater for a wide variety of needs. Supporters of the cause quoted the results obtained in Sweden where, after controlled experiments, selection had been abolished.

The arguments for and against comprehensive schools soon developed a political flavour and the achievement of comprehensive education became an avowed aim of the Labour Party. Previously, the 'social ladder' effect of the grammar schools had appealed to the Labour Party, whose Ministers even in 1945 had been opposed to comprehensive schools, at the same time subscribing to the common Labour belief that the social ladder drained away their children to become 'traitors to their class'. However, in 1965, shortly after the Labour Government took office, the House of Commons passed a motion which said

That this House, conscious of the need to raise educational standards at all levels, and regretting that the realisation of this objective is impeded by the separation of children into different types of secondary schools, notes with approval the efforts of local authorities to reorganize secondary education on comprehensive lines . . .

In July of that year the Department for Education and Science sent out a circular to local authorities saying that the Secretary of State† 'requests local education authorities, if they have not already done so, to prepare and submit to him plans for reorganizing secondary education in their areas on comprehensive lines'.

The implications of comprehensive education for curriculum

---

† The political content of the 'comprehensive' issue was further shown when this circular was rescinded by the Conservative Secretary of State when she took office in 1970. The social climate had, however, by then moved in favour of comprehensive schools even in Conservative districts.

developers are enormous. If by 'comprehensive' we mean 'multi-lateral' in the sense of the Spens Report, then the 'grammar school stream' would adopt the old 'grammar school syllabus' and so on. But what was to be done about the teachers who had been trained to deal with only one type of pupil?

Graduate teachers know how to teach what I will call the grammar school curriculum, and while the pupils will follow that if they must, there is every reason to suppose that comprehensive schools will land the community into a slough of despond if they take the grammar school curriculum as their yardstick for more than a minority of their students. There must be other ways of stimulating the spiritual, moral, mental and physical growth of adolescents, for whom the grammar school tradition or any of its derivatives are quite meaningless. (R. Wilson [1])

If a school is to merit the title 'comprehensive' it is not sufficient, though, for it to be merely a multilateral school. Indeed, the main objections to the tripartite system centred on selection and separation. A multilateral school would only serve to perpetuate both under one roof. Can a school, however, dispense with internal selection? How is it to handle an intake of, say, 180 pupils? The method adopted in many schools is to divide the pupils into classes by 'some measure of ability'; thus the brightest are grouped together, as are the slowest. The arguments for this solution, known as 'streaming', are that forms consisting of pupils of mixed ability handicap the teacher and are unfair to many children. Streaming of this nature, however, does seem to preserve the multilateral approach. A more refined means of avoiding mixed ability groups is to 'set', i.e. to superimpose upon the basic division into forms further divisions for individual subjects. Thus a pupil might be included in the brightest group for mathematics, that is, be ranked amongst the top 30 out of 180 for that subject, but would be with a group of average ability for English. Unfortunately, the advantages of setting are offset by the administrative problems it causes.

Many, though, would feel that if comprehensive education is to mean anything, then – at least in the early years – it must mean the end of any attempt to group by ability. The problem of teaching unstreamed, mixed ability groups now arises. The solutions proposed for this problem involve a breaking down of the class from a large unit to many smaller ones – groups of pupils working together or pupils working individually. This calls for new techniques of teaching and places emphasis on the use of work cards, unit packages and other products of educational technology to which we shall refer later.

Such techniques would be natural extensions of those now used in many primary schools, which have traditionally been 'comprehensive' though small.

Exercises

1 The opponents of the comprehensive system have used various kinds of arguments to support their case.

(a) List some of these arguments and discuss their merits. Which are academic and which political? Can they all be dismissed as selfish or reactionary?

(b) To create a good comprehensive school in a district may involve ending a good local grammar school. What arguments, if any, might justify this?

2 It has been argued that the American High School system has been ineffective in producing an intellectual elite. Indeed, special schemes have been developed for talented children, to enable them to have more specialised training than is customary in the USA. Is there any *necessary* incompatibility between the notions of comprehensive and specialised educations?

3 Discuss the argument that, since the values of the educational system are middle-class, then middle-class children obtain an advantage in 'selection examinations' such as the 11+. To what extent might mathematical ability, in particular, be independent of social class?

3. PROBLEMS OF CURRICULUM CHOICE

The provision of two types of examination – the General Certificate of Education intended for 'the top quarter of the ability range',† and the Certificate of Secondary Education intended for those between 'the 40th and 80th percentiles'† – has meant, however, that unstreamed classes can in general only be found in the first two or three years of secondary education. After that, the school has effectively carried out its own selection procedure and is faced with the problem posed by Professor Wilson (see p. 58). If the GCE classes are to follow a typically grammar school curriculum (in either a 'traditional' or a 'modern' form), it is necessary to work out curricula to be followed by the CSE classes and by those who do not aspire to CSE.

In the past the answer to the expansion of educational opportunities has almost invariably been the cheapest and most amateur, to offer newcomers what was already in existence and to adapt this gradually over the years as it became increasingly evident that what was designed for the old was not suited to the new. This process, unfortunately, has often served almost to nullify the benefits expected to result from expansion.

† Whatever these terms may mean!

More recently, machinery for promoting more rapid, and better, planned curriculum development has been established in many countries (see Part 4). So, for example, the recent plans to raise the school-leaving age in England resulted in an unprecedented outlay of thought and money spent in designing a curriculum, not just for an 'extra year' but for a five-year course to replace a four-year one. Even so, few people would claim that schools are fully prepared for such a step, but they are better prepared than when similar steps were taken in the past.

Other projects are still trying to remedy the oversights of the past and to hasten the provision of suitable curricula for, say, the pupil of average or less than average ability, and the culturally disadvantaged child. One project which has worked in the former area is that established in 1967 by the Schools Council (see Case Study 4 in Chapter 14). Many projects have concerned themselves with the disadvantaged child, including the School Mathematics Study Group, and some have concentrated their work entirely on providing for the socially deprived child (for example, Project SEED).†

The curriculum developer working in these particular areas is, however, beset by many problems of a theoretical and philosophical nature as well as the more obvious practical ones. To what extent should the 'aims and objectives' differ for differently-labelled groups, and to what extent is such labelling possible? Certainly the classification of schoolchildren as described in the Norwood Report (see p. 55) now seems unbelievably naïve, but to attempt to deny the existence of differences, and thus the need to cater for them, seems no more promising as a philosophy. Clearly, there is much of truth and value in the arguments both of the supporters of comprehensive education and of those who, for example, would wish to establish special schools on the Russian model for the mathematically gifted. It is by no means obvious which philosophy will and should prevail.

Exercises

1 'What is to be the course prescribed for the slower mathematician in the grammar school? He may be provided with any of these alternatives:

(a) the same syllabus as his more able contemporaries ... but covered more superficially and omitting anything unsuitable for minimum examination purposes;

(b) a similar syllabus but with a year longer before the examination;

(c) a substantially different syllabus.

† Special Elementary Education for the Disadvantaged.

The adoption of the first alternative involves some grave risks. Subject matter which pupils with some mathematical insight will understand and will fit into its proper context may well have little meaning for others, who are likely to find it tedious and probably frightening ... The second alternative ... may offer a solution to the problem of the pupil whose mathematical weakness ... comes from retardation (owing, for instance, to illness or frequent changes of school) or from late development, but not from an absence of mathematical ability ... a third alternative (c) which may differ substantially from that offered to the abler pupils in its content, its sequence and its method of presentation can bring new light and unexpected enjoyment to their learning of mathematics.' (HMSO [16], pp. 103–4)

Discuss the three alternatives mentioned above. To what extent do these views contradict those of the writer of the *Times Educational Supplement* leader of 21 May 1971, who wrote:

'To identify some definable group within the system as being slow-learners, and then find aims and objectives for them different to those of the rest of the school population sounds like a highly dangerous exercise, and certainly not in the interests of the children. The next step, after all, would be to suggest that they would do far better in different types of school – and look where that philosophy has led.'

What are the implications for the mathematical curriculum in schools?

2 Discuss the pros and cons of segregating highly gifted children in special schools. What is likely to happen to the children of such children?

3 Investigate the special problems of immigrant children with respect to the learning of mathematics. Devise a teaching unit for them, taking into account their special needs and limitations, such as those of vocabulary.

SOME FURTHER READING

Burgess [1], HMSO [4–6, 8, 10], Holt [1], Hooper [1], OECD [4], Pedley [1], Schools Council [2, 13], Vaizey [1], Vogeli [2]

5
Teachers

1. THE BEEBY MODEL

The potential of an educational system is directly related to the ability of its teachers. Hence, the more qualified and better trained teachers are, the easier it is to effect curriculum development. No matter how distinguished the members of a project team are, how carefully structured a new course is, how brilliantly the various educational media have been exploited, the success or failure of any innovation ultimately hinges on the receptiveness and flexibility of the classroom teacher. As Cecil Beeby has pointed out (Beeby [1]):

> There is one thing that distinguishes teaching from all other professions, except perhaps the Church – no change in practice, no change in the curriculum has any meaning unless the teacher understands it and accepts it. This is a simple and fundamental truth that no curriculum builder can ever afford to forget. If a young doctor gives an injection under instruction, or if an architect as a member of a team designs a roof truss, the efficiency of the injection or the strength of the roof does not depend on his faith in the formula he has used. With the teacher it does. If he does not understand the new method, or if he refuses to accept it other than superficially, instructions are of no avail. At the best, he will go on doing in effect what he has always done, and at the worst he will produce some travesty of modern teaching.

The teacher, then, plays a vital role in curriculum development whether the educational system is centralised, as, for example, in Scotland with its one mathematical syllabus for the School Leaving Certificate, or whether the individual school has considerable freedom, as in England. In the former case, new courses can, through ignorance, or malice, be taught in a way which is directly at variance with the ideas of those who designed the curriculum; whilst in the latter case the teacher is free not only to travesty change but to reject all initiatives, either good or bad.

The rate at which change can be assimilated into an educational system depends then on the teacher. Dr Beeby, writing of the problems of education in developing countries (Beeby [2]), postulated that an educational system must of necessity pass through four stages, described, by the following model:

Stage I. *The 'Dame School' stage*: at which the teachers are neither educated nor trained.

Stage II. *The stage of formalism*: at which the teachers are trained but poorly educated. This stage is characterised by the highly organised state of the classroom, the rigid syllabus, the fixed text-book and the emphasis placed on inspection. It is the stage found in England around 1900.

Stage III. *The stage of transition*: at which teachers are trained and better educated but still lack full professional competence. The aims are little different from those of Stage II but the syllabus and textbooks are less restrictive. Teaching is still 'formal' and 'there is little in the classroom to cater for the emotional and creative life of the child'.

Stage IV. *The stage of meaning*: at which teachers are well-trained and well-educated. Meaning and understanding are now stressed, individual differences are catered for and the teacher is involved in the assessment of his pupils. He may now be so confident as to reject any curriculum but his own.

It is Beeby's hypothesis that it is impossible to omit any of these stages in any educational system. If this is true, then the implications are far-reaching for the developing countries, where teaching is still at the stage of formalism (particularly in the primary schools). Attempts have been made to hustle a system from Stage II to Stage IV, omitting Stage III *en route*, but it must be admitted that so far no counter-examples have emerged to disprove Beeby's hypothesis.

The position in some countries is considerably easier and, for example, one could claim that all the schools in England had reached Stage III and some had attained Stage IV. Upgrading is, however, a long and continuing process. Steps are continually being taken in an attempt to ensure that teachers enter the profession better trained and better educated than their predecessors. Thus all would-be teachers in England are now required to have had some professional training in teaching, in addition to orthodox training in their academic speciality.† Again, students in Colleges of Education now have the opportunity to strengthen their academic training by means of a further year's study for the Bachelor of Education Degree. Regrettably, the tendency noted earlier to meet new demands in schools by

† Mathematicians have been 'temporarily' exempted from this requirement.

fitting newcomers into existing moulds has not been resisted at College of Education level; thus it cannot yet be claimed that the curricula designed for the new BEd courses are, in general, entirely apt. Doubtless these will improve and be made more relevant to the student's needs.

Exercises

1 In which of Beeby's stages would you place the educational system in which you grew up? Give reasons.

2 In which stage would you place a system (such as the French or Italian) in which teachers are frequently (over) educated in their subject but are not trained to teach?

2. IN-SERVICE EDUCATION

It is now realised, however, that it is not sufficient to ensure that teachers enter schools well-trained and well-educated at the beginning of their working lives. Steps must also be taken to ensure that they remain so as conditions change. Thus there is an evident need for continuing *in-service* education. To date, such training in Britain has been mainly of an informal nature – teachers have attended courses organised by Local Education Authorities, University Institutes of Education, Curriculum Development Projects, etc., on an optional basis and usually with no benefit to themselves other than that of knowing it would enable them to do a better job. That such informal education is insufficient to meet the needs posed by the rapid rate of development of both subjects and teaching methods was clearly recognised by the James Committee (HMSO [12]) which recommended that a more formal in-service education structure should be established. An indication of the form that in-service training might take in the future was contained in the White Paper on Education published in December 1972. Clearly, there is a need for greater cooperation between the various agencies now engaged on this task. For example, in England these include, in the field of mathematical education, the Department of Education and Science, Local Education Authorities, the universities (a recent innovation in this sector has been the establishment of posts and centres specifically concerned with in-service education), the subject associations and private trusts (such as SMP). Each of these bodies is making a small contribution to a massive task.

A powerful force in the improvement of teachers in a system can be the inspectorate of the system. In England, from the example set by Matthew Arnold in the mid-nineteenth century, the inspectors have frequently been able by subtle encouragement – rather than by official edict – to coax teachers into adopting new ways. They have also been able to shelter eccentric and original teachers from disapproval that might stifle their pioneering attempts. In other countries, for example Scotland, the inspectorate has played a more positive role and has led rather than merely encouraged (cf. p. 142). Indeed, the role of the inspectorate and the methods whereby they are chosen tell us much about a country's educational system. On this latter point, it is interesting to add that one of England's leading HMIs has said that if Picasso had applied for the post of Inspector, he would have got it, but on the grounds that he showed himself as a young man to be a thoroughly competent draughtsman.

Exercises

1 Discuss the advantages and disadvantages of the various means employed for the in-service training of teachers.
2 What would you infer about the inspectors in a certain European system, of which one reported that 'We told the teachers in the primary schools to start teaching about the system $\mathbb{Z}$ and the construction of the number system. People told us beforehand that it would be impossible, but no trouble was reported at all.'?

3. INITIAL TRAINING AND RECRUITMENT

The assembling of a competently trained force of mathematics teachers is still a long-term objective. The shortage of mathematics teachers is a problem that has been discussed on many occasions. Indeed, the Clarendon Commission of 1864 (HMSO [2]) is probably unique in reporting that 'it is easy to obtain mathematical masters of high ability who have had a university education', but then the commission was only concerned with the nine great public schools! The creation of the vast computer industry, accompanied by an enormous expansion of higher education, certainly had the effect of reducing the number of mathematics graduates 'with good honours degrees' who were attracted to teaching. The financial depression which hit most countries in the late 1960s and early 1970s has, however, had the effect of sending more good graduates into teaching – as did the depression of the 1930s – and this has done something to ameliorate what was becoming a desperate situation. For example, according to the *Times Educational Supplement* of 30 May 1969, half the mathematics teaching of 11–16-year-olds in English schools at that time was carried out by teachers having no 'mathematical qualifications', a position unlikely to be changed overnight or even within a decade.

Granted sufficient recruits, one must still decide how a mathematics teacher should be trained. Various bodies and committees have attempted to draw up curricula for the initial training of mathematics teachers – and here it must be remembered that all primary school teachers are automatically teachers of mathematics.

One famous set of recommendations (MAA [1]) was made by the Panel of Teacher Training of the (USA) Committee on the Undergraduate Program in Mathematics. The Panel attempted to define the mathematical education a teacher required, competently to teach at the five levels of teaching responsibility it identified: namely (with their English equivalents in parentheses)

(I) elementary (primary) school mathematics,
(II) elements of algebra and geometry (middle school mathematics),
(III) high school mathematics (O-level mathematics),
(IV) elements of calculus, linear algebra, probability, and so forth (A-level mathematics),
(V) college (university and College of Education) mathematics.

A more detailed look at the needs of teachers at the first level was taken by the Cambridge Conference on School Mathematics. Its findings are given and discussed in Chapter 15 of this book.

Amongst the several British publications on this matter, we should like to draw special attention to the publication of the Association of Teachers in Colleges and Departments of Education (ATCDE), *Teaching Mathematics* (ATCDE [1]). This report discusses not only what mathematics should be known by teachers, but – and this is equally important – how they should learn it, and what attitudes towards mathematics should be encouraged. If the student does not acquire such attitudes in his initial training, it is one of the purposes of in-service training to ensure that he does so as he matures.

The need to use to best effect what little time is available for professional and other training is, of course, obvious. Our success or otherwise in solving this problem will largely govern the rate at which mathematical education can be improved.

We began this chapter by quoting Dr Beeby on the crucial role of the teacher in curriculum development, and we end it by recalling the final 'Thesis' in Alexander Wittenberg's 'axiomatic' development [1], of the principles that should underlie the reform of mathematical education.

Thesis 20

No single issue is more important, in the teaching of mathematics as in education generally, than that teachers should be men and women of genuine intellectual distinction. An over-riding concern, in the shaping of the teaching of mathematics, must therefore be whether any proposals made will tend to attract or to repel, to challenge or to stifle, to stimulate or to discourage, to reward or to punish truly distinguished and creative teachers. Every one of the parameters which influence the recruitment of such teachers – including selection, training, salary scales, teaching loads, class sizes, independence . . . – must be a legitimate concern of those who care for excellence in the teaching of mathematics.

It is the role, then, of those concerned with the training of teachers – either in-service or initial – to develop this intellectual distinction in teachers. All those who design curricula must ensure that Wittenberg's other conditions are met. Apart from technical competence, a good teacher will also have dedication, unselfishness and a wish to make his pupils *better*. How to obtain these qualities is a further difficult problem: we should not be satisfied, however, with the production of competent dullards.

Exercises

1 'All the efforts of teachers in the Schools Council and elsewhere to develop new curricula may be abortive if curriculum development is taken to exclude examination of the part played by teachers in the curriculum, which is after all not a thing, but an activity.' (Barnes *et al.* [1], p. 11)

Discuss this statement and list factors in the teacher's behaviour which affect the curriculum and, in particular, that of mathematics.

2 'As an untrained teacher, I have even persuaded myself that the chief benefit of training is to enable a teacher to keep a class occupied without having to interest them.' (Rollett [1])

The late A. P. Rollett was an influential HMI, and his humorous asides normally contained more than a grain of truth. Discuss the significance of this remark.

3 'Many teachers believe that the decision about what is possible should be based not on the student's characteristics but on the teacher's – his teaching style, his ability, experience and personality. It is our considered judgement that most teachers can learn new ways of teaching students and that most can, if they will make the appropriate effort, help their students attain a great variety of educational objectives. The *teacher* does not, in our thinking, represent the major factor in determining the objectives which are possible.

It is the *teaching* which determines the objectives ...'   (Bloom [1], p. 11)
   Discuss this extract. Does it conflict with the views expressed by Beeby
(p. 62) and by Barnes (see Exercise 1)? To what extent should teachers
determine objectives and, if objectives are to be 'teacher determined',
can they be free of 'teacher characteristics'?

4 What could Wittenberg mean in his Thesis 20 (p. 67) by 'intellectual
   distinction'? Should this refer to mathematical ability only, or to other
   abilities? How can intellectual distinction (in your sense) be cultivated?

5 Investigate the origins and growth of teacher training colleges in your
   country.

6 'The research mathematician is loyal first to his profession and second to
   his employer. I expect the college mathematician to be loyal first to his
   college and second to the mathematics profession.'
                                   (H. Flanders, *Amer. Math. Monthly* **78**, 1971, 291–6)

   'I hold every man a debtor to his profession, from the which as men of
course do seek to receive countenance and profit, so ought they of duty to
endeavour themselves by way of amends to be a help and an ornament
thereunto.'   (Bacon)
   Discuss.

SOME FURTHER READING

Cane [1], Castle [1], Eraut [1], Gosden [1], HMSO [12], NSSE [1],
OECD [3], UNESCO [3]

6
Money

1. COSTS

Few, if any, statements concerning education have universal validity but one maxim that contains a considerable amount of truth is 'good education costs more than bad'. Certainly, if one considers Dr Beeby's four stages of education (p. 63) one notes that each stage is correspondingly more expensive than its predecessor. A well-trained teacher costs more than a badly-trained one – not only because his training will be more expensive but also because he will have to be paid a higher salary throughout his teaching career. If one wishes to cater for individual differences and to use activity methods, then this often necessitates bigger and better planned classrooms with additional items of furniture. A wider curriculum can mean more laboratories and equipment. Well-produced textbooks cost more than badly-produced ones. Visual aids cost money. A planned curriculum developed by a team working for years costs more than one handed down from some other sector of the educational system. Examinations meant to cater for the needs of individual schools and pupils cost considerably more to administer than do those of the traditional type.

Impoverished education will, therefore, almost certainly be bad education. What is not true, however, is that a greater expenditure of money will alone ensure a rise in the quality of education. A teacher does not automatically become a better teacher when given an overhead projector, nor does the class necessarily benefit when each pupil is linked, at vast expense, to a computer. Expenditure on buildings and equipment, unless accompanied by increased expenditure on teacher training, is rarely fruitful; and a good education has often been given in poor surroundings – but by good teachers.

The availability of money then, will have a great effect on the development of a curriculum. For example, in the field of mathematical education, improvements will depend upon the increased availability of money for such items as: initial training of teachers;

the in-service training of teachers; teachers' salaries; the provision of mathematics laboratories, of up-to-date textbooks, of audio-visual aids and of computing facilities; educational research and the financing of curriculum development projects. Should there be insufficient for all these items, then decisions must be made as to which to foster and which to delay.

Small wonder then that the educational future of some of the developing countries gives rise for concern. In these countries there is not sufficient money available to ensure that the vast increase in the quantity of education offered is not accompanied by a decline in the quality. Indeed, even money could hardly have guaranteed the preservation of quality, for the education which a country can offer at any time is ultimately limited by the number and quality of its teachers – factors which in turn are governed by the state of the educational system in the recent past. Thus the quality and quantity of education which can be offered tomorrow depend very much on the quality and quantity of that being offered today and even yesterday.

One of the key problems of education is that it is such a 'labour-intensive' industry – its costs are governed to a great extent by teachers' salaries. This means that any attempt to improve the quality of teachers either by increasing their pay or by up-grading their qualifications (and so entitling the teachers to pay-differentials) has a major effect on a country's educational budget. Whereas in a modern industry, new techniques and rising productivity permit steady wage increases without corresponding increases in the real cost of production, a 10 per cent salary increase for teachers usually translates to a 7 or 8 per cent increase in total 'costs of production'. And, of course, in direct contrast to industry's 'rising productivity', there is a constant demand in educational systems for smaller classes – that is, for *less* output per teacher, in quantitative terms. In some developing countries where buildings are poor and equipment virtually non-existent, teachers' salaries amount to a fantastic percentage of the total educational budget. For example, in 1964 over 93 per cent of the total public expenditure on education in Cambodia went on teachers' salaries (UNESCO [2]).

This problem is also encountered in the developed countries where any cutback on educational costs must be made from the spare left when the teachers' salaries have been met. In fact, this recently led in England to a reduction in the annual budget of the Schools Council.

It is unfortunate that research and development should be the first to suffer in such cases, particularly since they are not well endowed financially – Professor W. D. Wall pointed out in the 1960s that the British Government at that time earmarked more money for research into the development of new glues than for research in education.

Some idea of the way in which the educational budget is dispensed, and of the influence of teachers' salaries, can be obtained from the figures shown in Table 1, extracted from Vaizey and Sheehan [4]. (All amounts are given in millions of pounds relative to 1948 prices – allowing for inflation the actual amounts spent are considerably greater.)

TABLE 1   *Secondary education: England and Wales*

	1946	1955	1965
School population (millions)	1.1	2.0	2.9
Total current expenditure	38.7	66.7	118.5
Teachers' salaries	28.3	45.1	71.2
Salaries of non-teaching staff	0.4	1.1	10.5
Heat, light, upkeep of buildings	5.1	11.3	24.7
Books, stationery and materials	1.7	3.7⎫	12.0
Furniture, equipment and apparatus	0.9	1.8⎭	

(The ratio of teachers' salaries to total outgoings is slightly greater in primary schools than in secondary schools.)

Exercises

1 In the example above about research into glues, the reason could be that the Government might not have known that educational research even existed. Write a report designed to convince a politician that educational research is worth financing. (As a preliminary, write his secretary's report purporting to prove that it is *not* worth financing.)

2 What proportion of the taxes in your district is spent on education?

3 According to Trevelyan [1], the public money spent, per head of population, on secondary education in England in 1899 was 1/17th of that spent in Switzerland. What reasons would you give for this discrepancy?

2. NATIONAL CONTROL

In most countries, education is primarily financed and controlled by the State; and accordingly national policies on education can have a marked effect upon the curriculum.

'T' Minister knows nowt about t' curriculum' is a saying attributed to George Tomlinson, the Lancastrian ex-miner who was Minister of Education in the years following the Second World War. This ingenuous remark reveals the traditional attitude of the British Government to the school curriculum: that it was a matter to be decided in the schools without Government interference. For this there are historical reasons of religion and of high-minded liberal ideas about freedom. It was recognised, however, that such a system could be inefficient. By a slightly devious route this led to the establishment of the Schools Council for the Curriculum and Examinations (see p. 144), in which Governmental views on the curriculum could be expressed. Even so, the representatives of the teachers were still in the majority, and the liberty of individual schools to reject direction on particular issues was preserved. It must be noted, though, that an individual school must still respond to national policies on larger curricular issues. Thus, for example, if it is decided to fight premature specialisation by replacing the present pre-university examination system by a new scheme which insists upon candidates demonstrating that they have received a more general education, then all schools will have to fall into line. However, individual schools still have a part to play in the design of such a system and indeed, as has been demonstrated, in the torpedoing of suggested schemes.

Such a disinterested attitude is not typical of all Governments, however:

(The school in the German Democratic Republic) has the task of preparing the younger generation for life and work in a socialist Society.

It also aims at training general, well-developed, balanced personalities, who through a well-planned and systematic training and education process have gained thorough, exact and sure knowledge and ability and who are distinguished by socialist consciousness and socialist behaviour.

(*Education in the GDR*, 1962)

The implications of this statement for mathematical education are not immediately obvious. One example, however, is provided by the steps taken in the GDR to meet a foreseen shortage of personnel trained for high-level work in the computer industry. To help meet the need, in 1962 the top streams of 12-year-olds in four secondary schools were assigned a special curriculum with extra mathematics periods and a syllabus intended to extend over the remainder of their school careers. It was designed to train the pupils in aspects of numerical analysis and computing, which could be readily applied when they left school. This is clearly 'well-planned and systematic training and education', but carried out in a manner that would not be tolerated in other societies.

Other examples of the effect on the mathematics curriculum of a national education policy can be found in Vogeli [1], a report of the effects of the great Cultural Revolution in Communist China on mathematical education.

One must not, however, reject State interference in the curriculum out of hand. Too many of the developing countries, for example, have been handed down educational systems and curricula totally at variance with their needs. As President Nyerere [1] of Tanzania has pointed out:

'At the present time our curriculum and syllabus are geared to the examinations set . . . And the examinations our children at present sit are themselves geared to an international standard and practice which has developed regardless of our particular problems and needs. What we need to do now is to think first about the education we want to provide.'

Any such rethinking would almost certainly result in a desire to reform the curriculum of the secondary schools of Tanzania and, in particular, the mathematics curriculum. (See Phythian [1].)

Local authorities, too, can exercise a control over the curriculum. In some countries this is done by stipulating which textbooks are to be used (cf. p. 94). In Britain, local control is less obvious. However, curriculum development depends upon such factors as the availability of money to purchase new texts† and apparatus, and to finance attendance by teachers at in-service training courses; and here local authorities differ greatly in what they are willing (or able) to spend on these items and in the general encouragement which they give to schools wishing to participate in curriculum development projects.

In the USA, the disparity between the money available to schools in different localities is even more marked, for in 1970 only 7 per cent of schools' finances came from Federal funds, the remainder being collected locally. According to *The Observer* (14 Nov. 1971) this resulted in one school in Missouri having £89 per year per pupil to spend, whereas, at the other end of the affluence range, one school in Wyoming had £6064 per year per pupil.

3. RETURNS: EDUCATION AS AN INVESTMENT

We have seen how teachers can be regarded not only as an 'input' of an educational system, but also as one of its 'outputs'. In the

† According to the Education Statistics for 1970–1, the amount spent per primary school pupil on class and library books ranged from 46p in Swansea to £3.32 in Montgomery. At secondary level the extremes were Barnsley, £1.43, and Radnor, £4.59.

same way, we can think of education as producing money as well as needing it. Now, within a national economy there will be competition for money between such different parts of it as its education system, its health and welfare service, its military budget, etc. Clearly, if it could be shown that education *produced* wealth at a certain rate, then it would be easier for the educators in the country to argue for more funds to improve their system.

The preamble to the First Education Code (1871) of Japan contains the maxim that 'Knowledge may be regarded as the capital for raising oneself'. This notion, that human beings can be treated as capital, and that the acquisition of useful skills and knowledge can be seen as an increase in capital, has a long history. In the seventeenth century the idea was expressed by Sir William Petty and a hundred years later it was promulgated by Adam Smith.

Smith had a particular interest in education. He was a product of the Scottish educational system and was willing to credit that system with responsibility for 'the superior intelligence, and the providential, orderly habits of her people'. Education was seen by him, therefore, as the basis both of good civil Government and of economic growth. During the nineteenth century, the former view of education received most emphasis and John Stuart Mill was able to argue that 'It may be asserted without scruple, that the aim of all intellectual training for the mass of the people should be to cultivate common sense' (Mill [1]). The idea of education as a basis for economic growth was not, however, forgotten; and by the end of the century, references could be found to 'education as a national investment'.

Despite this long history of the involvement of economists in educational thinking, it was not until the 1950s that the subject 'economics of education' came into being. Since then it has become one of the most rapidly growing branches of economics. To quote Mark Blaug [1]:

The sort of questions that are asked in the economic analysis of education are: how much should a country spend on education and how should the expenditure be financed? Is education mainly 'investment' or mainly 'consumption'?† If investment, how large is its yield compared to other forms of investment in people and material equipment? If consumption, what are the determinants of the private demand for more or better education? What is the optimum combination of pupils' time, teachers, buildings, and equipment

---

† Roughly, goods and services from which consumers derive immediate benefit, or which are valued for their own sakes, are referred to as *consumption*; those which are used to improve and produce over a long term are known as *investment*.

embodied in schooling? What is the optimum structure of the educational pyramid, that is, the number in the different levels and channels of the educational system? What is the optimum mix of formal education within schools and colleges and informal education outside them? Lastly, what contribution does education make to the overall development of human resources and how far can we accelerate economic growth particularly in low-income countries, by controlling the expansion of educational systems?

Attempts to answer some of these questions have given rise to great controversy. For example, it has long been obvious that the earnings of people with different educational backgrounds have differed accordingly. Attempts have recently been made, therefore, treating expenditure on education as an investment, to measure the return earned on that investment. Thus it was found that 'In 1958 the average elementary school graduate could expect a lifetime income of about $182,000 as compared with about $258,000 for the average high school graduate . . . a college graduate could expect to receive $435,000' (Miller [1], p. 44).

Findings such as these led to slogans such as 'a college education is worth $100,000' and to calculations which showed that money spent on education produced a rate of return of 13 per cent (or even higher, according to the accounting system used), thus setting a quantitative seal of approval on George Eliot's argument that one had 'better spend an extra hundred or two on your son's education, than leave it him in your will'.†

Other economists, however, were not so easily persuaded by such arguments. How much of the increased earnings could really be attributed to 'education'? Might it not be due to greater intelligence, to parental wealth, education and influence? The most important factor determining the level of education achieved by the head of a household in the USA was shown by the survey to be not IQ nor ability, nor the income of parents, but the educational attainment of his father (Blaug [1]). Might the increase be due to restrictive practices in such professions as the law and medicine? Any arguments which ignore such factors are likely to be fallacious. Yet such 'research' is still given credence. Results purporting to evaluate 'the rate of return to society on its investment in human capital', obtained in Britain by a Government-sponsored team, were published in summer 1971 and received a warm welcome from *The Times*, amongst others (7 June). What the research showed was that the rate of return was higher on an HNC (a non-graduate qualification taken part-time) than on a PhD, a result which, granted the assumptions

† Mr Riley in Chapter III of *The Mill on the Floss* (1860).

of the project team, was pretty obvious. Whether such assumptions provide us with a satisfactory model of the situation is disputed, for example, in an article by Vaizey [2].

To date, the theories of educational economists have had little effect on the school curriculum. In certain countries, educational planning has led to attempts to divert students from arts to science – one such example from Tanzania is quoted in Skorov [1, p. 43] – but in general the models used have been too crude to provide specific guidance to curriculum planners. Thus 'a college education is worth $100,000' cannot hold regardless of what subjects are studied.

Doubtless, findings will be published by economists concerning the curriculum in the coming years. Experience to date, though, shows that such findings must be examined with great circumspection in order to ensure that the models used are realistic and that models developed for one system, say in an advanced Western country, are not applied unthinkingly to totally different systems, such as that of an underdeveloped country. Nevertheless, for reasons stated at the beginning of this section, it is important that, within a given national economy, economists should try seriously to cost the benefits of an education system.

Exercises

1 Is Mr Riley's advice (p. 75) still valid? Discuss some of the social issues it raises.

2 'T' Minister knows nowt about t' curriculum' (p. 71). Is this something to be deplored, or to rejoice over?

3 (a) The numbers of live births in the United Kingdom in the quarters of the years 1965, 1966, 1967, 1968, recorded in thousands, were as follows:

	1965	1966	1967	1968
1st quarter	257.5	251.7	253.1	247.7
2nd quarter	253.9	246.6	248.2	239.5
3rd quarter	252.5	251.0	237.4	239.6
4th quarter	236.2	229.1	221.9	220.0

(*Monthly Digest of Statistics*)

Predict likely figures for the four quarters of 1969, explaining clearly the method you are using.

(b) You are an official of the Department of Education and Science, concerned with the supply of primary teachers. The actual figures of live births in the United Kingdom in 1969 are not yet available; suggest purposes for which you might need to use predicted figures in your official work.   (Hull BEd, 1971)

SOME FURTHER READING

Anderson and Bowman [1], Blaug [2], Vaizey [3–5]

7

Educational Research and Theories

1. EDUCATIONAL THEORIES

At the time that Professor Wall drew his comparisons (p. 71) between the amount of money the British Government allocated to the development of new glues and to educational research, Professor John Vaizey remarked that, at least, the former outlay led to an increase in the stickiness of glue, whereas research on education, so far as he knew, did not lead to increases in the efficiency of education. It must be admitted that, to date, the design of secondary school curricula – the choice of syllabuses and of teaching methods – has not been influenced greatly by the products of educational research or by the thoughts and findings of philosophers and psychologists.

The traditional relationship between English teachers and educational theorists is well established by John Blackie [1], who writes:

I said just now that the English are pragmatists in religion, that they believe in what worked. This has really always been their approach to education. They have distrusted theories and the experts who thought of them. They were slow to be influenced by educational thinkers, few of whom were British. Those who were, Rachel and Margaret Macmillan, Susan Isaacs, Nancy Catty and Dorothy Gardner. were all practising teachers, and all, it may be observed, women. Rousseau (French), Pestalozzi (Swiss), Froebel (German), Dewey (American), Piaget (Swiss) have perhaps been the most powerful sources of influence, but it was an influence much disguised by being mediated through interpreters. It is probable that few English teachers have actually read any of the books of which those men were the authors.

Blackie is writing here of the English primary school, but what he has to say is equally valid for the secondary school – indeed, the theorist has had even less influence in that sector.

Of course, the position is beginning to change and should change even more quickly as the psychologists are able to tell us more about the processes of learning and teaching. Yet, for the moment, the developer has considerably more influence on British education than does the conventional research worker.

77

2. THE DISCOVERY METHOD

Just how slowly the educational system responds to the ideas of innovators, and some of the reasons for this tardiness, are shown by the history of the so-called 'discovery' or 'activity' method.

When the Nuffield Mathematics Project commenced its work in 1964, it took as one of its themes the old Chinese Proverb

'I hear and I forget,
I see and I remember,
I do and I understand.'

In this way, the Project emphasised the long history of the educational philosophy that one learns better by 'doing'. What was not recognised by all at that time, though, was how this particular belief has constantly recurred in educational history. Thus, to quote only a few examples taken from the last 200 years or so:

Teach your scholar to observe the phenomena of nature, you will soon raise his curiosity, but if you would have it grow do not be in too great a hurry to satisfy the curiosity. Put the problems before him and let him solve them himself. Let him know nothing because you have told him, but because he has learnt it for himself.

Undoubtedly the notions of things thus acquired for oneself are clearer and much more convincing that those acquired from the teaching of others . . .
(Rousseau, *Emile*, 1762)

The best way of cultivating the mental faculties is to *do ourselves* all that we wish to accomplish . . . The best way to understand is to do.
(Kant, *On Education*, 1802)

Do not, therefore, always answer your children's questions at once and directly but. *as soon as they have gathered sufficient strength and experience*, furnish them with the means to find the answers.
(Froebel, *Education of Man*, 1826)

Scientific education is not arrived at, and never can be arrived at, by young people crammed at School and College with ready-made knowledge . . . As preparation for learning electricity, do not be satisfied with once showing the child that sealing wax rubbed on flannel will attract bits of paper, but let him have a stick of wax, or better, a common vulcanite comb and a piece of flannel, and keep them, and try all the experiments he wants to try. Let him learn by experience . . . Do not attempt to explain why . . ., but let him find out that so it is.
(M. E. Boole, *The Preparation of a Child for Science*, 1904)

Running through all the work is the central notion that the children must be set free to make their own discoveries and think for themselves, and so achieve understanding, instead of learning off mysterious drills.
(*Mathematics Begins*, Nuffield Project, 1967)

The effect of this pleading can now be seen in many primary schools in Britain, but by no means in all. At the secondary school level even less progress has been made in getting these beliefs accepted and acted upon.

What has been responsible for this delay?

The answer that contributors to the *Black Papers* (Cox and Dyson [1–3]) would give is 'the commonsense of the British teacher'. This is, perhaps, a somewhat over-naïve view! Certainly, it is not one which those who have seen such methods working well in schools would take. Yet in many schools such methods do not 'work well'. There are still many teachers who, to quote a Nuffield Guide, think that 'setting the children free' means 'starting a riot with a roomful of junk for ammunition'. The real reasons which underlie the non-acceptance of discovery methods are constraints within the educational system – external examinations and the lack of suitably qualified teachers. Discovery methods make considerably greater demands upon teachers than do conventional ones. The teacher is no longer the director of the classroom situation, proceeding along well-worn paths, but is now the manager of an educational enterprise which will never work in exactly the same way two years running. Not only are the recurrent demands upon the teacher greater, but he will require different and more extensive initial training to ensure that he is capable of becoming the master rather than the victim of the method. No wonder that past attempts to encourage such teaching techniques have often failed. Recently the constraint of external examinations has been largely removed from the primary schools, but it still remains in secondary schools, where the difficulties of teaching by discovery methods against the schedules of the GCE examinations are too readily apparent.

We see then how external constraints – the professional competence of the teacher, the time he is allowed for preparation, and the examination system – can completely negate the arguments of generations of educational theorists and innovators.

Exercises

1 The quotations printed above all contain dogmatic assertions. Do you agree with them?

2 Read and criticise the paper on discovery methods by G. H. Bantock in Cox and Dyson [2].

3. THEORIES OF LEARNING

It would be wrong to suggest that the work of educational researchers has been entirely ignored by curriculum planners and teachers. In particular, in the field of mathematics, great interest has been shown in the work of Piaget and his school in Geneva. Piaget's researches are now 'well-known' in the sense that most teachers have heard of them – but usually through interpreters and in a general sense rather than by first-hand knowledge. Recently, the Nuffield Mathematics Project has cooperated with the Piaget Institute to devise methods of checking a child's learning, and Piaget's work has provided a stimulus for educational research throughout the world.† Piaget's main researches have, of course, been directed at the primary and pre-primary levels of education. If we consider his 'four stages in the development of a child's thinking':

1. The sensory-motor stage: up to 18 months of age.
2. The pre-operative stage: $1\frac{1}{2}$ to 6 or 7 years.
3. The concrete operations stage: 6 or 7 years to 11 or 12 years.
4. The stage of formal operations: above 12 years.

we see that it is only the last that influences the planning of the secondary school (Grades 7 and upward) curriculum.

It is important to add that Piaget's researches and theories are basically epistemological, i.e. connected with learning and knowing, and have nothing to say about teaching methods. His theories do, however, provide a framework around which a curriculum developer can build.

The theory of learning is still, however, in its infancy and a precise science of learning has yet to emerge. What has emerged during the century that psychologists have studied these problems, is a number of theories. One of these, the *stimulus–response* theory of Skinner, is briefly described in a later chapter, but in general the theories do not lend themselves to brief descriptions, and the interested reader is referred to specialist books on the topic, for example Hilgard [2] or Stones [1].

As Blackie observes above, though, the teacher 'usually believes in what worked'. If one applies this philosophy to questions of learning, it is unlikely that one would disagree too violently with fourteen points on learning advanced by Hilgard [1]:‡

† For example, between 1965 and 1971 more than a dozen doctoral candidates at Teachers College, Columbia University, completed or worked on theses inspired by Piaget's work (Rosskopf *et al.* [1]). Nevertheless, it must not be believed that Piaget's theories have been universally accepted. There has been some opposition to them in, for example, Russia, the USA and Britain.
‡ A slightly modified version of these fourteen points can be found in Hilgard [2]. We have reprinted the earlier version since it is more easily followed out of context.

1. In deciding who should learn what, the capacities of the learner are very important. Brighter people can learn things less bright ones cannot learn; in general, older children can learn more rapidly than younger ones; the decline of ability with age, in the adult years, depends upon what it is that is being learned.

2. A motivated learner acquires what he learns more readily than one who is not motivated. The relevant motives include both general and specific ones, for example, desire to learn, need for achievement (general), desire for a certain reward or to avoid a threatened punishment (specific).

3. Motivation that is too intense (especially pain, fear, anxiety) may be accompanied by distracting emotional states, so that excessive motivation may be less effective than moderate motivation for learning some kinds of tasks, especially those involving difficult discriminations.

4. Learning under the control of reward is usually preferable to learning under the control of punishment. Correspondingly, learning motivated by success is preferable to learning motivated by failure. Even though the theoretical issue is still unresolved, the practical outcome must take into account the social by-products, which tend to be more favorable under reward than under punishment.

5. Learning under intrinsic motivation is preferable to learning under extrinsic motivation.

6. Tolerance for failure is best taught through providing a backlog of success that compensates for experienced failure.

7. Individuals need practice in setting realistic goals for themselves, goals neither so low as to elicit little effort nor so high as to foreordain to failure. Realistic goal-setting leads to more satisfactory improvement than unrealistic goal-setting.

8. The personal history of the individual, for example, his reaction to authority may hamper or enhance his ability to learn from a given teacher.

9. Active participation by a learner is preferable to passive reception when learning, for example, from a lecture or a motion picture.

10. Meaningful materials and meaningful tasks are learned more readily than tasks not understood by the learner.

11. There is no substitute for repetitive practice in the overlearning of skills (for instance, the performance of a concert pianist), or in the memorization of unrelated facts that have to be automatized.

12. Information about the nature of a good performance, knowledge of his own mistakes, and knowledge of successful results, aid learning.

13. Transfer to new tasks will be better if, in learning, the learner can discover relationships for himself, and if he has experience during learning of applying the principles within a variety of tasks.

14. Spaced or distributed recalls are advantageous in fixing material that is to be long retained.

In noting the absence of a science of learning, and thus, *a fortiori*, of a science of mathematical education, we might, however, claim that our prime aim at the present time is to develop a sound technology of education. As consolation, we may remember that in many spheres of endeavour, technology is ahead of pure science; so often 'technique precedes science' (cf. Ellul [1]).

4. WHAT DO WE WANT TO KNOW?

As we remarked in the opening paragraphs of this chapter, educational research has not greatly influenced curriculum planners. It is relevant, then, to look at some of the questions which one would like educational researchers to answer – questions which lie at the heart of any attempt to devise a mathematical curriculum.

The absence of a science of learning has already been mentioned. In default of this, it would still be valuable to know more about the answers to such questions as, for example, in what ways does 'discovery' learning help, and how do individual differences between students affect their learning of mathematical concepts?

Piaget has written on concept formation, but how do we tell whether someone 'has a concept', and what exactly does this phrase mean?

How do children acquire the art of problem solving or, as Pólya [2] terms it, heuristics? To what extent do the environment and the culture affect a child's capacity to learn mathematics?

Learning, however, is only one facet of the problem, for, in addition to a theory of learning, we should also like a theory of teaching. What can be said about the way in which the characteristics of teachers affect learning, about methodology and about the most suitable role for the teacher in the classroom? What is the connection between what can be learned and what can be taught? Is the traditional assumption justified, that these are equal?

Again, many other questions arise in connection with curriculum design. Some computational skill is clearly needed by all children, but how much? What are the significant factors to be borne in mind when planning a terminal course for the less able 16-year-old? How are courses and students to be assessed (see Chapter 3, Section 11)?

More 'local' problems might refer, say, to the best method of introducing matrices, granted that it had to be done with students at a given age level and with given mathematical experience, and cultural background.

Clearly, there are many such problems (see, for further examples, Begle [1] and Long et al. [1]), and few of them are amenable to simple solution. Certainly, the answers are unlikely to be found in PhD theses. If the questions are to be answered adequately, then they will have to be tackled in a cooperative manner by full-time research workers. One suggestion as to how this might be done is contained in the SMSG Newsletter No. 39 (August 1972), *Final Report of the SMSG Panel on Research* (which also includes a most valuable list of research papers in mathematical education).

Even then, the findings, if they are to be of value, will have to be unequivocal and applicable within well-defined limits. Far too frequently current educational research is seen by teachers as consisting of statements about 'teaching in the large', that is, what goes on in everyone else's classroom, which lack relevance to their own classes' needs and to their own teaching.

The problems of educational research are, therefore, obvious and many. Until greater effort and thought are applied to their solution, however, mathematical education will have to remain a technology rather than a science.

Exercises

1 Discuss 'What do we want to know?' further.
2 Comment on Hilgard's fourteen points (p. 81).
3 Investigate the literature concerning mathematical models of learning.

SOME FURTHER READING

Beth and Piaget [1], Dienes [2], Hooper [1], Peel [1], Philp [1], Pidgeon [1], Shulman [1], Skemp [1]

8
Educational Technology and the New Media

1. THE SYSTEMS APPROACH

One of the terms used by educators with great enthusiasm during the late 1960s was 'educational technology'. Anyone using it was doubly blessed, for it was both a fashionable phrase, and one which had no generally accepted meaning. Thus little that could be claimed on its behalf could logically be denied.

To some writers it meant equipment:

> When we speak of the 'new educational technology' . . . we refer to instructional television, radio, films, programmed learning, and language laboratories.
> (Wilbur Schramm [1])

to others, more:

> Educational technology . . . implies an applied science, or if you like, the application of sciences to education. Precisely which sciences are involved is a matter for experiment – certainly one would include sociology and psychology, lighting, printing and photography, acoustics and electronics, but perhaps increasingly we must consider management, architecture and building science, ergonomics, systems analysis and communication. (K. Austwick [1])

Clearly, it is essential that one should distinguish between the two different aspects of 'educational technology' which emerge from these quotations. On the one hand there is what one might describe as 'technology *in* education', i.e. the use of electro-mechanical and other teaching aids. On the other, there is the 'technology *of* education' – an attempt to bring into education experiences gained in other sciences, in particular, management, ergonomics and systems analysis. These experiences will enable the educator, hopefully, to match available facilities to defined needs in a systematic manner, a method which is now often referred to as a 'systems approach'. Thus, for example, it has been suggested that when planning a learning system, that is, some section of the curriculum, the curriculum developer should ensure that his design includes:

1. A thorough analysis of the objectives of the learning area in question.

2. A clear specification of the initial requirements demanded of the learner before embarking on this.

3. Adequate means of assessing that both final objectives and initial requirements have in fact been attained to the requisite level.

4. A careful ordering of the different processes and sequential stages in which progress is most likely to be effectively made by the learner from the initial starting point to the final objectives.

5. A suitable deployment of all available equipment, materials and personnel to meet the various requirements identified under 4, paying particular attention to the interaction between the individual learner and these resources.

6. An organisational structure by means of which the system of learning materials so developed may be continuously tested, revised and improved in the light of 'feedback'. (NCET [1], p. 9)

These aims are hardly of breathtaking novelty or likely to lead to dissent if considered in their widest terms. A university department when demanding two A-levels from each entrant is clearly satisfying point 2, and it might well argue that in providing suitable degree examinations it is also satisfying the third aim. This, however, is taking the aims outlined above in their broadest sense. It is only when one concentrates attention upon limited areas of the curriculum that one sees how often many of the above requirements are not met. Students may generally be assumed to be equipped to tackle the first week of a course; regrettably, many emerge from that first week ill-equipped to begin the second.

A systems approach, therefore, is helpful not so much in that it contributes to the armoury of a teacher, but in that it provides him with an inventory against which he can check that he is fully armed.

Even when one accepts the principles outlined above, many points of interpretation can give rise to argument. How are we to express the objectives of a learning area? Is it always possible to ensure that such objectives have been attained? Is there a danger that important objectives, not easily measured, will be neglected? How do we cater for the individual differences of students within a learning system?

Exercises

1 R. M. Gagné has suggested that one should attempt to distinguish the hierarchy of knowledge required for any learning task and then plan the relevant teaching routines with the hierarchical structure in mind.

Gagné's presentation of the hierarchy of knowledge required to find formulae for the sum of a finite series is reproduced in Figure 1.

(a) Take a specific series and interpret Gagné's subtasks in terms of your example.

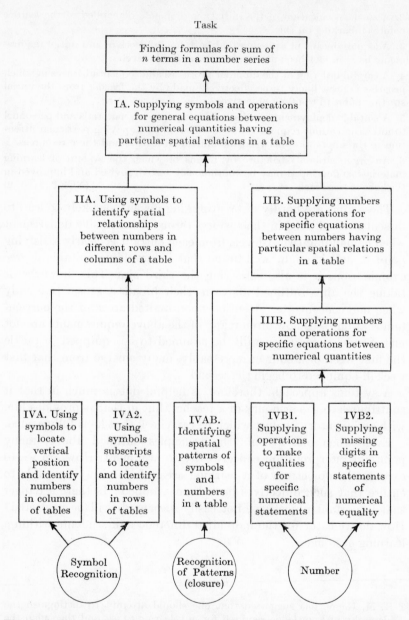

Fig. 1. A diagram illustrating the hierarchy of knowledge required for the task of finding formulas for the sum of *n* terms in a number series. Taken from Robert M. Gagné, 'The acquisition of knowledge', *Psychological Review* 1969 (1962), pp. 355–65. Copyright 1962 by the American Psychological Association. Reprinted by permission.

(*b*) Either construct your own diagram for this particular task (if you disagree with that given by Gagné) or take some other mathematical task and provide a hierarchical model for that.

2 Figure 2 (p. 88) is reproduced from Howson [2] and illustrates the structure of a possible 'unit' on probability. Before commencing the unit each student would have to show a grasp of certain prerequisites (A). The first aim of the unit would be to assess the student's knowledge of probability. This would be done by means of a diagnostic test (B) taking up to 15 minutes. The results of this test would determine whether the student proceeded to (C), (F), (I) or (J).

Either take this example and
(i) having decided upon the content of (C), (E), (F) and (H), list carefully the prerequisites (A) required,
(ii) devise a diagnostic test (B) to satisfy the requirements of your unit,
or select another topic, draw a comparable flow diagram and answer questions (i) and (ii) above for your proposed unit.

2. PROGRAMMED LEARNING

One of the ways in which it was hoped to answer the last question of Section 1 was by the technique of programmed learning. Some sixty years ago, E. L. Thorndike wrote that personal instruction for all would be guaranteed if only one could produce a book arranged so that page two was invisible until the directions outlined on page one were successfully carried out. In the 1920s, a machine equivalent to such a book was designed by S. L. Pressey. But Pressey's work attracted little attention and it was not until almost thirty years later that B. F. Skinner successfully launched the great 'programmed learning movement'. Skinner, a Harvard psychologist, had carried out research work on rats and pigeons, just as Thorndike had worked with cats, and suggested that some of his findings on the way animals learn could be generalised with benefit to the classroom situation. Thus pupils were to be encouraged by being frequently 'rewarded' for supplying correct 'responses'. In this case the 'reward' was to be the immediate knowledge that they had obtained the correct answer to a question posed. Thus learning was to be broken down into a number of small steps, each demanding a response so graded as to ensure almost 100 per cent correct answers. Pupils would then build up a sequence of successful responses (cf. Hilgard, point 6, p. 81) and this would provide the motivation required to make them persevere with the learning system. This was the underlying philosophy of the linear (Skinner)-type programme. Its supporters claimed that it catered for individual differences by allowing each pupil to

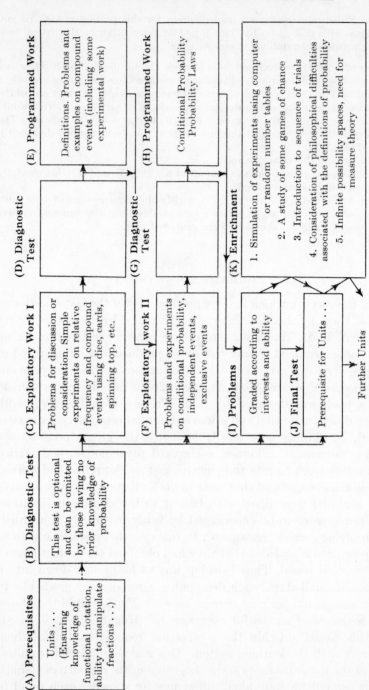

(A) Prerequisites

Units... (Ensuring knowledge of functional notation, ability to manipulate fractions...)

(B) Diagnostic Test

This test is optional and can be omitted by those having no prior knowledge of probability

(C) Exploratory Work I

Problems for discussion or consideration. Simple experiments on relative frequency and compound events using dice, cards, spinning top, etc.

(D) Diagnostic Test

(E) Programmed Work

Definitions. Problems and examples on compound events (including some experimental work)

(F) Exploratory work II

Problems and experiments on conditional probability, independent events, exclusive events

(G) Diagnostic Test

(H) Programmed Work

Conditional Probability Probability Laws

(I) Problems

Graded according to interests and ability

(J) Final Test

Prerequisite for Units...

(K) Enrichment

1. Simulation of experiments using computer or random number tables
2. A study of some games of chance
3. Introduction to sequence of trials
4. Consideration of philosophical difficulties associated with the definitions of probability
5. Infinite possibility spaces, need for measure theory

Further Units

Fig. 2. A flow diagram for a unit on probability.

work his way through a programme at his own rate. Sceptics pointed out that the clever would be bored to tears by a programme consisting of a sequence of steps designed to be easily attainable by weaker pupils; and should a pupil fail at any one stage, then all, or nearly all, was lost.

One attempt to meet these deficiencies was the branching (Crowder) programme which offered alternative routes and saw the 'response', not as something which was to be arranged to be correct, but as supplying information which could be used to determine whether the next point in the programme was to be presented or whether additional material was required. The increased availability of the computer and its power to handle complicated branching programmes made Crowder-type programmes considerably more attractive. However, programmed learning failed initially to achieve any noticeable breakthrough; and this failure could be blamed partly on the energy expended on rivalry between the two factions, led by Skinner and Crowder. More recently, attempts have been made to free programmed learning from its self-imposed straitjacket. For example, programmed systems have been designed using media other than the printed word, group learning (as opposed to each individual being isolated in his own study booth), discussion, practical work, and both linear and branching programmes. The choice depends upon which technique seems best suited for the task in hand.

Exercises

1 Work through a programmed text and criticise it.
2 Construct a programmed unit (either linear or branching) on a topic of your choice.

3. BOOKS

It is not surprising that early programmers should concentrate upon the written word as their medium of communication, for the textbook has for centuries been the teacher's most valuable, and sometimes only, aid. Some of the first textbooks, written in the fifteenth century, were manuals in the art of courtly behaviour – odd predecessors of modern books on etiquette. Later there followed textbooks for the teaching of Latin and Grammar.

The first great writer of mathematical textbooks in English was Robert Recorde (1510?–58), a Welshman of wide interests whose career encompassed a Fellowship of All Souls and death in a debtors' prison. Recorde's picturesquely named books, *The Grounde of Artes* (Arithmetic) 1540, *The Pathway of Knowledge* (Geometry) 1551, *The Whetstone of Whitte* (Algebra) 1557, *The Urinal of Physick* (Medicine) 1547, had a great influence, and indeed introduced to England such symbols as =, + and −. The success of these books naturally gave rise to competition, and by 1656 another author, Thomas Willsford, was complaining that 'The Stationers' Shops seem oppressed with Arithmetics'. This did not, however, prevent him from publishing his own − just as is done today.

Recorde's books were cast in the form of a dialogue between teacher and pupil, a form which has found favour with many authors from Plato onwards. For example, the most popular arithmetic text in America during the eighteenth century − Thomas Dilworth's *The Schoolmaster's Assistant*, a reprint of an English text − was cast in 'a *Catechetical Form* . . .; for *Children* can better judge the Force of an *Answer*, than follow *Reason* through a Chain of *Consequences*. Hence also it proves a very good *examining Book*; for at any time in what Place soever the *Scholar* appears to be defective, he can immediately be put back to that *Place* again.'

Thus the opening lines of Dilworth's book are:

Of Arithmetic in general

Q. What *is* Arithmetic ?
A. *Arithmetic* is the Art or Science of computing by Numbers, either whole or in Fractions.
Q. *What is* Number ?
A. *Number* is one or more Quantities, answering to the Question *How many* ?
Q. *What is Arithmetic in* Whole Numbers ?
A. Arithmetic *in whole Numbers* or *Integers*, supposes its Numbers to be entire Quantities, and not divided into Parts.

Here we have the beginnings of a linear programme or, if we offer a choice of answers and ask the student to select the correct one, of a branching programme! Like so many other programmes, it is riddled with ambiguities, many of which in this case need a twentieth-century technique to resolve them.

The development of the textbook quickened as more emphasis was placed on the utilisation of knowledge rather than on its acquisition. It must be remembered that the translations of Euclid's *Elements* in use in British schools of the nineteenth century, and those of Legendre's *Eléments de Géometrie* used in the United States, contained few problems or exercises for the student and proved most of the propositions they stated. Thus these books accomplished Legendre's avowed aim 'to accustom the student to great strictness in reasoning' but in so doing failed to develop his powers of reasoning. Books which did provide exercises followed the pattern – state a rule, give an example, and set exercises on the rule. Authors of textbooks were happy to follow this pattern for many years. Fortunately, as in all fields and all periods of education, there were those who were ahead of their time. One such was A. C. Clairaut, who in 1741 published his *Eléments de Géometrie*. His aims were very different from those of Euclid, Legendre and the writers of the catechetical-form texts: 'I wish to employ my readers constantly in solving problems.' This he did in the hope that they would thereby acquire the spirit of discovery. To achieve this objective he had to transform the traditional approach to geometry, for

to avoid the dryness normally belonging to the study of Geometry some authors have added after every important theorem the practical application, but while they thus show the utility of Geometry they do not facilitate its study; for as every theorem comes before the application the mind does not return to concrete objects until after it has become tired with abstract ideas.

Clairaut's different views on the aims of mathematical education and, in particular, his emphasis on the need for the student to acquire the spirit of discovery, caused him to recast traditional teaching methods and the form of the textbook.

Unfortunately, though, 'problems' and 'exercises' (or 'examples' as the latter were often called) cannot be considered to be synonyms, for while the former represent a challenge, the latter too often mean a test of techniques; and the latter came to predominate in textbooks. 'The English system of cramming a book with examples is simply vicious, and a positive hindrance to mental progress', wrote F. S. Macaulay in 1909 (*Math. Gazette*), for 'a misguided youth may attack hundreds of easy questions, and never a hard one, without discovering his folly'. One wonders what Macaulay would have said to the publisher's representative who complained to the *Mathematical Gazette* in 1948 about a review, printed in that journal, of an algebra textbook published by his firm. He protested that 'K. S. Snell singles out

$$\frac{a^{2b}\,d^{2e}}{b^{2c}} \quad \text{when} \quad a = 5, \text{etc.}$$

as an example of the manipulative exercises of Chapter I: he does not say that this is the 279th example out of 289.'

Books so overloaded with tedious exercises seem to have disappeared. It still remains true, though, that authors are expected to supply not only many problems but also consolidatory exercises. That this expectation was not fulfilled was responsible for many of the criticisms of *Mathematics: A New Approach* by D. E. Mansfield and D. Thompson (1962). This influential book was the forerunner of the modern mathematics texts to be published in Britain during that decade. The need for problems has long been coupled with the influence of public examinations and the consequent drilling for them; and in England its result has been that the only persons who experience the creative activity of making up problems are examiners and authors – who are in least need of the practice.

Perhaps even more significant than the need to provide examples is the apparent onus on curriculum development projects to provide books. It is true that some projects, for example, the University of Illinois Arithmetic Project and the BBC 'Maths Today' series, have been based upon the use of the media of film and television respectively. Nevertheless, they have both had to fall back upon the book to supplement their work. In the case of many primary school projects the book has played its part as a teachers' guide rather than a pupil's text, but nevertheless it has still had a central role.

With all this emphasis on books it is perhaps surprising that university students should find it a new (and difficult) experience to read a mathematics text. The reason for this is that teachers themselves have often had little training in the use of books and that there is a long tradition in British schools of not reading the text, but simply of using the lists of exercises to illuminate what is stated verbally by the teacher. Clearly pupils need training to read textbooks for themselves, for many reasons. The task is not easy, for the art of mathematical authorship is most demanding, and it must be admitted that few mathematicians seem capable of describing mathematics in a way that can be followed by those who are not themselves professional mathematicians. Attempts have recently been made to cope with these problems at a school level. The initial aim of the SMP, for example, was to provide a set of books which could be read and understood by the pupil. To some extent the Project succeeded in its aim, for the GCE books it produced are readable. Unfortunately, though, the standard of literacy which they demand from pupils is extremely high — this probably reflects the strong public-school, middle-class leanings of the Project in its early days. An attempt to get children to start reading mathematics at school even earlier was made by Minnemast. In 1964 this project was planning to have reading taught in parallel with its kindergarten mathematics courses so that the children could use the mathematics texts to increase their vocabulary.

It is perhaps surprising that the textbook should have continued to be used in traditional form by reformers; for by its very nature – its cost and its comprehensiveness – it militates against changes in the curriculum. Even its appearance, particularly when in hardback form, proclaims authority and permanence. Yet financial and administrative constraints, together with the ease of use of a textbook with its inherent structure, have helped to preserve its supremacy in the face of opposition from other written forms. Of these we may mention work cards (see, for example, Nuffield Mathematics Project – *Problems, Green Set*, 1969), topic books (see, for example, the series *Topics from Mathematics* by Fielker and Mold), or mixed packages similar to the Jackdaw packs used in the teaching of history. And, in response to requests from teachers, the Contemporary School Mathematics Series (St Dunstan's Project) evolved from a topic book form to that of a series of conventional textbooks.

Modern textbooks can differ in some respects from their predecessors – there is the 'scrambled text' used in programmed learning, and the inspired use of colour by Papy in his *Mathématiques modernes* series (but even Papy does not use colour in a manner essentially different from that of Oliver Byrne in his book *Euclid in Colour* published in 1847).† In the main, though, there have not been any breakthroughs in the design of textbooks, although the series *Experiences in Mathematics* (NCTM) does contain examples of textbook writing cast in the form of a running dialogue between two boys – an interesting adaptation of Recorde's method – whilst Mme Lucienne Félix makes use of *three* characters in one of her books; Dessi, who draws and makes deductions, Mati, who does the mathematical thinking, and Logi, who reduces all to abstract ideas.

The respect that is by tradition accorded the written word makes books an interest of the state. Thus, while textbooks are used in every educational system, not every country allows the individual teacher to select texts at will. The requisitions list, on which books had to be placed before they could be used in schools, disappeared in London some ninety years ago, but it still lives on in many systems. Many English teachers who take for granted the freedom to select

† Perhaps more surprisingly, the first English translation of Euclid's *Elements* (H. Billingsley, 1570) contained figures made of paper and so pasted in the book that they could be opened up to make actual models of the space figures. (See Shenton [1].)

their own texts, would be surprised to find that such freedom is far from universal, and that rules such as the following are still to be found.

(1) A teacher shall not use or permit to be used as a textbook ... any book that is not approved by the Minister or the regulations.

(2) Where a teacher uses as a textbook, or negligently or wilfully permits to be used as a textbook by the pupils of his school ... a book that is not approved ..., the Minister, on the report of the inspector of the school, may suspend the teacher and the board that operates the school ...

Exercises

1 Take a 'classical' mathematics textbook such as Hall and Knight, Godfrey and Siddons, or Durell, and a present-day text intended for students of similar age and ability. What differences in educational and mathematical objectives are evident? What new techniques in textbook writing and design are apparent?

2 What should be the aim of 'Exercises' in a mathematics text? Relate your answer to the target population of the text.

3 Give arguments for and against the establishment of a 'requisitions list'.

4 In 1905, Mrs G. C. Young (see p. 15) wrote a book [1] to explain geometry to children via their parents. According to Dame Mary Cartwright [1], 'the examination system killed it'. The book has recently been reprinted. Read it, consider why initially it failed to achieve its objectives, and judge whether it could be remodelled so as to be of use today.

4. RADIO, FILMS AND TV

Although the textbook is still the principal aid to the teacher, other important aids have been developed during the last fifty years. The first of these, radio, has had less effect on the teaching of mathematics than on other subjects. But radio can play a part in a multi-media, mathematical package, as has been demonstrated in Britain by the Open University. Such packages are, however, still novel and their use has not been fully explored. It is possible that a mixture of radio and books could be useful in those developing countries where transistor radios are commonplace and television sets rare; but most mathematicians use their visual sense as a help in mathematical thought, and this severely limits the part which radio can play in mathematical education.

Film is a more appealing mathematical medium, for it can be not only a substitute for the presence of the informed teacher in the classroom, but the provider of a unique mathematical experience. 'If the dynamic nature of films is properly used', wrote Caleb Gattegno [1], 'they may display in front of the pupils the whole family of situations which in a textbook are illustrated by perhaps only one example, and may produce a theorem which expresses the *invariant* property embedded in the dynamic pattern.'

The validity of this statement will be accepted by anyone who has seen the pioneering geometry films made by Nicolet or later films such as T. J. Fletcher's *Simson Line* and *Four-Point Conics*. Such films have much to offer. For use in the classroom, though, they have their drawbacks. Showing films can be a laborious and cumbersome business and the setting up of apparatus, rearrangement of furniture, and possible change of classroom often impose an air of occasion on the class. Such formalities counter the response demanded by the film. The fullscale film is also expensive, and apt to contain too many mathematical ideas for the viewer to absorb. These objections do not apply to the film loop which can be home-made, short (i.e. up to five minutes in duration), viewed informally by groups or individuals, and rerun quickly as many times as the viewer wishes. The use of film loops, whether home-made or commercially produced, has not been widespread in schools, in spite of their demonstrable value. Certainly their use can greatly assist in the teaching of geometry, for example.

Loops, however, must be well produced; otherwise they suffer from the comparisions made by pupils who are used to polished films and television.

Films have important uses in curriculum development in addition to that of enriching the teaching of subjects, because they can demonstrate a new curriculum in action. Thus several projects, for example, the Madison Project and the Illinois Arithmetic Project, have shown films of the 'reformed' classroom as a technique of effecting change. In the notes it prepared for teachers attending its 1968 Institute for Elementary School Teachers, the latter project points out: 'You will be watching the teaching of classes of children of various grades and backgrounds. Watching classes being taught (even when all of them are not entirely successful) will help you learn both the subject matter and the variety of ways it can be pursued in the classroom.'

It is in the primary sector that film has been most effectively used in this way in Britain. Films such as *I Do . . . and I Understand* (Nuffield Mathematics Project) and *Maths Alive* have shown primary school teachers how their mathematics teaching can be organised in a manner totally at variance with traditional routine (and, it must be added, at variance with the methods used in the Illinois Project referred to above). These films demonstrated how the philosophies of Geoffrey Matthews (of Nuffield) and Edith Biggs could be put into practice; and also attracted the attention of uncommitted teachers to the activities of the reformers.

Television introduces a new flexibility to the classroom – a flexibility which will be enormously increased by the advent of cassettes which the teacher can obtain pre-recorded or can record himself.

This is indeed a contribution from technology which may erode the grip on educational progress at present exerted by the teacher. In several developing countries attempts have been made to make television a central component of the educational system. In American Samoa, for example, six open-circuit television channels were used to transmit primary and secondary school lessons to every state school on the islands. Similar, if smaller-scale, schemes have been used in Niger, Mexico and Peru to counter the shortage of schools and teachers. Television has been used in these countries and elsewhere, not only for teaching students but also for the in-service training of teachers.

In Glasgow the introduction of the Scottish Mathematics Group's texts to the city's schools was facilitated by means of lessons given over a closed-circuit television link. Each year's work was repeated twice, and then dropped, on the assumption that by that time teachers would be familiar with the work and with one way in which it could be taught. It was hoped that they would then feel confident to embark on its teaching unaided.

A less rigid use of television was demonstrated by the BBC, which provided a two year secondary school course, *Maths Today*. Although it had television as its central component, this course also used other media such as film strips, work books and teacher's guides.

One major drawback of television and films as educational media is, however, the fact that the individual pupil is unable to adjust the speed of presentation of material to his rate of learning. With a book he can always re-read, in a classroom he can generally question, but the television and film go inexorably on, regardless of whether or not the pupil can follow.

Exercises

1 Give an example of a mathematical topic that could serve as the theme of a film intended to justify Gattegno's claim (p. 94).

2 Design, and if possible make, a film loop to illustrate some aspect of a topic of your choice.

3 Watch a TV programme on mathematics and write a criticism of it, paying especial attention to those occasions on which particular advantage was taken of the medium, and to those where the limitations of the medium were most apparent.

4 'I do not think we have begun to scratch the surface of training in visualisation – whether related to the arts, to science or simply to the pleasures of viewing our environment more richly.'   (J. Bruner)

  Describe materials that a mathematics teacher might provide and the measures he might take to allow learners to cultivate their visual functioning.   (Exeter MA, 1971)

5. INDIVIDUALISED INSTRUCTION

Individualisation of televised work will be possible with the advent of cassettes. In the meantime, a useful substitute – although lacking the visual element – is the cassette tape recorder. L. C. Taylor [1] in his book *Resources for Learning* suggests that the tape recorder is the most valuable piece of equipment in any attempt to individualise learning. Taylor is here speaking of the school curriculum in general and, in fact, mentions that mathematics learning can be individualised by means of printed materials alone. That tape recorders can play a part in school mathematics teaching has been demonstrated, however, by many teachers (see, for example, Banks [1]) who point out that children find that the use of a tape recorder helps to break the monotony of continual reading.

  The possibility of individualised instruction leads naturally to consideration of Computer Assisted Instruction (CAI). In a famous article, Patrick Suppes [1] claimed that 'The computer makes the individualization of instruction easier because it can be programmed to follow each student's history of learning successes and failures and to use his past performance as a basis for selecting the new problems and new concepts to which he should be exposed next.' Here is an example of understatement relative to the claims usually made on behalf of CAI. Certainly, in the development of skills by means of drill and practice, and in the testing of such skills, the computer can be of great (if expensive) assistance. However, in other modes of operation (see, for example, *Computers for Education*, NCET [2]), enormous problems are posed by the programmer's having to anticipate the students' responses and questions. The uses of the computer in a specifically mathematical context are, of course, many and we shall refer to these again in a later chapter. But, in the near future, learning systems built about a computer are unlikely to become common; although experiments like those carried out by Suppes in Stanford and Sylvia Charp in Philadelphia, among others, are likely to proliferate. In Britain such experimentation has been encouraged by the Government's decision in 1972 to set aside substantial funds for research and development in the field of computer assisted learning.

The computer has caught educators unprepared, for CAI demands a knowledge of techniques of programmed learning in advance of those which have so far been developed. Nor is it obvious how the vast powers of the computer can best be used. It is to be hoped that experiments in the design of multi-media instructional packages, such as that proposed in *Continuing Mathematics* (Howson [2]), will fill gaps in our knowledge of the techniques of programmed learning, and in the design of learning systems. Perhaps they will also demonstrate the particular roles in instruction which the computer is best fitted to play.

The potential of educational technology is generally accepted, but there is little evidence, as yet, of its making a significant contribution to education above the 'educational aid' level. Neither can one see many examples of the 'systems approach' in action. At the present time one can point to the multi-media packages of the Open University and the BBC schools programmes, or to the range of programmed materials developed in Sweden by Hermods (the Individualised Maths Instruction (IMU) course for 13–16-year-olds (see, for further details, L. C. Taylor [1])) as examples of educational technology. Yet the former examples are hardly individualised; and while the latter is planned, compiled and tested according to the tenets of the 'systems approach', it is restricted in its use of media (books and tapes). Also, some educators would argue that it is too limited in its educational objectives. (Even more, the IMU course cost £750,000, spread over seven years, to develop (Becher *et al.* [1]), and is reported to be heavily criticised and little-used by teachers who find it too complex to organise.)

Clearly, the implications of educational technology are unlikely to be appreciated, and to be accounted for in educational planning, until a number of such systems are in operational use.

Certainly, the development of learning systems capable of use on an independent basis would have enormous effects on curriculum development. For, besides making education less dependent on the supply of suitably qualified teachers, the range of options and alternative paths through a topic could be greatly extended. Thus 'individualised instruction' would come to mean not only that the student could proceed at his own rate but that he could choose different approaches according to his particular style of learning.

Let us conclude this section with an emphasis: no 'package' is of value unless the material, written to support it, is well thought out, and produced imaginatively. The heavy cash investment associated with the production of 'packages' can lead to over-optimistic sales campaigns, and a resulting diversion of valuable funds to buy things of little value. Like all educational materials and aids, 'packages' should be approached with a healthy scepticism.

Exercises

1 Is it good to allow a student to choose an approach 'suited to his particular style of learning'? Should one of the aims of education be to develop *all* styles of learning in a pupil? Discuss these questions and the problems to which they give rise, and attempt to describe some different learning styles.

2 Criticise some of the claims made by publishers, manufacturers, project directors, etc., on behalf of their mathematical wares.

3 'The principal apparatus of mathematical teaching is a blackboard and a box of chalks. But the chalks should be of many colours . . . Strictly speaking, nothing else is necessary . . .' (Mathematical Association [2])
   To what extent is this comment (made in 1939) still valid?

4 Write an account of *one* of the following mathematical materials: Cuisenaire; Geo-Boards; Circuit Boards. Though your account should not attempt to be a comprehensive survey, you should include some discussion of the following aspects:
   (i) the properties of the material itself: multivalence, imagery, structures, etc.
   (ii) some ways in which the material is used with children and the range of work they respond with.
   (iii) ways in which such work may be seen to generate a syllabus for the mathematical education of some pupils.
   (iv) implications for the ways in which a teacher works with children.
   (Exeter MA, 1969)

SOME FURTHER READING

Atkinson and Wilson [1], Brown *et al.* [1], Glaser [1], Hooper [1], Leith [1], Math. Assn [8], NCTM [3], Nickson [1], OECD [5], Skinner [1]

PART 3
DETERMINANTS OF CHANGE: INTERNAL

Mathematics is a constantly growing organism with a history of over 4000 years. Its growth has become exponentially rapid within the last century or so and it is therefore not surprising that this growth should be reflected in changes in mathematical curricula. Below the university level, there is a social reason for these changes: pupils trained on traditional school syllabuses flounder on the new-style university mathematics that their secondary teachers never knew. With the rapidly increasing numbers flowing into the universities, the high wastage-rate becomes *politically* important and some *rapprochement* between secondary and tertiary level has to be devised.

Mathematics is also being used in many new ways by commerce and industry. Thus it is desirable that many pupils who do not aspire to a university education are introduced to these new applications, for example in connection with computers, before they leave school.

We therefore now look at the changes in mathematics itself, that have led to these social changes. The reader is assumed to know what is meant by such terms as 'linear algebra' etc.; for further detail he should consult Bell's *Development of Mathematics* [3] or Griffiths [1].

9
New Material and New Light on Old Topics

1. THE NINETEENTH-CENTURY LEGACY

The 'traditional' mathematics taught at secondary level usually stops with calculus of an eighteenth-century kind and possibly with a little projective geometry done in the style of an extension to Euclid. This material came into the syllabus around 1900, and reflected the manner of university examinations of the 1850s. It was a reasonable basis on which to build a university course of the type found in 1900, when no pupil would find an upsetting mathematical change in the transition from school to university.

But profound changes in the growth of mathematics had taken place during the nineteenth century, and glimmerings of them began to show in the more forward-looking university courses by 1910. Let us look briefly at these mathematical changes.

Throughout the nineteenth century, commercial arithmetic, algebra, geometry in the style of Euclid, and Newtonian calculus were fairly common knowledge. There were few creative mathematicians, but among these, ideas of great potential were evolving. First, there was mathematical physics, the next extension of theoretical physics after the analytical mechanics of Laplace and Lagrange. In 1828, Green had published his essay on *The application of mathematical analysis to the theories of electricity and magnetism*. This began modern mathematical physics in Britain, and by the 1870s it was in the hands of men like Maxwell, Kelvin, Tait, Rayleigh, Helmholtz, Kirchhoff, Gibbs (in the USA) and, by the 1890s, Poincaré. Much of the physics had later to be revised in the light of relativity and quantum physics, but the associated mathematical ideas are still very much alive. Briefly, the attempts to understand Nature led to descriptions of it that we now call *mathematical models* (see p. 290); and an important feature of those models was the occurrence of differential and integral equations. These, we are still learning to solve. Questions of probability and statistics were around too, because statistical thermodynamics and actuarial

102

mathematics were beginning, with mathematical genetics not far behind, based on the probability theory of the eighteenth century. In order to solve the equations and construct the models, we need pure mathematics. By the end of the nineteenth century there was a rich legacy of finished mathematics, together with some half-emergent ideas that seem to us now of even greater importance. Let us enumerate them:

(A) *The notion of an algorithm*. Gauss, Wentzel, Abel, Galois and others had discussed such questions as the possibility or otherwise of constructing (with ruler and compass) the regular polygons, of trisecting angles and duplicating the cube, and the solution of equations by radicals. The notion of a 'procedure' or 'algorithm' for carrying out the proposed constructions had to be analysed. Taken with the work of Babbage (1792–1871) on his 'analytical engine', together with the later, descendent ideas of such logicians of the 1930s as A. M. Turing, these notions were the right ones for developing the electronic digital computer, which has had such an effect on contemporary life.

(B) *The notion of algebraic structure*. Gauss and Galois, especially, had needed to develop the notions of *field* and *group* to do their work. Although some German universities had lecture courses on groups from the 1850s, Jordan's book *Traité des Substitutions* was the first to publicise the notions widely; it appeared in 1871. Quite independently, Hamilton in 1852 developed his algebra of quaternions to apply to physics, while Boole, in 1854, published *An Investigation of the Laws of Thought* in which he developed the algebra of sets – Boolean algebra. From the complex numbers and the quaternions it was a natural step to investigate 'hypercomplex numbers', and by 1870 enough was known for Benjamin Peirce of Harvard to write his book *Linear Associative Algebras*. Cayley and Sylvester, too, had investigated matrices and their algebraic structure, partly with a view to developing a language for geometry, which we will mention below.

(C) *The notion of function*. The theory of Fourier series had led Dirichlet in the 1830s to formulate a satisfactory notion of function which was more or less the modern one. It was a basic tool for the development of complex variable theory, and above all, for Riemann's theory of functions. Here, one important idea was that of considering, not just special functions like $\sin x$, $\ln x$, $J_0(x)$, $\Gamma(x)$, in the eighteenth-century spirit, but rather the set of *all* (differentiable) functions from one domain of complex numbers to another, and asking for the 'typical', 'generic' structure of the functions of the entire set. This is a broadening of our horizons comparable to asking for a study of Man, rather than 'my neighbour'. Even more, Riemann replaced the domains and ranges of his families of functions by his 'Riemann surfaces': he had taken the first step from the flat plane† of complex numbers, to the 'curved' n-dimensional manifolds which Poincaré later found to be the proper models for studying differential equations.

(D) *The development of geometry.* In the early nineteenth century, projective geometry had flourished with Monge and Poncelet. Grassmann had invented his calculus of linear subspaces of n-dimensional Euclidean space, and algebra had to be developed for describing these rich, new structures. From a philosophical point of view, however, the really liberating event was the discovery of non-Euclidean geometry by Gauss, Bolyai and Lobachewsky: it was at last realised – though Gauss lacked the courage to face the publicity – that Euclidean geometry was but one possible model by which to describe the external world. L. Bers has given an illuminating idea‡ of this philosophical difference by comparing the 'Euclidean' attitude of the Declaration of Independence ('We hold these truths to be self-evident . . .') with the 'non-Euclidean' attitude of Lincoln's Gettysburg Address of 1863 ('. . . a nation dedicated to the proposition that all men are created equal'). Perhaps the climate of the time was ripe for such ideas to generate in different places, but in mathematics, the first to see the implications of this relativistic point of view was Riemann, who seized on Gauss's differential geometry of surfaces to incorporate it into his n-manifolds, with a 'Riemannian metric'. He even hinted that the proper model of force should be curvature, as Einstein later taught us. By 1872, Klein had unified all these geometries within his Erlangen programme, wherein he stressed that the important mathematical feature of any geometry is its group of symmetries – here is the notion of function again. A wholly natural generalisation of Klein's ideas is the contemporary theory of categories and functors,§ and his message has been amply justified: *to study an object, observe how it maps into known objects and how they map into it, and, in particular, how it maps into itself.*

In the closing years of the nineteenth century the study of geometry led to yet another crisis in the philosophy of mathematics.

† Or rather, from the Gaussian sphere of complex numbers consisting of the complex plane together with the 'point at infinity'.
‡ Discussed in Moise [1], p. 383.
§ For an elementary introduction, see Griffiths [1], Ch. 38.

Until that time geometers had believed that geometry dealt with real space. Thus, although Moritz Pasch in his attempt to reorganise Euclidean geometry (1882) had realised that it was impossible to do correctly what Euclid had attempted, namely to provide non-circular definitions of such terms as point, line and plane, he still thought of these in physical terms. Giuseppe Peano (1889) and David Hilbert (1899) (in his *Grundlagen der Geometrie* (*Foundations of Geometry*)) were, however, soon to remove geometry from sensory space. 'Let us consider, three kinds of things . . . called points . . . called lines . . . called planes' began Hilbert, thereby making geometry an abstract formalistic study – pure mathematics rather than physics. Such thoughts about geometry were one part of the motivation for the whole philosophy and logic of mathematics at the turn of the century, and the way was prepared for such aphorisms as Bertrand Russell's [1]: 'Mathematics may be defined as the subject in which we never know what we are talking about nor whether what we are saying is true' and Einstein's [1] 'As far as mathematical theorems refer to reality, they are not sure, and as far as they are sure, they do not refer to reality.'

(E) *The arithmetisation of analysis*. By the mid-nineteenth century, Weierstrass, Heine and others had invented the language of rigorous analysis. The older, vague ideas of 'moving to a limit' and of 'infinitesimal' had made it difficult to give convincing proofs of the rules required for doing the increasingly more complicated calculus. Simple inequalities, the fearsome 'epsilonology' of contemporary undergraduates, removed the vagueness and made acceptable proofs possible. Once the language of analysis was invented, however, it at once showed gaps in our understanding of the number system; and Dedekind and Cantor filled them in. Cantor's theory of sets provided us with a serviceable working language which, nevertheless, was not precise enough to avoid the famous contradictions concerning the Infinite. These led to the work of Peano, Bertrand Russell, A. N. Whitehead, K. Goedel and the later logicians.

2. FURTHER DEVELOPMENTS IN MATHEMATICS

These mathematical ideas were digested and extended in a way that has greatly affected not only mathematics itself but also our contemporary society. First, the improvement in precision of language made it possible (granted some ingenuity also) to prove many suspected results, like the transcendence of e (Hermite, 1873) and π

(Lindemann, 1882), and the Prime Number Theorem (Hadamard and de la Vallée Poussin, 1896); the Dirichlet problem in potential theory could be dealt with properly, whereas Riemann had had to assume it; and such facts as the denumerability of the algebraic numbers and the non-denumerability of the reals could be established. At last, too, the vital notion of *isomorphism* could be expressed objectively.

Secondly, the language, and the multiplicity of structures, led to the *axiomatic method*, which we shall discuss in greater detail in Chapter 16. The way in which Hilbert used this method to remove the blemishes in Euclid has already been mentioned. Its use was also supported by the dictum of E. H. Moore that when two theories possessed similar theorems, it was a mathematician's duty to uncover a common underlying structure (which is rather different from looking for an isomorphism between, say, groups; one looks for an isomorphism between languages). Thus, following the classical success of potential theory – the common structure of several theories of mathematical physics such as electrostatics, ideal fluids, etc. – more *applications* of the common abstract theory could be sought. This is the practical justification of the 'abstract' axiomatic approach. Its greatest exponent has been Nicolas Bourbaki. His enormous and still growing book, begun in the 1930s (see p. 109), has effected a great unification of mathematics by the use of the axiomatic method. As a host of specialities in separate compartments, mathematics was getting too much to comprehend, but Bourbaki unifies it by expressing the different parts in the language of set theory, looking at them all from the point of view, 'Here is a set with a special kind of structure, let us study that structure and its relations with others we already know.' In particular, the axiomatic method helps to separate the details of a complicated construction of a mathematical tool, from the use to which the tool is put. For an example, see the proof of Brouwer's Fixed Point Theorem in Griffiths ([1], p. 441). The method also forces a revision in our attitude to what an acceptable answer to a question should be: Dedekind's answer to the question 'What is a number?' was to formulate a set of axioms that any good *system* of numbers should satisfy, demonstrate the existence of such a system, and leave it at that. And an engineer's emphasis creeps in, because of the question 'Can your axioms and hypotheses be verified in a finite number of steps – *are they computable?*'. Classification theorems arise, to say how many things satisfy a given set of axioms. All this is a far cry from the widely held contemporary misconception that axioms are simply things one varies at will: one concentrates on significant sets of axioms, significant because they describe abstractly some interesting concrete situation.

Thirdly, the language and the structures were available for the creation after 1900 of the branch of geometry called algebraic topology and its many offshoots. Following ideas of Euler, Listing, and Tait (who invented knot theory), Poincaré had created it to deal with problems of celestial mechanics when Riemann's two-dimensional methods became inadequate; but it took half a century of the efforts of many mathematicians, using the language of set theory, logic, analysis and algebra, before it was sufficiently organised and flexible enough for a strong attack to be made on the problems envisaged by Poincaré. The resulting weapon of attack is called 'global analysis': for an example of the influence of this approach in mechanics, see Abraham [1].

3. THE DISSEMINATION OF MATHEMATICAL KNOWLEDGE

At a research level, mathematics is communicated verbally and through the medium of research papers in technical journals. Eventually, the significant ideas are digested and written into books, and these are further simplified, before being included in more elementary books, and so on. This process has been hastened in many cases by writers of popular expository texts.

In mathematical physics, Thomson and Tait's *Treatise on Natural Philosophy* (1879) and Maxwell's *Treatise on Electricity and Magnetism* (1873) had a unifying, enthusiastic view; and these books were followed by a line of semi-popular works by Einstein, Weyl, Eddington, Schrödinger, Jeans and others. The Christmas Lectures at the Royal Institution (going back to Faraday) yielded several good popular monographs, and the Darwinian controversies of the late nineteenth century generated popular expositions that were followed by other works by biologists and geologists. A taste was created that was later to be met by the science fiction of H. G. Wells and his successors.

In mathematics proper, however, perhaps the first significant book of this kind was the two-volume work *Elementary Mathematics from an Advanced Standpoint* (1908) by Felix Klein. It was a record of a course of lectures to Prussian schoolmasters, given by Klein in 1895–6 at a time when it was becoming clear that mathematics was already so vast that there could be few universalists again (in spite of, or rather because of, Klein's contemporary, Poincaré). Klein's knowledge of mathematics was enormous, and he seemed to be able to keep it in his mind in an organised way, through the imposition of general unifying principles such as his Erlangen programme (see p. 104). Consequently, although the subject matter of his lectures encompasses little more than that of a good traditional English sixth form course, he treats it in a coherent way. True, he splits it into the distinct compartments of arithmetic, algebra, analysis and geometry, but he stresses their mutual links. Also he shows how a knowledge of the historical development of a piece of mathematics can be an aid in explaining it clearly. He has nothing to say about the actual teaching of the material, and in fact simply lectures on it in the authoritarian way of a Teutonic professor. One particular foresight of Klein's was his recognition of the Brunsviga calculating machine as an aid for teaching arithmetic (Vol. 1, p. 17). Klein's great moral is that it is possible and essential to have a comprehensive view of mathematics, which helps to break it down ready for teachers to work on it, for inclusion into a secondary curriculum. His approach influenced the Mathematical Association in Britain, especially with regard to calculus.

Hilbert also had this unifying view, and embodied it in the advanced work *Methods of Mathematical Physics* (1931), written with Courant (physics being, as they said, 'too beautiful to leave to physicists'). With Cohn-Vossen, Hilbert wrote a more elementary, and very beautiful, book on geometry, *Geometry and the Imagination* (1932), in which there appeared several topics from 'research mathematics', like the topology of surfaces, and the crystallographic groups. Courant continued this process with the classic *What is Mathematics?* (1941), written with H. Robbins, and there have been several other works of the same sort. A more eccentric book, without the Germanic 'World View', was that of W. W. Rouse Ball (1892) entitled *Mathematical Essays and Recreations*, which brought many combinatorial problems to the public eye, such as the four-colour conjecture, problems in knot-theory, and the Kirkman problem about schoolgirls (see, for example, *SMP Book 2*, p. 282, Question 14). All these works showed diligent teachers that what they taught in their official syllabuses was not all that was known, nor all that was best chosen.

At university level, the rigorous analysis of the late nineteenth century had first been put into textbook form by such eminent mathematicians as Picard and de la Vallée Poussin, each of whom wrote a *Traité d'Analyse*, which expounded the new ideas with great clarity. These were a revelation to British mathematicians of the time, and were eagerly absorbed by some of the Cambridge School, like Russell, Whitehead, Forsyth, Hobson and others. The classic English undergraduate text emerged in 1908, written by G. H. Hardy and entitled *Pure Mathematics*. With its aid, rigorous analysis was taught to British undergraduates by the many enthusiasts for this new kind of mathematics. That work could then be followed by more advanced material in the theory of functions, to be found in classics like Whittaker and Watson's *Modern Analysis* and successive texts on complex variable theory. 'Analysis' became the prestige part of the courses for mathematicians; failure in analysis frequently meant exclusion from any but the most elementary mathematics courses.

4. NEW DIRECTIONS IN UNDERGRADUATE MATHEMATICS

Hardy's book was to have a marked effect on the relationship between schools and universities in Britain, for it ushered in a new type of university mathematics which no longer was a mere extension of school mathematics. It also had the effect of introducing students to a part of mathematics and to a style of mathematical thought which they found far from easy. One result of this was that many graduates returned to schoolteaching not only incapable of preparing their pupils for the hurdles to come, but disillusioned by higher mathematics and intent on perpetuating that school mathematics with which they felt at ease. The rapid expansion of the universities in the 1950s and 1960s only exacerbated this problem.

By this time, analysis was not the sole culprit in increasing the gulf between university mathematics and that of the unchanging school curriculum. Many of the younger university teachers now found analysis entirely natural, in a 'third generation' way: their PhD work had taught them modern algebra, the axiomatic method, and other great ideas of the nineteenth and twentieth centuries. At the very least, this made them want to teach even old-fashioned sylla-buses (where these were still laid down) in a different, more rigorous way. In France, the same urge led several mathematicians around 1938 to combine to write a textbook that would replace the classical, but now out-dated, *Traités*. As they worked, however, they found that only a *thorough-going,* unifying treatment of all mathematics would suffice to explain everything clearly and accurately. Their one 'textbook' has now grown to a shelfful of separate chapters and still continues to grow. The writers use a joint name, that of Nicolas Bourbaki, a somewhat comically incompetent General of the Franco-Prussian war; and some of the older men have retired for younger men to take their place. For a detailed account of this unique mathematical phenomenon, see Halmos [1] and Dieudonné [3].

In various universities, then, the younger teachers read the authors we have mentioned – especially Bourbaki and books like Courant and Robbins – and began to supplement the old material of the syllabus with material they found it more congenial to teach. Sometimes also they were more narrowly specialised than their colleagues of an earlier generation. The result of all this was to make university mathematics more different from secondary school mathematics than ever.

Some idea of the difficulties which followed these changes in mathematics can be seen by considering one particular aspect of the calculus and observing how the implications of its generality have varied at different times and levels.

In traditional calculus, pupils learn to find derivatives by looking at specific expressions like

$$\frac{d}{dx}\,(3 \sin x + 5x^4) = 3 \cos x + 20x^3$$

and then being told 'you do all other sums of functions in the same way'. The climate of Hardy's *Pure Mathematics* would make the rule explicit in the form

'If $f(x)$ and $g(x)$ are differentiable at x, so is $uf(x) + vg(x)$ for any real constants u and v, and its derivative is $uf'(x) + vg'(x)$.'

By the 1950s this had become:

'The set of real functions differentiable at x is a real vector space A, a subspace of the vector space B of all real functions, and the differential operator $D\colon A \to B$ is linear.'

Mathematics gains enormously by our being able to make general explicit formulations of this kind (not least as a preliminary to writing computer proofs), but there is a severe problem in getting pupils to comprehend this generality, even if proofs are omitted. Even the intermediate stage, the Hardy formulation, is found difficult by many university students, particularly engineers.

It was partly to cope with this situation that such projects as the

110

NEW MATERIAL AND NEW LIGHT ON OLD TOPICS

SMP and the SMSG came into being. The reasons why people cared about the situation were, of course, mixed. There was the reason of state, that the output of university mathematics graduates should increase rather than diminish; there was also the enthusiast's reason that the ideas A–E of Section 2 were so important that pupils should be made ready to receive them as soon as possible – and at the secondary stage more time is available for getting pupils over hard conceptual thresholds than there is later (see Quadling [1]).

Other reasons which prompted the reformers sprang from the new and increased uses of mathematics in science, industry and commerce. It is to these applications of mathematics that we now turn.

Exercises

1 '. . . without mathematical infinity, there would be no science at all, because there would be nothing general'.   (H. Poincaré)
   'From time immemorial the infinite has stirred men's emotions more than any other question. Hardly any other idea has stimulated the mind so fruitfully. Yet no other concept needs clarification more than it does.'
                                                                 (D. Hilbert)
   Discuss these and your own thoughts on the matter of infinity, particularly with reference to children.   (Exeter MA, 1971)

2 Discuss the thesis 'Practically every important development in science and mathematics from 1600 to 1900 was connected . . . with differential and integral methods.'   (Margaret Baron)
                                                          (Southampton BSc, 1970)

3 'The whole form of modern mathematical thinking was created by Euler. It is only with the greatest difficulty that one is able to follow the writings of any author preceding Euler, because it was not yet known how to let the formulas speak for themselves. This art Euler was the first one to teach.'
                                                                 (F. Rudio)
   Support this thesis.   (Southampton BSc, 1971)

4 Write essays on the great ideas A–E described in this chapter.

5 Explain briefly the importance for mathematical education of any one of the following books:
   (i)   Hilbert's *Grundlagen der Geometrie* (1899),
   (ii) Hardy's *Pure Mathematics* (1908),
   (iii) Birkhoff and MacLane's *A Survey of Modern Algebra* (1941).

6 'Preferring conceptual to algorithmic approaches is one of the most conspicuous features of what is really modern in modern mathematics.'
                                                              (Freudenthal [2])
   Give examples of the 'conceptual approaches' to which Freudenthal refers.

7 Bourbaki has said that one should not change established notation without grave reasons. Find examples of necessary, and of unnecessary, changes in mathematical notation by authors. Discuss the role played by notation, and

111

find examples from the history of mathematics where the invention of a new notation has cleared the way for progress. (See, for example, Ball [1], MacLane [3].)

SOME FURTHER READING

Bourbaki [1], Dubbey [1], Félix [1], Kline [4], Manheim [1], MAA [3], Newman [1], Newsom [1]

10

New Applications of Mathematics

1. TECHNOLOGY, COMMERCE AND WAR

The enormous growth of technology in the last 100 years, spurred on by wars of the twentieth century, has seen a host of applications of mathematics. These applications have given mathematics a pervasive influence in modern life, thus increasing its educational importance. We can give here only a very brief survey; for further details we refer the reader to such books as Kline [1] and the SMSG course for teachers (M. S. Bell [1]).

Elementary Newtonian mechanics had, of course, allowed nineteenth-century engineers to calculate the stresses in structures and to design machinery. But good intuitive guessing, based on experience, was to many engineers a more significant tool. Apart from astronomy, perhaps the first place where mathematics could probe where 'common sense' could not, was in electrical matters, with Lord Kelvin's analysis (in the late nineteenth century) of transmission lines, in connection with the Atlantic cables. And it was Maxwell's mathematical analysis of electromagnetic theory in the 1850s that suggested the possibility of radio communication, and led Hertz to demonstrate experimentally, *a generation later*, the existence of electromagnetic propagation. From these beginnings grew the electrical industry, which made considerable use of mathematics from the areas of integral transforms, complex variable and conformal mapping, tensors, Boolean algebra and algebraic topology (for the analysis of networks and circuits); and the telephone led to the creation by Shannon and Weaver [1] and others, of the mathematical theory of communication, using ideas of probability and statistics.

Aeroplanes are a twentieth-century invention, and a mathematical theory of aerodynamics is required to help design them reliably; this can involve the use of penetrating tools of mathematical analysis to discuss convergence of series, etc. With very high speeds, any 'common sense' an engineer acquired on low-speed aircraft can be dangerously misleading; mathematics is again needed to probe. Such vehicles as aircraft, ships and missiles need automatic guidance to replace an inadequate human steersman; and the guidance machinery needs for its design the mathematics of non-linear differential equations and the calculus of variations. This type of mathematics seems best attacked by 'qualitative methods', using the language of topology and geometry.

To anyone with an idea of the physical principles involved, it is not surprising that the mathematics we have mentioned should have been needed. It is, after all, basically an extension of the calculus. More esoteric extensions were needed for general relativity theory (tensor calculus, differential geometry), developed from Einstein's work around 1910; and also for quantum mechanics (matrices, linear operators, Hilbert space and functional analysis) – the physics of the atom that led to the atomic energy industry.

Outside physics, experimental workers had become great users of methods of statistical design and inference, especially in biology; and then came World War II. During its time, the science of operational research and the related 'systems analysis' was established, using particularly the techniques of linear algebra and inequalities, as well as new statistical methods. Out of this have grown such branches of mathematics as linear programming and optimisation theory. These methods have also been useful in the control of warehouse stocks and for making various commercial decisions. In the latter field, the basic theory is that of von Neumann and Morgenstern's book *The Theory of Games and Economic Behaviour* (1944) which combined linear algebra, theory of convexity, and even topological fixed-point theorems. The necessity for scheduling vast construction projects, and minimising expensive delays, led to the development in the 1950s of 'critical path analysis', a branch of the theory of linear graphs. Other parts of this graph theory have useful applications to such problems as methods of cheapest transport along road networks. Wartime conditions (in 'hot' and 'cold' periods) have involved a considerable mathematical effort in the making and cracking of codes, and in the related problems of translating from one language to another. These fields are examples of the application, and consequent growth, of 'combinatorial (or 'discrete') mathematics', as distinct from the 'continuous mathematics' of the calculus. The necessity to make important military or commercial decisions has led to the new science of conflict studies, which uses many kinds of mathematics to manufacture models of situations that can be analysed with a view to predicting their equilibrium. As further examples of these trends we may mention the way in which the quantitative aspects of subjects such as geography, economics, etc. are now being investigated. Thus, for example, *Geography* (1969) contains 'A selective bibliography on quantitative methods in geography', where papers are described which use a knowledge of statistics, probability and operations research. Perhaps more surprisingly, we find papers by geographers entitled 'Topology and geography', 'Connectivity of the interstate highway system', 'Network models in geography' and 'A graph theory interpretation of nodal regions'.

Exercises

1 Find out how mathematics is used in some of the fields mentioned above.
2 Write essays on various branches of applied mathematics such as cybernetics, electromagnetic theory, conflict studies, etc.
3 Discuss, in a manner suitable for a first course in mechanics, the main mechanical principles involved in the flight and propulsion of a helicopter, a jet airliner, an artificial satellite and a free space ship. (Bristol BEd, 1970)

2. THE COMPUTER

Many of the applications of mathematics described above would have been impossible without the development of the electronic computer. Until the 1940s, computing equipment was mechanical, not electrical – from the abacus to hand calculators, to the complicated 'differential analyser' of the 1930s (see Crank [1]). But these would solve only the specific type of problem for which they were designed – hand calculators for simple arithmetic, differential analysers for a differential equation – and for each new problem within the type, they had to be reset. Being mechanical they were essentially slow. In the mid-nineteenth century, Babbage had designed an all-purpose computer, his 'Analytical Engine', described with the help of Lady Lovelace, daughter of Byron (see, for example, Dubbey [2]). Unfortunately, since it depended on complicated cogwheel mechanisms, the friction was too great to allow the engine to work. His efforts did raise the standard of high precision metalwork in England (see Kline [1], p. 53) and his theoretical ideas were of great interest once electronic equipment could replace the cogwheels. By the late 1940s, the first electronic 'digital' computers were working, and from then on their technical development has been rapid. Consequently, vast calculations could be performed that would have been impossible before, even though methods were known for doing the calculations 'in principle'. For example, large amounts of statistical information could now be processed by computer, as could the tedious calculations of combinatorial mathematics. Code-cracking and language translation became feasible. Differential equations could now be solved, in the sense that numerical 'tables' of the solutions could be computed. Weather prediction was improved, because the equations could be solved by computer although the speed is still not yet great enough for high accuracy of prediction.

These new possibilities for solving problems in applied mathematics are changing the attitudes of mathematicians to the idea of what a solution is. Before the computer came, the ideal was to 'solve' a differential equation in the form of a 'closed' formula involving familiar functions, or infinite series. Now, it is often more informative to have the computer print the solution in graphical form, or to display it visually to be modified by a light-pen, or even to make a film to show how solutions change with time. This change in the form of a solution leads to changes in the questions asked; now one often deals with 'discrete' mathematics rather than the 'continuous' model of classical mathematical physics. For a fuller discussion see Thwaites [2].

Intentionally, electronic computers are designed to have an 'all purpose' nature. Thus their tasks need not be limited to numerical calculation as if they are mere glorified slide-rules. Hence they can perform such non-numerical work as data processing, and many simple clerical jobs like sending out bills or calculating net wages after tax. Naturally, therefore, computers have become essential tools in commercial organisations; also they can be used for controlling manufacturing processes, for example, to steer the cutter of a machine tool, to control the working of an oil refinery or the switching operations in power distribution, or to work out a 'critical path' through a vast network.

Once again, then, we have an influence within mathematics, which is related to the social activities involving mathematics. The computer is changing mathematicians' attitudes to mathematics, as mentioned above; and its large-scale introduction into industry and commerce creates a high demand for personnel to programme and operate the computer installations. Repercussions therefore follow in the teaching of mathematics throughout the educational system of any country with a high investment in computing. In particular, university courses in cybernetics, numerical analysis, data processing, computer software, computer architecture and computer science (to name but a few) have been designed to train graduates for careers with computer firms or installations. Such courses are changing rapidly with the growth of knowledge about computer applications and design; and they have not as yet been the subject of much analysis by curriculum designers. There have also been approaches to teaching mathematics via computing, as for instance in Dorn and Greenberg [1]. The way in which this forces one to rethink calculus is well explained by G. S. Young [1].

At primary and secondary level, however, the situation is stable enough for it to be feasible to give large numbers of pupils some acquaintance with basic notions about computing and the related mathematical attitudes concerning a logical approach to the planning of work. Flow-charts and iterative procedures are the most obvious aspects that can be treated in an elementary way, while simple programming languages can be taught if access to a computer is possible (this last is an expensive constraint which will surely improve with time). For these reasons, various projects have attempted in different countries to spread computer education.

The objectives and treatment offered by such projects differ considerably. Some projects have concentrated upon the social implications of computer technology. The increasing use of computers has profound sociological effects: patterns of employment, modes of education, opportunities for leisure, man's health and life span are amongst the aspects of society which will be affected. These social implications are too great to be ignored, and projects such as the Computer Concepts Course (Philadelphia) and that organised by the National Computing Centre (England) have sought to ensure that pupils are made aware of them.

Other projects have concentrated more on an attempt to explain the way in which computers work. The depth of treatment has varied considerably. For example, the SMP O-level texts include a description of the various parts of a computer and an introduction to the theory of programming. The St Dunstan's (CSM) Project took such considerations further than the SMP and included questions on these topics in its O-level papers from 1964 onwards. A 'sixth-form' course of this nature is provided by the Engineering Concepts Curriculum Project in its Senior High School Course, *The Man-Made World*.

The SMP and St Dunstan's projects did not specifically require

117

pupils to have 'hands on' experience of computers, although obviously such experience was desirable. It is, however, much easier to gain access to a computer nowadays than it was when these projects originally planned their syllabuses. In the meantime, several projects have included computing as part of their mathematics course and have used the computer to solve problems in mathematics. Thus one project was established by the Massachusetts Department of Education in 1965. This project used remote terminals and had time-sharing facilities – a means of computer access to which, it seems, most schools will soon aspire. In order to prepare for the increased availability of this facility, several projects are now busy attempting to find out how it can best be used, for it is clear that the presence of computer terminals in the classroom could provide us with an opportunity to introduce new mathematical topics and to approach traditional topics in a new way.

The interest in computer education has manifested itself in many reports. As examples, we mention here *Computers and the Schools* (HMSO [18]), a report prepared for the Scottish Education Department, *Computer Science in Secondary Education* (OECD [6]) and *Computers and the Teaching of Numerical Mathematics in the Upper Secondary School* (Math. Assn [3]). In Britain much useful work in this field has been achieved by the Schools Committee of the British Computer Society.

From the point of view of the organisation of curriculum change (see Part 4), it is interesting that in Britain, computer education has also led to the first industry-controlled curriculum development project. The project known as Computer Education in Schools (CES) is run by International Computers Limited (it was originally launched by the Hoskyns Group in 1968).

3. APPLIED MATHEMATICS AND MATHEMATICAL EDUCATION:
AN OPEN RESEARCH PROBLEM

We have already shown how the demand for computer personnel has affected educational programmes. It is not so easy to show a similar influence produced by the enormous increase in applications of mathematics outlined in Section 1, although the demand for applied mathematicians is obvious.† Certainly, the British tradition in mathematical physics, and the arguments of such men as Perry

† There are, however, indications that a different form of training is desirable (see McLone [1]).

(see Chapter 2), were responsible for having the British mathematical curriculum firmly related to classical applied mathematics from the early 1900s. But this has not been true of curricula in many other countries, which have left 'applications' to the physics teacher. Even in Britain, 'applied mathematics' normally means mechanics – a difficult and not-too-popular subject – with a shortage of good teachers of it that leads to an even greater shortage as the next generation of pupils avoids teaching the subject. Consequently, statistics has come into the curriculum within the last decade or so, not so much as a rational choice for its own merits, but simply because teachers who are competent at pure mathematics are more willing to teach statistics than mechanics. Because the tradition in statistics is so short, it is often treated in a less artificial way than mechanics, with more 'real' and appealing applications. But there is considerable dispute among statisticians about what should be taught, and criticism of present syllabuses.

Clearly, it is essential that techniques be devised for producing pupils who will be good at applying mathematics in their chosen professions. This is a completely open problem to which teachers should apply their minds. It is certainly far more difficult than producing reforms in the teaching of pure mathematics, perhaps because we seem to have such a small stock of good problems in applications of mathematics that can be attacked by beginners. For these reasons, perhaps, and the curious conservatism of applied mathematicians as a class, no projects are known to us that are based on the teaching of applied mathematics.

However, several projects have attempted to find simple applications of mathematics which can be introduced to pupils at an early age. We take up the details in Chapter 19.

In the USA and certain other countries, engineering has great prestige, but in Britain this has not been the case, for reasons of nineteenth-century snobbery. Consequently, perhaps, this explains why the US project most concerned with applied mathematics should be called the Engineering Concepts Curriculum Project. One English attempt at improvement is a new and interesting A-level examination called 'Engineering Science' (offered by the JMB) designed by engineers who wish to attract pupils into the study of engineering: they are not mathematicians, and they had to use the word 'science' because of its prestige in English schools. Their approach is well worth the attention of mathematics teachers. Further examples of attempts at this problem, are discussed in Chapter 19.

119

Exercise

'Computer technology is now so complicated that students have no time to worry about general matters like its impact on society.'

Discuss. Is the situation analogous to that of other technologies such as chemical- and weapon-technology?

4. PURE VERSUS APPLIED MATHEMATICS

We have been stressing that the role of pure mathematics – or 'mathematics' as we should prefer to call it – in applications is to provide models of aspects of the world, with techniques for handling the models. This implies that the models are to hand, having been already thought of within mathematics itself. Mathematical history shows many cases where this convenient state of affairs has occurred, but in many other cases, the mathematics has had to be developed in step with the applications. Indeed, almost all the mathematics of the nineteenth century grew from applications: even the logic and rigour were developed for showing that mathematical arguments were reliable tools, and that mathematics contained no implicit errors. But a new intellectual universe was then uncovered for the twentieth century to investigate. Marshall Stone [2], the Chicago mathematician who made significant contributions to this abstract world, could say

When we stop to compare the mathematics of today with mathematics as it was at the close of the nineteenth century, we may well be amazed to note how rapidly our mathematical knowledge has grown in quantity and complexity, but we should also not fail to observe how closely this development has been involved with an emphasis on abstraction and an increasing concern with the perception and analysis of broad mathematical patterns. Indeed, upon closer examination we see that this new orientation, made possible only by the divorce of mathematics from its applications, has been the true source of its tremendous growth during the present century.

His point of view would certainly be echoed by Bourbaki (see p. 109), who insists that the true intuitive understanding of mathematics comes from a knowledge of many mathematical structures, not from physical models which may be too limited in scope. But others have maintained that this attitude is too inward-looking and can lead to degenerate forms of mathematics. They would maintain that most good mathematics can only be inspired by contemplating an application of it; for then the application may suggest the questions to ask and intuitive reasons for their answers. An eminent exponent of this second point of view was Richard Courant, who embodied his beliefs educationally in the way he moulded the growth of the Courant Institute of Mathematical Sciences, New York University. The criticism of Kline (see p. 139) is on similar lines: he was on Courant's staff at the time that Courant had gone to New York from Göttingen, when that world centre of mathematics was broken up in the 1930s as a result of the Nazi persecution of the Jews. Since the line of professors in Göttingen reached back from Courant via Riemann to Gauss, Courant had understandably absorbed a tradition of getting inspiration from Nature; and for years the Institute in New York was the leading American centre for that kind of mathematics – differential equations, fluid dynamics, magneto-hydrodynamics and electromagnetic theory. Few other American mathematicians – as distinct from engineers – worked in these fields because of the strong American predilection for 'abstract' mathematics. Paradoxically, this 'abstract' strand also has a Göttingen influence, because the early American research mathematicians were largely trained there. They were reinforced in the 1930s by an influx of refugees from Germany, to the great benefit of such centres of mathematics as the Institute for Advanced Study in Princeton (which once had Einstein, von Neumann and H. Weyl simultaneously on its staff). The paradox is resolved if we accept the fact that mathematics has always the dual characteristics of being *both* the Queen *and* the handmaiden, of having aesthetic *and* utilitarian attractions – and both aspects are present in the work of Gauss, Riemann and many other great mathematicians.

Some curriculum designers have gone very strongly for one aspect or the other; Courant himself in designing a curriculum for graduate students at his Institute neglected the 'abstract' aspect of their education, although this neglect was often repaired by W. Magnus, another member of the staff – a leading group theorist who ran a seminar in electromagnetic theory! Papy in Belgium neglects applications of mathematics entirely in his curriculum; as did the early SMSG curricula. Such neglect has led to accusations that pupils are being trained solely as future pure mathematics PhDs, and that employment opportunities for these will be severely restricted.

Within the British tradition, it would be uncharacteristic to expect anything but a compromise position to be taken, in view of the successes of British mathematical physicists and the national distrust of general philosophical theories. True, some criticism has been heard, that pupils who have taken the SMP and other modern curricula have 'less technique' than the others. But such criticism has never been other than amateur, frequently with no other requirement for 'technique' than as something traditionally associated with examinations – one teaches engineers the technique of integration, not because engineers request it but because mathematicians have always asked such questions in examinations they imagine to be suitable for engineers. More solid criticism has come from Hammersley [1] who suggests certain problems as tests of technique. He does not appear, however, to have given the same thought to the teaching difficulties as he has to the mathematics.

Ignorance concerning what users of mathematics actually need, as distinct from what academic mathematicians think they may need, is beginning to be combated by the publication of certain surveys. These attempt to find out what mathematics is actually used by graduate engineers and mathematicians in their later employment. Two such are a Swedish survey (Hastad [1]) of engineers and a British survey (McLone [1]) of mathematicians. These indicate that only fairly elementary mathematics is actually used; but in the case of engineers at any rate, it is not surprising that they use the tools they are happiest with and they traditionally tend only to learn elementary mathematics. This view is supported by a survey (Nuffield [1]) carried out at Salford in the 1960s, which showed that US electrical engineers used more mathematics in their published papers than did their British counterparts – presumably because they knew more. However, interpretation of such statements requires care, because of the 'back of the envelope' school of thought, which prides itself on having such a good physical insight that only simple mathematical models are needed. This technique has been fruitful in the hands of such men as G. I. Taylor, but less brilliant men may profit from the help of more powerful mathematical tools.

Thus, there are genuine problems facing curriculum designers when they try to supply a mathematical basis which will serve pupils for a long period of their future lives. The alternative is to depend on adults returning to schools for retraining when necessary, assuming they can learn new concepts as they grow older.

To sum up this section, we wish to emphasise that curriculum designers must be aware of the two aspects, 'pure' and 'applied', of mathematics. There may be reasons for emphasising one aspect rather than the other, but if either is neglected there will be trouble. Neglect the 'pure' aspect, and potential applied mathematicians are excluded from powerful tools; neglect the 'applied', and pupils may be cut off from sources of inspiration, employment and the support of society's taxpayers.

Exercises

1 In many countries (e.g. France) mathematics is taught, divorced from applications. What justifications can you find for doing this (cf. the quotation of Stone on p. 120)?

2 It has been said that 'The sole English contribution to mathematical education has been to over-emphasise its applications.' Discuss.

3 Too much fear of the taxpayer can endanger creativity and critical thinking. How can independence be encouraged?

SOME FURTHER READING

Boehm [1], Holt and Marjoram [1], Newman [1]

PART 4
INSTRUMENTS OF CHANGE

The last Part showed how the continual growth of mathematics, and of its applications, results in pressures for changes within the teaching of mathematics. The previous Part showed that social changes can also serve to initiate developments in mathematics teaching. It is however, one thing to perceive that changes are necessary, and it is quite another to decide what should be done or how best to do it. We now consider ways in which change has been brought about, through individual, and through collective action. This will set the stage for Part 5, where we shall consider the details of some of the mathematics necessary for writing the 'scenarios' underlying some of the changes that we now describe only from without.

11
Individuals and Associations

1. THE INDIVIDUAL INNOVATOR

The most obvious instrument for change in any educational system is the individual. From Plato on, the history of education reflects the part played by individuals, through Comenius, Rousseau, Pestalozzi, Froebel, Dewey, Montessori, down to A. S. Neill and Piaget. Yet, as was mentioned earlier, the effect which an individual produces upon an educational system as a whole is often ephemeral and usually extremely small. Nevertheless, the passion and compassion moving such individuals occurs to a lesser extent in many teachers who will often be moved and influenced to do things once change is put to them.

Consider for example the work of Arthur Hill and his brothers at Hazelwood School, Birmingham, in the 1820s; for a fascinating account see, for instance, Stewart [1], or Lawrence [1]. Their use of discovery methods, the establishment of an elected committee of boys to help govern the working of the school, and the abolition of corporal punishment, fully justify their place in a list of progressive educationalists (though whether confinement in the dark is more acceptable than corporal punishment is a moot point). The curriculum of Hazelwood, too, although more conventional than the organisational side of the school – a characteristic shared by many 'progressive schools' – was wider than usual at that time. It encompassed gymnastics and physical exercise, and anticipated recent trends in the teaching of modern languages by emphasising ability to communicate, rather than a knowledge of grammar. Further, the teaching of geography to the senior class was carried out entirely in French (a practice similar to that obtaining nowadays, in, for example, certain special Russian schools). Also, the curriculum gave particular emphasis to science.

What, however, was the effect of the Hills' work? What influence did it have on contemporary education and on the English educational system in the century that followed?

126

Frequently the innovator working in 'his' experimental school, whether it be the Hill brothers, Mayo, Russell, Neill or Hahn, has been peculiarly free from external constraints. For example, he has been free of parental pressure to produce good examination results, and pressure from educational authorities or governing boards. Cecil Beeby [3] has written that 'The reformer like the golfer, must play the ball from where it lies' and such innovators have often been seen as demonstrating how to make wood shots from the fairway – an exhilarating pastime but not one which the man stuck in the rough can emulate. Moreover, just as in golf one obtains greater length at the cost of accuracy, so in education the greater the reforms the less predictable are their outcomes. Thus the Hills' achievements were not without cost:

By juries and committees, by marks, and by appeals to a sense of honour, discipline was maintained. But this was done, I think, at too great a sacrifice: the thoughtlessness, the spring, the elation of childhood were taken from us; we were premature men ... the school was in truth a moral hotbed, which forced us into a precocious imitation of maturity.
(W. L. Sargent, *Essays of a Birmingham Manufacturer*, quoted in Stewart [1])

A less impressive shot than the well-hit wood is the hack out of the rough, back on to the fairway. Yet this can be extremely valuable to the golfer; and the worth of its educational analogue is shown by the way in which the name of Arnold of Rugby is known to many who know nothing of the Hill brothers (with the exception of Rowland who is remembered for his development of the British postal system, rather than for educational reasons). Arnold's reforms were less revolutionary than those of the Hills, but they possessed the great advantage of being readily applicable and directly transferable to a section of the educational system. Indeed, Arnold must share the credit for the reform of the public schools with others, such as Butler and Kennedy of Shrewsbury. Their influence remains today in the secondary school systems of many countries. Their success reminds us of what still remains the case, namely that developments which can be adopted easily by large sections of the educational system are the ones that are at a premium, and that attempts to make too radical a change can result in a loss of confidence by both teachers and pupils, causing unhappiness to them.

Pioneering work by individuals, therefore, although a necessary condition for educational progress, does not serve to guarantee it. Indeed there are often educational forces barring progress against which a single teacher is relatively powerless, particularly when that

127

force is exerted from a different part of the educational system. For example, the first school textbook on geometry to be published in Britain which departed radically from Euclid's *Elements* was written by J. M. Wilson,† the senior mathematics master at Rugby, and appeared in 1867. The book aroused great interest and was adopted by certain schools; but the great difficulty for a teacher was to reconcile the use of such a book with his job of preparing candidates for external examinations – for these allowed no geometrical proofs other than those of Euclid!

Clearly no single schoolmaster could persuade universities and examination boards to change their ways and permit alternative approaches. It was with this problem in mind that Rawdon Levett, a young schoolmaster from Birmingham, wrote the following letter to *Nature* (26 May 1870) and thereby indirectly produced one of the first potent instruments for curriculum change – the professional subject association:

Euclid as a Textbook

There are many engaged in the work of education in this country, besides those who have come prominently forward in the matter, who feel strongly that geometry as now taught falls far short of being that powerful means of education in the highest sense which it might easily be made. They find themselves, in the majority of cases, compelled to use in their classes a textbook which should long ago have become obsolete.

We have lately had instances in abundance of the power of combined action. If the leaders of the agitation for the reform of our geometrical teaching would organise an Anti-Euclid Association, I feel sure they would meet with considerable and daily-increasing support. We of the rank and file do not feel strong enough to act alone, and yet think we might do something to help forward the good cause by cooperating with others.

The immediate object of such an association should be in my opinion:

(1) to collect and distribute information connected with the subject;
(2) to induce examining bodies to frame their questions in Geometry without reference to any particular textbook.

King Edward's School   RAWDON LEVETT
Birmingham

Exercises

1 What were the 'instances of the power of combined action' mentioned by Levett? Investigate.

2 G. M. Minchin (*Nature* **69**, 1903) wrote how in 1895 he was 'assured by many friends that if (the attempt to get rid of Euclid's order and language)

† For an interesting picture of the good mathematics graduate of his period, see the Presidential Address given by Wilson to the Mathematical Association in 1921 (*Math. Gazette* **10**, p. 239).

were successful the foundations of all logical thought in England would be destroyed'. Were these warnings in any way justified? Investigate and discuss.

2. SUBJECT ASSOCIATIONS

The publication of Levett's letter was the first step in the foundation of the Association for the Improvement of Geometrical Teaching, a body of schoolteachers, fortified with university dons, who sought to produce a new syllabus for geometry and to persuade examination boards to accept it. The Association soon widened its interest. In 1894 it published the first number of the *Mathematical Gazette* and in 1897 changed its name to 'The Mathematical Association', by which time it had fellow subject associations concerned with science and geography, amongst other topics. The early history of the Association is recounted in the special issues of the *Gazette* published in 1948 and 1971. The flavour of its early meetings can also be judged from various reports in the *Gazette*: fierce arguments, formal dinners and *Flatland*† done as a play.

The Mathematical association of America, and the National Council of Teachers of Mathematics, came into being a little later – in 1915 and 1920 respectively. The relatively late establishment of these two bodies is, however, somewhat misleading, for regional associations of mathematics teachers had been at work earlier. For example, the Association of Teachers of Mathematics in the Middle States and Maryland was established in 1903; and its journal, *The Mathematics Teacher*, was taken over in 1921 by the NCTM. Similarly, the *American Mathematical Monthly* pre-dates the MAA.

Since their institution, the subject associations have played a great part in promoting curriculum reform, through their journals, reports on particular aspects of mathematical education, and by providing platforms for the individual innovator. In England the years 1910–40 especially were almost a golden age of mathematics teaching at sixth form level, and it was carried on by mathematics teachers who were influenced by – and often contributed to – the publications of the Mathematical Association.

Subject associations have been created on occasions by a desire to

---

† E. A. Abbot's *Flatland* (1884) is a piece of 'mathematical fiction' set in a two-dimensional world. Flatlanders are plane figures whose shape depends on their social status, for example, 'Professional Men and Gentlemen are Squares or Pentagons . . . The Nobility . . . beginning at Six-sided figures'.

prevent changes taking place. Thus, in an account of the establishment of the Glasgow Mathematical Association (Math. Assn [4]), we read:

In 1927, some changes in the teaching of Geometry were threatened by the Scottish Education Department, such as the dropping of Playfair's Axiom and the proofs of congruence. Euclid, it was felt, was being eroded, and so a group of teachers of Mathematics met along with the Glasgow University Mathematics Department and formed the Euclidean Society, to defend the Euclidean tradition.†

It is unfortunate, however, that the potency of subject associations tends to decrease considerably with age and with increase of membership. This, of course, is understandable. The original members of the AIGT were drawn together by a common wish to reform the teaching of geometry. A teacher nowadays joins the Mathematical Association because it is a representative body of mathematics teachers and its journals discuss things of interest to him. Its very 'responsibility' leads it to number more 'benevolent conservatives' amongst its members than it does 'actionaries'. The classic answer to this problem is for a splinter group of reformers to detach itself from the main body and this is indeed what happened in Britain in the early 1950s. At that time, and led by Caleb Gattegno, a number of educators came together to form the 'Association for Teaching Aids in Mathematics'. As with the AIGT, members soon found the title of the association too limiting, and in 1962 it became 'The Association of Teachers of Mathematics'. The title of its journal *Mathematics Teaching*, instituted in 1955, continued unchanged. Even though still young, the ATM has contributed much to mathematical education, particularly in the development of teaching aids and new approaches in the classroom. It has placed particular emphasis on forging links with other disciplines, such as psychology; and in coordinating the teaching of mathematics with that of other subjects, for example English.

Amongst the early members of the ATM were many teachers from the non-grammar, secondary schools whose interests were not catered for by the Mathematical Association. A few comprehensive schools were being established at that time, and some teachers were also trying to improve mathematics teaching in secondary modern schools. In such schools, the absence of external examination pressures meant

† One cannot believe that the SED literally intended to dispense with Playfair's Axiom (see p. 233). It is of interest, however, to attempt to plan a school geometry course which does not use this axiom or its equivalent. Coxeter [1 Ch. 15] will assist any reader wishing to take up the challenge.

that it was often possible to experiment freely. The ATM was able to attempt, therefore, to satisfy this clearly defined need.

The interests of the two organisations are now beginning to co-incide – as is shown by the large common membership. The coming of comprehensive schools has forced the Mathematical Association to extend its traditional range of interests, witness its new journal *Mathematics in School*. Also, in 1967 the ATM showed its concern with sixth-formers (but not only the cleverest and most academic) by inaugurating a project which aimed to develop work in mathematics in which a major part of the activity was to be the investigation by individual pupils of substantial open problems. By encouraging what might be termed 'research-type activity at the pupil's level' it was hoped to give pupils a deeper understanding of the nature of mathematics and to bring into play many of their own intellectual powers.

Of the several interesting ATM publications, we recommend two in particular to the reader, since so much of the early ATM philosophy – never explicitly stated – can be inferred from them. The books are *Some Lessons in Mathematics* (ATM [1]) and *Notes on Mathematics in Primary Schools* (ATM [2]).

As we stated earlier, a large, representative body is unlikely to move quickly on subjects which are of interest to all its members. However, it still remains true that even large associations can produce far-sighted and *avant-garde* proposals on topics which are to them of minority interest. Thus the most forward-looking report to be issued by the Mathematical Association in recent years was its *Report on the Teaching of Mathematics in Primary Schools* (1955). This was possible because the Association had in the past been almost entirely concerned with secondary education – and that mainly in grammar and public schools – and those few members of the Association who were interested in the primary sector were deeply-committed enthusiasts and not by any means representative delegates of teachers in the primary school.

3. THE PRIMARY SCHOOL

One drawback of the subject association as an agency for curriculum development is that it is usually very badly represented in primary schools. The number of members of the Mathematical Association and ATM combined – even ignoring the large overlap – is still far less than the number of primary schools in England. The teacher in a

131

primary school has to teach many subjects and accordingly is unlikely to belong to associations concerned with a single one. Some
enthusiasts will join, other teachers will attend association branch
meetings when they are specifically concerned with primary education, but, in general, it is difficult for subject associations to establish
close ties with primary school teachers.

In 1970, only about 60 per cent of all entrants to Colleges of
Education in England had obtained an O-level pass in Mathematics
and yet it is from these that the primary schools recruit the bulk of
their staff. Thus the particular difficulties of effecting changes in the
primary schools can be seen. Also, since they employ a majority of
women, primary schools are subject to frequent changes of staff.
Nevertheless, in recent years primary schools have still contrived
to be pace-setters in innovation!

Perhaps a principal reason for such innovation is that the primary
schools have long been influenced by the liberal ideas of Pestalozzi,
Froebel and others. In their classrooms, free activity and expression
have been encouraged as the teaching has become less authoritarian.
The simplest way of teaching, developed in the nineteenth-century
elementary schools, was 'father-centred' teaching, intended for
teaching the Bible, the three Rs and knowledge of one's place in
society. Under the influence of liberal ideas and the growth of
professional and psychological skills, the teaching moved to a more
'mother-centred' style, where the teacher plays more the role of a
mother in a family and may even teach different age-levels simultaneously, using the older children to guide the younger.†

Such an 'open' environment has a great effect on the teaching of
mathematics. Selection at 11 or so came into disfavour in England
during the 1960s, even though it has not yet been abolished in all
areas. In schools where it was abolished, there was no longer a need
to cram arithmetic in order to get pupils into secondary schools.
Consequently, more attractive subject matter could be chosen, and
the manner of learning could use, in the 'open' environment, such
methods as activity and discovery. Most important for scientific
subjects that cannot thrive without rational argument, dissent and
criticism, the primary school classroom had often become a place
where rational argument with a pupil's teacher *and his friends* was
encouraged. No need for quiet mice! Freed from the time-pressure
of examinations, the primary pupil could be involved in mathe-

† This may sound like the method used in the old 'Dame Schools', but is much more
  sophisticated for various reasons, including the education of the teacher.

matical project work, with similar effects in other subjects also. An important contemporary problem is: how to encourage a similar tendency in the secondary schools, whose concern to get pupils through public examinations cannot be lightly set aside.

In the past, the problems of curriculum development in primary schools were either 'solved' by administrative action – notably in England by the 'payments by result' scheme† of the nineteenth century – or were ignored.

Thus, if we consider, for example, the syllabus laid down for Grades 1–4 in the Northwest Territories of Canada in 1902, we find:

Addition, subtraction, multiplication and division of numbers up to 100
Addition, subtraction, multiplication and division of fractions (halves to twenty-fifths)
Use and meaning of fractions to hundredths
Percentage
Use and meaning of measurement units
Reading Roman numerals to C
In Grades 5 and 6 this work was extended to include common fractions, decimal fractions, applications of percentage, and use of weights and measures.
(NCTM [2])

This 1902 syllabus differed in no essential way from that which the authors of this book were taught in England in the 1930s.

True, a Consultative Committee on the Primary School reported (HMSO [15]) in 1931 and headed its mathematical section 'Arithmetic and Simple Geometry'; it began by remarking that 'There is strong agreement . . . that too much time is given to arithmetic in primary schools, and a general regret that too little attention is given to the study of simple geometrical form.' Unfortunately, the section went on for over 300 further lines, yet only nineteen of them were devoted to non-arithmetical topics!

'Mathematics' was, until the 1960s, not a word to be used in primary schools, as a glance at library shelves will reveal.

When the Plowden Report, *Children and their Primary Schools*, was published in 1967, it was still true to say that

Until comparatively recently a typical 'scheme of work' in a primary school could have been summarised somewhat as follows:
Composition and decomposition of 10.
The four rules of addition, subtraction, multiplication and division.
The four rules in money.
Tables of length, weight, capacity and time.

† Details of the working of this almost universally condemned scheme can be found in any text on the history of English education.

The four rules with these.
Vulgar fractions.
Simple problems.

Thankfully, opinions were changing, and the Inspectorate, to its credit, played its part in encouraging primary schools to think of mathematics rather than arithmetic. The Introduction to ATM's *Notes on Mathematics in Primary Schools* (also published in 1967) expressed the view that

The mathematical experiences of a child before the age of 11, and the responses he has been encouraged to make to them, largely determine his potential mathematical development. It is no longer possible to believe that the learning of mathematics properly begins in the secondary school, and that the only essential preparation for this stage is a certain minimum of computational skill in arithmetic.

Why had it taken so long for it to be appreciated that mathematics in the primary school could, and should, mean more than arithmetic? What instruments were to be used to effect changes in the primary school curriculum?

The answer to the first question was that for too long mathematicians had been happy to acquiesce with the grammar school headmaster who replied 'when asked what knowledge he required from the boys who came to him from the Junior Schools. "I should be happy if they were on friendly terms with numbers up to 100" ' (quoted approvingly by A. P. Rollett [1]).

Yet this headmaster did not really seek the little for which he asked. For the standard drill with multiplication tables never enabled the pupils to get on 'friendly terms' with the prime numbers, for example, or to investigate intriguing number patterns. Certainly much of the blame for the stagnation of the primary school syllabus must be placed on mathematicians from other sectors of education who failed to give advice, encouragement and a clear lead.

We have mentioned the interest aroused by the Mathematical Association's Report in 1955. About the same time, various pieces of invaluable apparatus, such as Cuisenaire Rods and Dienes' Multi-Base Arithmetic Blocks (MAB blocks), began to find their way into primary schools. Some of this material had been introduced through the enthusiasm of ATM (recall that its original name was concerned with teaching aids). Z. P. Dienes was a mathematician who became interested in mathematical education, and who had ready access to the schools of Leicestershire which from the mid-1950s have been very experimentally-minded.

Even more importantly, because it demonstrated the commitment of the Ministry of Education, Edith Biggs (an HMI) began to mount short courses throughout the country at which she demonstrated what mathematics could mean in primary schools. Miss Biggs, by her work in Britain, Canada and many other countries, and by her writings (see, for example, *Mathematics in Primary Schools* (HMSO [17]) and Biggs [1, 2]), showed once again the power and the limitations of an individual. For, although an individual can arouse enthusiasm, there must be a great deal of organisation, and the commitment of a group of disciples to ensure that, once aroused, it is not quickly dissipated. That is to say, *reform must be made practical and given an administrative framework*, a point often unpalatable to revolutionaries!

It is always essential, therefore, that work of this kind should be consolidated by some means or other. In the late 1950s innovators in countries throughout the world began developing a new instrument for consolidating change – the 'project'.

Exercises

1 Write essays on the individual innovators mentioned in Section 1.

2 Apparatus has been used in Froebel schools for many years. Find out the reasons for its use, and the ideas underlying the *mathematical* apparatus. What other tactile materials were used in primary schools before Cuisenaire Rods, Dienes material, etc.?

3 The very logical minds of young children often interpret the mathematical commands of a teacher in a way he did not expect, because he did not notice the ambiguity of his language. The child is often correct relative to *his* interpretation, but the teacher marks him wrong. Discuss the likely effects of such marking, and techniques for avoiding undesirable consequences.

4 Why do so many intelligent people dislike what they consider to be mathematics?

5 Traditional texts are often ambiguous or unclear in their explanations. For example, numbers are introduced ostensively – '1, 2, 3, . . . are numbers', '$\frac{1}{3}, \frac{5}{8}, \ldots$ are fractions', but from these special cases the books infer general rules of arithmetic. Also, they may not distinguish between the x's in 'Solve $3 + x = 1 + 2x$' and '$(x + y)^2 = x^2 + y^2 + 2xy$', nor between the types of equality. Take a standard 'classic' and try to explain in the language of mathematics what the author is saying. Now try to explain it to the age-group for which it was intended.

6 Consult issues of the *Mathematical Gazette, American Mathematical Monthly*, etc., published during the first 25 years of this century, and list ideas and suggestions about mathematical education which are to be found in them and which were not to be generally accepted until some 50 years later. Try

to account for the lack of support they received initially and for their long period of neglect.

7 Charles Godfrey (a noted reformer of the first two decades of this century and a member of the mathematical 'G and S' partnership) wrote in 1920 [1], concerning the desirability of producing reports on teaching practices, that 'an Association cannot easily do more than register the average opinion of sound teachers'. Discuss this statement.

8 Is it surprising that the least well-trained should be amongst those most willing to experiment? Is this a necessarily desirable trait?

9 Discuss the changes in classroom organisation that are needed if one accepts that it is mathematically necessary that children be expected to cooperate with friends in a mathematics lesson, and to have mathematical arguments with them.

10 There has recently been criticism from secondary school teachers of the primary school pupils they meet who do not know their tables. Investigate whether this criticism is valid. What was the situation in the days before 'secondary education for all'? Discuss.

SOME FURTHER READING

ATM [2], Biggs [1, 2], Clegg [1], HMSO [15, 17], Hooper [1], Howson [1, 3, 7], Math. Assn. [6], MAA [2], J. M. Wilson [1]

12
Curriculum Development Projects

1. FIRST MOVES

One of the 'classical' projects began in Massachusetts, USA.

> In an effort to improve the teaching of high school physics I want to propose
> an experiment involving the preparation of a large number of moving picture
> shorts . . . complete with text books, problem books, question cards and answer
> cards . . . but before taking up the detailed mechanism it is necessary first to
> look at the subject matter.

This quotation comes from a memorandum written by Professor
Jerrold R. Zacharias, a physicist at the Massachusetts Institute of
Technology, in March 1956. His memorandum is often taken as the
starting point of one of the most remarkable educational phenomena
of recent times.

Even in the brief extract we quote, we note three distinct features
which were later to characterise curriculum development projects:

(1) it was to be an 'experiment' which to Zacharias implied the
need for trials, feedback, adjustments, and the possibility of failure;

(2) emphasis was to be placed on the selection of suitable subject
matter – no black boxes without careful, prior thought about their
contents;

(3) the outcome of the project would be materials – films, books,
etc.

The project to which Zacharias's memorandum gave rise was the
Physical Science Study Committee (PSSC). In 1957 the National
Science Foundation (NSF) granted PSSC the sum of $245,000 for the
purpose of planning and writing materials, and later sponsored
summer institutes at which teachers were trained to use the PSSC
materials. By then, two other characteristics of a project had
emerged: the establishment of a broadly-based group of authors,

and the acceptance of the need for a project not only to prepare materials, but also to train teachers in their use.

Whenever one marks out a particular person or group as having made a breakthrough, it is certain that claims will be advanced on behalf of others. (For example, how much mathematics attributed to later mathematicians was known by Gauss?) It is possible then to claim that the honour for mounting the first curriculum development project should go to the late Max Beberman and his colleagues at Urbana, who in 1952 founded the University of Illinois Committee on School Mathematics (UICSM). Certainly, Beberman's project was not so broadly based, nor as large as the PSSC, but it did have all the characteristics of later projects: it produced materials which were tested in pilot schools and later revised, and it conducted institutes at which teachers were trained to use the materials. Moreover, it refused to let its units be used by teachers who had not been so trained. Nevertheless, the PSSC was the first project to command national interest and to stimulate emulators.†

2. SPUTNIK

One reason for this interest was that it happened to be getting under way just at the time that the satellite Sputnik I was launched (November 1957) by the Russians. This event had an enormous effect on American complacency about their undoubted engineering capacity and its superiority over that of the Russians. In their preoccupation with engineering consumer goods, the Americans had grown contemptuous of clumsier Russian rivals; but now they had to ask themselves whether first things had really come first. Since education produces scientists and technologists, it was argued that funds must be pumped into education, especially to improve its scientific and mathematical quality. The 'hawks' of Congress were quick to produce cash – education for them was now 'Defense' – and educators were naturally quick to express their arguments for education in the hawks' terms, even though their own aims may have been entirely different. Certainly, past history did not encourage them to argue on liberal grounds. Indeed, in the new 'Defense' climate, it was possible to obtain funds for special educational programmes for underprivileged children, a notion that five years earlier would probably have been labelled 'Communist' and un-American, owing to the influence of that infamous Senator, the late Joseph McCarthy.

† See Beberman [1] for a fuller account of the aims of UICSM.

3. SMSG AND OTHERS

A multitude of projects followed in the wake of PSSC. Of these, the biggest and best-known was the School Mathematics Study Group (SMSG), which grew out of two conferences of mathematicians sponsored by the NSF in February 1958, one in Chicago, the other in Boston. As a result of these meetings, Richard Brauer, the President of the American Mathematical Society, appointed a committee of eight mathematicians to implement the conference findings. This committee invited Edward G. Begle (then of Yale University) to direct the work of an organisation – to be called the School Mathematics Study Group – which would be responsible for preparing new mathematics courses and training teachers to work with them. Since its establishment, the SMSG has ranged far and wide over mathematical education – it has produced texts for average and above-average High School pupils (including programmed texts and alternative approaches), for Junior High School pupils (including special texts for slower students who experience difficulty in reading), for elementary school pupils and for culturally disadvantaged children. The project has also produced a great deal of supplementary material, including a series of monographs for teachers and a set of thirty half-hour films intended for the in-service training of elementary school teachers. In 1962, it initiated a five-year study, the reports of which began to be published in 1968; they form the National Longitudinal Study of Mathematical Abilities. Details of the formation and early work of the SMSG can be found in one of its publications *SMSG: The Making of a Curriculum* (1965). Criticism of the work of this and other projects can be found in articles by Wittenberg [2], Morris Kline [2–3] and in one (Ahlfors [1]) which appeared in 1962 in *The Mathematics Teacher* and the *American Mathematical Monthly* over the signature of seventy-five American mathematicians, including such men as Birkhoff, Courant and Coxeter (to proceed no further down the list). Kline's criticisms concerned the choice of subject matter for the new courses; indeed he saw the main problem of the teaching of mathematics as being not the choice of new mathematical topics, but the *way* in which mathematics was taught. For he was reasonably happy with the old topics, and the materials prepared by SMSG appeared to him to be more likely to result in a deterioration of mathematics teaching than in an improvement. The seventy-five mathematicians were concerned more about the way that reform was being dominated by university mathematicians, who appeared more eager to train a new generation of professional mathematicians than to provide a mathematics course suited for the average child. They deplored the premature abstraction of the SMSG material and the absence of links with science.

Such criticisms of modern mathematics courses were later to be heard in other countries.

4. EUROPEAN PROJECTS

About the same time that the SMSG was established, the first reforms in Europe were beginning. In Belgium, Lenger and Servais initiated a programme concerned with the mathematical education of young children; and 1959 saw the first of the Arlon seminars which were to produce a rich harvest of surveys and reports of experiments in teaching such topics as topology, analysis and vector spaces (see Papy [1]). The Scandinavian countries, too, began to plan new school programmes.

The Arlon seminar of 1959 was quickly followed by a seminar of even more importance: one held at Royaumont in November 1959 under the aegis of the Organisation for European Economic Cooperation, which was attended by mathematicians and educators from eighteen countries. The report of this seminar, *New Thinking in School Mathematics*, had a great effect in many countries, as did its sequel, *Synopses for Modern Secondary School Mathematics*, which was written the following year by a group of international experts meeting under the sponsorship of OEEC.

The effect was not always what the authors had intended; for both reports left themselves wide open to the criticisms which the US mathematicians were to make of SMSG in 1962. Indeed in writing of these reports in the *Mathematical Gazette* (1962) Professor R. L. Goodstein said:

It may well be that school courses are too narrow and that we should replace the depth we seek in the form of hard examples by a spread of topics, but many mathematicians hold that the heart of Mathematics lies in solving problems, not in learning concepts ... When we make changes in school syllabuses we must make them to add to the interest in the school course, and to deepen understanding, and not simply to advance the starting-point of university courses.

5. BRITISH CURRICULUM DEVELOPMENT

Britain lagged in the race to start projects; and the events leading to the creation of the various British projects, which began in the 1960s, followed a different pattern from those in other countries.

In England, the educational system is decentralised as far as matters of the curriculum are concerned, and each school has a considerable degree of freedom. This is not the case, for example, in Scotland or in many other countries. Moreover, throughout Britain 'Mathematics' in the twentieth century has always included 'Applied Mathematics', and it has always been the aim of schools to stress the applications of the subject. The secondary schoolteacher, too, in Britain and particularly in Scotland, has usually been well qualified. Also, partly as a result of the depression in the 1930s and the fact that the higher-educational system had not expanded as fast as that of North America, there were many academically outstanding teachers in the schools. One result of these characteristics was that the conferences which, as in the United States, led to the inauguration of projects, differed in constitution from those in Chicago and Boston. Thus the projects themselves were organised on different lines from the SMSG. The first of the conferences was held in Oxford in 1957 and was organised by J. M. Hammersley, a statistician. Its membership was drawn from schools, universities and industry; and it was financed by industry. Great attention was paid to modern industrial applications of mathematics. A participant noted later that the conference 'certainly emphasised the fact that "modern" mathematics does not consist entirely of abstract algebra' and that no mention was made at that time of the Illinois project since it had 'little relevance to conditions here' (Rollett [2]). A second conference took place in Liverpool in 1959 and a third in Southampton in 1961. The Southampton conference differed from the others in that it followed Royaumont and the establishment of projects in the USA and Europe. Indeed, Professor Bryan Thwaites, who organised this conference, had invited Professor Begle and Henry Swain of the SMSG, and Professor Papy of Brussels, to participate in its work. The record of the conference, *On Teaching Mathematics* (Thwaites [1]), still demonstrates, however, the enormous emphasis placed on the utility of mathematics – an objective of mathematical education which appeared to be given little attention by the projects directed by Begle and Papy.

The chief result of this conference was the establishment of the School Mathematics Project (SMP), which was to become the leading mathematics project in Britain. The significant features about the institution of the SMP were:

(a) responsibility for drawing up new syllabuses and preparing the necessary textbooks was in the hands of a group of teachers;

(b) finance was provided by industry and not by public funds;

(c) the schoolteachers involved remained in their schools, so that they could try out the new materials as they were written (writing did not take place at summer workshops);

(d) the project did not receive official recognition and support from the Department of Education and Science, but worked within the established bounds of freedom which allowed teachers to design their own curricula and to have examinations set on them.

The SMP was centred upon a university (first at Southampton, later in London) and received assistance and advice from many university teachers; but its activities were directed, and its texts written, by practising schoolteachers. In this as in many other ways it differed from SMSG and other American counterparts. It shared its 'teacher-domination', however, with the majority of other projects in England and Wales. (Indeed, those projects in Britain which were 'university-dominated' showed a remarkable inability to get off the ground.) These tended to differ from the SMP not so much in outlook but in scale. No public funds were available at that time for curriculum development; and other projects were able to attract only very limited support from industry and foundations, or did not choose to seek it. Thus shortage of money, and the resultant overworking of personnel, greatly hampered the work of the Midlands Mathematical Experiment; while the Contemporary School Mathematics Project (St Dunstan's) restricted its activities to the provision of texts and suitable examinations, and did not attempt to organise in-service training for teachers who might wish to use its materials.

If we summarise by saying that at that time curriculum development in the US was university-dominated, and in England teacher-dominated, then in Scotland (as later in many developing countries) it was 'inspector-dominated'. The Scottish educational system has always been more tightly controlled than the English and the inspectors have retained greater power and influence (cf. p. 65). Thus in Scotland the movement to reform curricula was led by HMIs. The result of this involvement was that changes were effected throughout the educational system rather than within pockets of the system.

Each type of 'domination' has its advantages and disadvantages. Criticisms of the 'university-dominated' projects have already been made – their output has a tendency to be orientated towards the future mathematician and to be unteachable. 'Teacher-dominated' projects automatically prepare more practicable materials, although a striving for academic respectability often results in their setting themselves over-ambitious targets; but teachers are apt to lack the overall view of the subject which a university don possesses, and are unlikely to be as aware of modern developments. Thus undue emphasis is sometimes placed on interesting but relatively unimportant facets of the subject. The great advantage of the 'teacher-dominated' project, though, is the fact that the person involved is seen by fellow teachers as 'one of us' rather than as someone from above trying to impose his views on those below. This latter has sometimes proved a drawback of the 'inspector-dominated' project. Another failing of such centralised projects is that reforms cannot be so sweeping if they are to be absorbed simultaneously by the whole system. (That this latter point can also work in favour of a project is demonstrated by the widespread, international use of the texts of the Scottish Mathematics Group.)

6. PROJECTS AND TEACHER TRAINING

The arguments for 'rolling reform' or 'continuous reform' are many, and it is obvious that curricula cannot be allowed to stagnate. However, the effort involved in making any change is so great and the immediate effects are often so unsettling, that a large number of small steps is not necessarily preferable to a smaller number of large ones. Reforms initiated by Ministries do, moreover, commit those Ministries involved to an extensive in-service training programme. Merely changing the textbooks and the syllabus is insufficient to bring about curriculum development. As Dr Beeby [1] has remarked, 'the average teacher has a very great capacity for going on doing the same thing under a different name'. The toughest part of any development work is the in-service training stage. In a system where an individual school adopts a project's work when its teachers wish to do so, and when they feel equipped to follow the new curriculum, then in-service training becomes the responsibility of the individual rather than that of the central authority.

It can be argued that the teacher is allowed too much freedom if he is permitted to keep on following a curriculum generally recognised to be outdated. These problems of freedom and the teacher are, however, extremely complicated. They have been discussed at length on occasion, but much more remains to be said, especially concerning those degrees of freedom which a teacher possesses in actuality and those which he can be safely allowed.

Exercises

1 Compare and contrast the amount of freedom allowed to individual schools and teachers in different educational systems.
2 What is 'reasonable' freedom for
   (i) a department of mathematics,
   (ii) an individual teacher within that department?
3 Discuss the role of the head of a school mathematics department.
4 Discuss the effects of putting pressure on a good teacher of 'traditional' mathematics to switch to new material which may be uncongenial to him.

7. THE SCHOOLS COUNCIL

The projects we have described in this chapter have in the main been concerned with mathematics. As such, they have sought to ensure that the time allocated in school to the teaching of mathematics is used to best effect. We mentioned in earlier chapters, however, that this is not sufficient – it is necessary constantly to re-examine the place of mathematics in the curriculum as a whole. Now this is not a task that can be undertaken by a subject association or by most projects. It is necessary to have some agency which can oversee the development of the curriculum in its entirety. This need led in 1964 to the establishment in England of the Schools Council for the Curriculum and Examinations, and in Scotland to that of the Consultative Committee on Curriculum.

The precursor of the Schools Council was a Curriculum Study Group, consisting of HMIs, administrators and co-opted experts, which was set up within the Ministry of Education in 1962. This was seen by some as a possible threat to the freedom of the teacher and the local education authorities; the upshot was that after the situation had been investigated by a working party under the chairmanship of Sir John Lockwood, it was decided to establish a Council which would be a coalition. The involved parties would include: all the associations of local government, the universities, teachers and the Department of Education and Science. The Council's duty was to seek

to uphold and interpret the principle that each school should have the fullest possible measure of responsibility for its own work, with its own curriculum and teaching methods based on the needs of its own pupils and evolved by its own staff, and ... through cooperative study of common problems, to assist

144

all who have individual or joint responsibilities for, or in connection with, the schools' curricula and examinations to co-ordinate their actions in harmony with this principle. (HMSO [9])

The Council is financed from central and local government funds. Among its duties it tries to ensure that teachers and others are informed about development work in progress. This it does by means of its publications, including a newsletter *Dialogue* distributed to all schools. Further, it finances and supports a large number of curriculum development projects, for example, that on mathematics for the less able (see p. 60); and it mounts a wide-ranging programme of research projects in different subjects which offer a promise of help to those wishing to devise new curricula. The Council has also interested itself in the improvement of examination methods and techniques, and in the reform of the external examination system. It has met least success in its efforts to reform sixth-form education and, in particular, to fight over-specialisation.

There is no obvious American counterpart to the Schools Council, but one organisation that merits special mention is the Education Development Center (EDC) of Boston. This was established in 1958 – as Educational Services Incorporated (ESI) – to handle the physics course developed by PSSC. Since that time it has become widely involved in many curriculum development projects. One of these is the University of Illinois Arithmetic Project (see p. 95); others include Mathematics Curriculum Study – responsible for the Cambridge Conferences, see CCSM [1, 2] – and the African Education Program which includes the widely known Entebbe Mathematics Project. Unlike the Schools Council, EDC received money from both public and private sources, and is a private though non-profit-making organisation.

8. SOME GENERAL FEATURES OF PROJECTS

In the last fifteen years the curriculum development project has emerged as a powerful instrument for effecting change. It is valuable therefore, to consider projects that have developed in different parts of the world and to see what characteristics they share. Considering the diversity of educational practice, it is perhaps surprising that they should have so many common features.

The principal feature of most projects is that they possess a clearly defined objective which is thought to be capable of attainment within a specified time. Thus a project might aim to develop materials for culturally disadvantaged children within a certain age range, and to train teachers in their use. Such a project might have a life span of, say, four years. A project aiming to devise a five-year course to O-level would probably require about seven years to carry out its task.

Projects cost money. If the materials are to be written and tested with a thoroughness that textbooks written by individual authors never are, and if in-service training is to be provided, then quite a large capital outlay is required. It may be that an outstandingly successful project will recoup the initial outlay from royalties. However, it would be wrong to expect all projects to do this, and if one were to attempt to budget curriculum development on the assumption that all development costs would be recovered, then minority interests would be neglected. The availability of money is no guarantee of success – as is shown by examples which we shall leave anonymous. But, as we remarked earlier, a shortage of money can result in excessive strain on the project team and will detract from the project's work.

The money to mount a project can come from either private or public funds. Many of the early projects in Britain were financed either by grants from industry or by the Nuffield Foundation, a charitable body established by Lord Nuffield, the founder of Morris Motors. Later projects were financed by the Schools Council from public funds. In the USA, too, projects received money from a variety of sources. For example, the Madison Project was given financial assistance by the National Science Foundation, the Marcel Holzer and Alfred P. Sloan Foundations, and a group of industries and trade unions.

Irrespective of the source of money – whether private or public – it is essential that the initial appeal for assistance should be accompanied by some plan of action outlining the aims of the project and the methods it would use to attain them. The writing of such a plan is now frequently treated as a separate little project – the prelude to the main work – and is termed a 'feasibility study'. Such a study is generally carried out by one or two persons who will spend up to a year visiting schools and other educational establishments. There they collect opinions on suitable objectives for the project and on appropriate methods, and, in particular, they look out for existing work which can be taken as a starting point for the work of the project. For, if the project can build on existing work, it is more easily seen to be a practical proposition and not just a theoretical exercise. The study will probably conclude with proposals for staffing and housing the project, together with estimates of the costs involved.

(Published examples of feasibility studies in the field of mathematical education are Schools Council [3] and Howson [2].)

Once money becomes available, the next and probably crucial decision is the appointment of the Project Director. It is significant that many projects are frequently described by reference to their director's name, for his is a vital role. He will not only have to be competent academically but will also have to recruit and guide his staff, manage the business and financial side of the project, and deal with publishers, press and the educational establishment. All evidence shows that the appointment of a first-class director is an essential step in the organisation of a project – alone it does not guarantee a project's success, but one knows of no project which has succeeded with a dud director.

The first task of the project director is to gather together his 'team'. Here, as we have mentioned earlier, there has been some divergence between British and American ways. Thus the teams of the British projects consist almost entirely of teachers, or lecturers from Colleges of Education, with schoolteaching experience; whereas American projects have tended to involve a much greater proportion of university teachers. Thus, selecting a representative text by each of several projects at random from the bookshelf and classifying named contributors by the part of the educational system in which they were then active (year of publication is quoted) we have:

	Schools	College of Education	University
Scottish Mathematics Group (Scotland) 1967	14	3	–
School Mathematics Project (England) 1967	19	3	1
Joint Schools Project (Ghana–British influenced) 1967	6	–	–
Entebbe Project (Africa–American influenced) 1965	4	–	10
School Mathematics Study Group (USA) 1960	3	–	3
Secondary School Mathematics Curriculum Improvement Study (USA) 1968	3	6	15

We have earlier stated arguments for and against 'teacher-' or 'university-' dominated projects. Here it is more relevant to note the number of people involved and to note that the writing of project materials is a cooperative effort. It should be added that in order to involve university, industrial and other interests, many projects now have consultative committees on which representatives

147

of these bodies serve. The SSMCIS university total is, one suspects, swelled by the inclusion of people who in Britain would have been thought of as consultants rather than contributors.

The result of having such large writing teams is that draft chapters are seen by many and are therefore unlikely to contain major mathematical mistakes. In certain cases the size of the team is reflected in the wealth of problems and examples which the materials contain – a range which could not be equalled by an individual author. Disadvantages are that frequently there is little consistency in writing style and level of presentation.

Writing teams work in different ways, from project to project. Some, like SMP and SMG, use authors who remain in their schools and divide their day between writing and teaching, with the greater part of their time devoted to the latter. Other projects, such as the Nuffield Mathematics Project, had a central team of writers seconded from their usual employment who, as part of their work, visited project schools, and taught in them. Some projects, for instance the Entebbe Mathematics Project, have prepared their materials at writing sessions held for six weeks or so in vacations. The authors in these instances have sometimes had only very limited opportunities to test their materials in schools.

All projects recognise that they are unlikely to construct a successful, revolutionary set of materials at their first attempt. For this reason, projects usually prepare draft materials which are then tested in selected 'pilot' schools. The number of pilot schools varies, but experience indicates that this number should not be more than thirty or so, owing to the need to keep a close watch on all the schools involved and to collect detailed criticisms from all the teachers. After this stage, when the rough corners have been smoothed off the draft materials, a second draft can be made available to a wider range of schools – wider both in a geographical sense and in their placing in the mathematical spectrum. For now, immediate communication is not so important; and during the first stage, the hazards of preparing an initial draft are so great that one is well advised to restrict experiments to schools with good staffs and in which the conditions are conducive to success.

Once the materials have been tested on this wider front, final versions can be prepared.†

---

† A flow-chart describing the steps in the preparation of one of the Comprehensive School Mathematics Program's packages (CSMP) can be found in Kaufman and Steiner [1].

Commitments to in-service training will grow with the number of schools involved in the project. The project team will have the task of training the teachers from the pilot schools, but whilst doing this it must also select teachers from these schools who, a year later, can train teachers from the 'second-wave' schools. In this way the number of potential course-leaders and tutors will grow with the project.

Various kinds of in-service training have been used. Teacher's guides have been used by many projects, occasionally supplemented by films. Television – both open and closed circuit – has also been utilised. Another frequently used method is the annual course to which teachers come to get their 'batteries charged' for the year ahead. This method, by itself, is rarely more successful than the annual charging of car batteries would be. It is necessary that a teacher should have more frequent contact with project team members or with other teachers who are using the project's materials. To ensure this, the Nuffield Mathematics Project established 'teachers' centres' in all the areas in which it had pilot schools, to which teachers could come for training and reassurance. These centres have grown and multiplied, and in so doing have cast off their original Nuffield labels and become general 'curriculum development centres' concerned with all aspects of the curriculum and with both primary and secondary education. They are supported financially by the local education authority.

Similar centres, often backed with collections of audio-visual resources, or even animals, which teachers can borrow, are to be found in many countries.

In many countries too, projects need to ensure that suitable external examinations are provided for pupils using their materials. In addition, it is necessary that such pupils are not unfairly penalised (or rewarded) for having followed the project's course. This will necessitate the project's cooperating with the examination boards concerned and with other sectors of education (for example, the universities) and with professional institutions whose entry requirements are likely to be affected.

The earlier individual innovators could only assess subjectively whether or not their teaching was 'better' for the pupils; also only by self-introspection could they usually have insight into pupil psychology, since they had no resources to do research on pupils. With greater resources, a good project will not only be concerned with assessing pupils, but also with examining and assessing *itself* – was all the effort worthwhile? Did it result in an improved curriculum,

149

or not? It is a common experience – known as the Hawthorne effect – that the enthusiasm of those conducting an experiment, and the increased sense of involvement of those taking part in one, will lead to an initial improvement on what was done before. Thus such effects must be allowed for when assessing a project; would the average teacher produce such an effect once the approach becomes standard? Projects are established to attain certain objectives; it is, therefore, reasonable to ask, once the project's life has run its course, whether or not those objectives have been achieved. We have already mentioned the testing and rewriting of materials – what is referred to as 'on-going evaluation'. There is also a need for 'final evaluation' – the checking of whether or not a project has achieved its aims; and for 'comparative evaluation' – the comparison of the effectiveness and desirability of alternative approaches.

The degree of difficulty of carrying out final evaluation depends upon the way in which the project's aims were initially stated. If a project were established to teach pupils to solve quadratic equations, then it would be easy to carry out an evaluation of its success. Most projects, however, have somewhat woollier aims: 'The principal purpose of the project is to contribute to the improvement of the teaching of mathematics' (University of Maryland Project), 'The application of on-going psychological research to the learning and teaching of mathematics' (Psychology and Mathematics Project), 'To give the children an awareness of the basic structure of mathematics and also to find and apply mathematics in the world in which the children live' (Contemporary School Mathematics).† A final evaluation of the last-named project would present many problems to the investigators, whereas the first two are so vague as to make talk of final evaluation meaningless – more detailed objectives would have to be sought. The second project would, however, lend itself to comparative evaluation. Does 'the application of on-going psychological research' lead to better learning and teaching of mathematics and, if so, in what ways?

Evaluation of mathematics projects has not engaged a great deal of interest in Britain, where it has mainly degenerated into a form of 'voting with one's feet'; it being interpreted that those projects with large followings have been successful, whereas those which attracted

† These quotations are taken from the *Sixth Report of the International Clearinghouse on Science and Mathematics Curricular Developments, 1968*, published by the University of Maryland. These annual reports contain a great deal of information about mathematics projects throughout the world.

few disciples, and failed to hold on to those which they had, were failures. This is a cheap method of evaluation and would certainly satisfy the cynic who believes that the principal aim of all projects is to attract customers for their wares. It is obvious, though, that if curriculum development is to proceed on healthy lines, then improved tools for evaluation will be needed and projects will have to take final and comparative evaluation more seriously.

The most unhealthy form of project can be that which is operated by some Governments. The 'reform' is laid down by fiat of an Education Minister, who then has reasons of prestige for not wishing to have an unfavourable evaluation. Like all such authoritarian modes of reform (however well-meaning), the technique is self-defeating. Rigorous criticism from all involved is necessary to avoid waste – most importantly of pupils' time – and the possibility of teachers seeming to teach new material while in effect they garble it. Mathematics and science, especially, depend on rational conviction and reasoned argument for their growth. Any reforms in their teaching must respect this fundamental part of their nature.

9. 'THE GOOD IS OFT INTERRED WITH THEIR BONES'

In Section 8 we stressed how projects are usually conceived with a specific life span. Thus in 1972, for example, both the SMSG and the Nuffield Mathematics Project concluded their official work. Occasionally a project's life has been lengthened to enable materials to be completed or, as in the case of the Mathematics for the Majority Project, a 'continuation' project has been set up to fill some of the gaps left – or revealed – by the original one. Clearly, public money cannot be promised indefinitely to any one project. Equally obviously, the preparation of materials and their adoption by a number of schools (be it large or small) marks only the first phase of an educational advance. Only constant use will reveal the deficiencies of materials and indicate just what training is required by teachers before they can use them with insight. Such constant use is also likely to have a stultifying effect; for example, the books of Godfrey and Siddons were revolutionary in their time, but they became a hindrance to progress. The spirit in which the books were written was forgotten – only the solutions to outmoded problems remained.

It is important, then, either that steps should be taken to keep a project in being so that its materials can be constantly revised and teachers helped to use them with care and imagination, or that project materials are withdrawn from publication after a number of years. Only if these steps are taken will the dangers of adherence to outdated dogma be avoided.

In any case, an important objective of a good project will be to educate teachers (as well as pupils) and bring them nearer to Stage 4 of Beeby's model. At that stage, hopefully, local groups of teachers, self- rather than project-disciplined, will have formed. Like Nelson's captains, the teachers will, by cooperation and readiness, be thoroughly familiar with 'each other's strengths and weaknesses; and with the standards expected of them as mathematicians and as teachers. If they persist in retaining an old book, it will not be through idleness or ignorance; if they reject an old technique, it will not be through fashion; if they select something new, they will do so for positive reasons.

Exercises

1 During the 1920s many hundreds of schools in England (and elsewhere) followed the Dalton Plan – an approach to learning based on individual assignments (see, for example, the article by Miss F. A. Yeldham reprinted in *Math. Gazette* 55 (1971), pp. 200–7, and L. C. Taylor [1]). By 1970 the scheme lingered on in a handful of schools. Find out more about the Dalton Plan and try to account for its disappearance from schools. What lessons can contemporary projects learn from the history of the Dalton Plan?

2 A review in the *Times Educational Supplement* (10 Nov. 1972) of Thwaites [3] said: 'Nothing in this country (Britain) matches the professionalism of the work in curriculum study being carried out by the Americans...' Assuming that the author was not referring to 'practising for money', what do you think she meant by the term 'professionalism'? What kinds of 'professionalism' are to be valued in curriculum study? With the meaning you attach to the quotation, is the statement a valid one?

3 Is it good that children should use draft texts which may contain errors (both mathematical and pedagogical)? Discuss some of the implications of the use of such texts.

4 Some textbooks are more popular than others because they do not stretch either the pupils or the teacher too much. What are the implications of this for projects?

5 How can a project best explain its aims to parents? Take a particular topic and devise an explanation of it and the reasons for teaching (or not teaching) it, suitable for parents.

6 The British projects in the early 1960s relied to a large extent on the part-time work of the enthusiastic, experienced teacher. Later projects have often used full-time workers who no longer teach in schools. What differences in approach would you expect? Investigate the methods of different projects and see if your expectations were justified.

152

7 Is the project 'model' as used in the 1960s necessarily applicable today? Give your reasons.

SOME FURTHER READING

Hooper [1], HMSO [9], Howson [4, 5], MME Director's Report 1963–5, Schools Council [11, 12], Thwaites [3]

PART 5
THE CURRICULUM IN THE LARGE

WHAT IS A CURRICULUM?

Throughout the opening chapters of this book we have frequently used the words 'curriculum' and 'curriculum development', without specifying exactly what we meant by the terms. We hope that it has always been reasonably clear from the context what meanings were to be ascribed to the words, and so we have, in one sense, defined them; for the reader has accustomed himself to accepting them, in much the same way that a child grows to understand the adjective 'red'. However, just as a mathematician constantly looks back on his work and attempts to redefine terms or to specify them more precisely, so it is valuable for us to study some of the meanings one can give to the word 'curriculum'.

Many attempts have been made in recent years to give definitions: thus, for example, we have

... all the learning which is planned and guided by the school. It should comprise several interrelated components – precise statements of objectives for each area, the knowledge and learning experiences most likely to achieve the stated objectives and the means of deciding the degree to which the objectives are achieved. (J. F. Kerr [1])

... the entire program of the school's work ... It is *everything* that the students and their teachers do. Thus it is twofold in nature, being made up of the activities, the things done, and of the materials with which they are done. (H. Rugg [1])

These definitions differ considerably in their style and usefulness. The former definition is clearly stated and at first sight has a scientific exactness. On deeper consideration, however, doubts begin to creep in. With what 'precision' are objectives to be stated? How are we to measure the extent to which objectives have been attained? What do we mean by 'most likely' – does this refer to the individual pupil, or pupils in general?

Perhaps the most conspicuous failing of this definition is its essentially statical nature. For, as anyone knows, who has given a lecture course or taken part in a curriculum development project, the situation is much more fluid than the definition would make it appear. The amount of 'feedback' is such as to be continually affecting one's objectives and methods for attaining them.

155

Nevertheless, despite its shortcomings this definition points the way – unlike the latter, which is so all-embracing and vague as to be of little practical help – to what we could take as a working definition of the term, 'curriculum'.

We recognise in it four statements that are always present – but which are constantly being modified as a result of feedback from the system together with changes in inputs: for it must be remembered that any curriculum must be planned *bearing in mind the system for which it is designed and the constraints operating within the system*. Thus we have statements of:

(*a*) *Purpose*: a statement, or an assumption (for purposes usually exist even when they are not explicitly stated) of the aims of one's teaching;

(*b*) *Content*: a statement of the content of what students are to learn and experience, and of the choices they will be offered;

(*c*) *Method*: a statement of the method or methods which are considered to be most likely to achieve the aims set out in (*a*);

(*d*) *Assessment*: a statement of how the course and the work of pupils are to be evaluated.

If we restrict our thinking to mathematics, then we see that *purpose* corresponds to the question 'Why teach mathematics?' which we considered in Part 1. *Content* and *method* both demand statements, but the types of statement required will depend greatly upon the teachers in the system concerned. A friend, on taking up his first teaching post, was told by the Head of Department, that he was to teach Form IV and that 'normally I start with Pythagoras and go on from there'. That, together with the syllabus of the external examination the boys were to take the following summer, was considered to be a sufficiently extensive statement of 'content'. Nothing whatsoever was said about 'method'. One may infer, however, that there was an unspoken assumption that the young teacher would give the course in the 'conventional manner', i.e. as he himself had been given it while a pupil. He was in a stable system where change was not expected. Such freedom is admirable if the teacher is competent enough to benefit from it, but with less intellectually sophisticated teachers a mere examination syllabus will not suffice. More detailed statements on content and method will be necessary. In the following chapters we shall consider some of the problems relating to *content*. The problems of *method* belong, however, more to the province of education than of mathematics, and are of a pedagogical nature lying outside the scope of this book. We have mentioned certain of these problems in Part 2 and would here only wish to emphasise again that 'method' is an essential ingredient of 'curriculum' and that it cannot be excluded from the considerations of individuals or projects. The *assessment* of a project itself was mentioned on page 150, and in Part 7 we shall consider some of the ways in which the work of *pupils* can be assessed and evaluated by their teachers, themselves, or others.

13
The Traditional Approach

1. THE SYLLABUS

In this chapter we discuss the mathematics curriculum as it evolved for British grammar and public schools in the early part of this century, following the pioneering work of the AIGT (p. 129) and such reformers as Perry (p. 17). After a rapid evolution it jelled and, with little further change, became what is known as the 'traditional syllabus'. Its contents are described in the syllabuses of any of the Examining Boards, which differ from each other only in minor variations.

Those responsible for the reforms that led to the 'traditional syllabus' would, no doubt, have been surprised by its longevity. For as the example we reproduce below in Section 3 shows, they designed a syllabus to meet the needs of their pupils and which made use of the mathematics available to them. They would surely have been amazed and dismayed that the syllabus should remain when the needs had changed and new possibilities had been opened.

2. STAGES A, B AND C

It will help if at this stage we explain some terminology used in the Mathematical Association's 1923 report on the teaching of geometry – the three 'stages' of learning and teaching.† In Stage A, mathematical material is introduced almost as a body of experimental fact, as for example when geometrical feeling is imparted via geometrical drawing. Sometimes this 'experimental' stage is called 'pre-mathematical' because no attempt is made to organise the facts in a mathematical way; it corresponds roughly to the attitude of the Egyptians before the Greeks imposed the Euclidean organisation of mathematics. Stage B – the 'deductive' stage – then begins to organise and

† Although nowadays only three 'stages' are remembered, it is worth pointing out that as originally envisaged (see Math. Assn. [1]) there were two other stages: Stage D in which one attempted to generalise, and Stage E in which one considered philosophy and the foundations of the subject.

enrich the material, but without too much pedantry, and with no attempt at a full axiomatic development. The underlying logical attitude is 'local' rather than 'global', because pupils passing through the stage are only being introduced to the ideas of proof and the algorithmic manipulation of symbols, etc. It corresponds to the eighteenth-century approach to calculus, before the critical build-up of analysis in the nineteenth century – which is the paradigm of a 'Stage C' development.

This model of 'mathematical development' is an extremely valuable one. However, it must be realised that although a pupil may be at Stage C in one branch of mathematics he may well be at Stage A or Stage B in another. Teachers have to recognise the appropriate times in a pupil's development, when it is right to switch from one stage to another. Considerable stress can be caused when a pupil with Stage C notions in one topic meets Stage B notions in another; and the pupil needs help to cope with the stress. This problem also arises when a teacher, having coped well with his own Stage C development, is expected to give a Stage B (or even A) treatment. Sometimes he cannot return to the lower stages, so great was his own effort at jumping the thresholds: and sometimes he is ignorant of the existence of conceptual thresholds in others. Consequently he communicates badly with his pupils.

In the case studies of this chapter and the next, the reader may well ask himself at which stage the approach under discussion is aimed (by the composer of the curriculum). Some composers do not, of course, think in terms of these stages: for example the very abstract and formal approaches of some projects are very much in the style of Stage C and their authors do not recognise Stages A and B as mathematics – or as stages in which a professional mathematician should display interest!

We begin with an example of curriculum design from pre-World War I days – which shows that there is nothing essentially novel about the principles outlined above – and then move on to more recent examples.

3. CASE STUDY 1

Designing a course for naval cadets at Osborne and Dartmouth, England

The account which follows is an edited version of a paper prepared by J. W. Mercer for the International Commission on the Teaching of Mathematics, shortly after that body was established following the 1908 International Congress of Mathematicians. It was one of a series of papers written by mathematical educators in Britain which were published by the Board of Education in 1912 under the title *The Teaching of Mathematics in the United Kingdom* (HMSO [14]).

The course at the two colleges, which trained naval officers, was for boys from 13 to 17, and was taken by all the cadets. About six hours a week were set aside for mathematics. It will be seen that the work has a strong scientific and engineering bent. The difficulty of finding teachers who could teach such a course without overstressing the academic mathematical aspects was as great then as it remains today. In a preliminary note to Mercer's paper, his headmaster, C. E. (later Sir Cyril) Ashford, explains that:

A man must have received an essentially mathematical training if he is to realise the importance of sound foundations, and to lay them properly; success in this training demands a concentration of effort incompatible with the practical study of science and engineering which is essential for the teacher of practical mathematics. The plan adopted at Osborne and Dartmouth is to secure trained mathematicians (as teachers) and make all possible opportunities whereby they can develop a scientific habit of mind. It is not sufficient to maintain close contact outside the classroom between mathematician and engineer or scientist, nor to provide for courses, however extended, of 'looking on' at science teaching, but a most marked effect is produced by arranging for mathematical masters to give science teaching themselves for a few terms both in lecture room and laboratory.

Ashford also has pertinent remarks to make on the subject of 'accuracy':

The fear has often been expressed that a reduction of the traditional drill in manipulation, by which a knowledge of these standard types (processes and results) was instilled, would lead to a decrease in accuracy . . . It is not easy to bring forward proof or disproof of this view, but experience goes to show that other factors are more important in producing accurate work. First of these is age, or maturity of mind; it is largely a matter of the growth of the feeling of responsibility . . .

Obviously, 'accuracy' was never what it used to be! As remarked in Section 1, Mercer's way of introducing the various topics is what is now known as 'traditional' (Stages A and B) in Britain, and even nowadays there are few variations from it. The content is usually taken for granted as something that all pupils who study mathematics should know, even if their experience is nowadays enriched by the material of a more modern syllabus. This is why we quote Mercer at such length. It should be remembered that Mercer was able to plan this curriculum, and follow it, because the Naval Colleges did their own examining: he was not preparing pupils for public examinations and was working very much in the manner of the present-day CSE Mode 3 (see p. 41). In most schools, however, this 'traditional' material was adopted without Mercer's reasons, perhaps because it might be useful for the pupils, and eventually because the examinations were based on it. Once the reasons were forgotten, degeneracy set in.

MATHEMATICS AT OSBORNE AND DARTMOUTH

The Course of Instruction

The colleges at Osborne and Dartmouth have this great advantage over other public schools, from the point of view of curriculum-framing, that all the boys are being trained for the same career, and it is fairly clear what use they are likely to make of their subjects of study in the future. Let it be clearly understood at the outset that Mathematics is taught to naval cadets mainly as a useful tool which can be employed in physics, navigation, engineering, and, later, in special gunnery and torpedo work. This does not mean that the subject is reduced to a number of rules of thumb – far from it. It is true that information must be gained and the memory cultivated, but these are of little use unless at the same time the boy acquires a logical way of thinking, and the power to apply his stock of knowledge to a new problem. We cannot spare the time to teach Mathematics merely as a 'mental gymnastic' without regard to its usefulness, but there is no reason why Mathematics which is intended to be of use in the future should not be an effective instrument of mental discipline; properly taught it is probably more powerful in this direction than the old-fashioned type of Mathematics, which, in the earlier stages at any rate, was much too abstract and artificial, and to the ordinary boy seemed to lead nowhere and to have no point of contact with actual experience. We try to steer a middle course between two extremes – one in which Mathematics is treated solely as an instrument for getting results, the other in which it is looked upon merely as a means of mental discipline and no interest is taken in its practical applications.

The statement that all boys are being trained for the same career must be modified, in that cadets will eventually be either general service officers, or specialists in Engineering, Gunnery etc. The standard of Mathematics expected from those who take specialist courses will be high. Now it is found here, as everywhere else, that boys of the same age differ very widely in their aptitude for Mathematics, and before cadets reach their last year at Dartmouth we can mark out some of each group as being those from whom specialists will be drawn. The syllabus provides, first, for a minimum course to be taken by everyone; secondly, for some extensions of this course to be taken by better boys in the lower half; thirdly, for higher work to be done by the upper half, and fourthly, for still higher work for a very small number of cadets at the top.

The problem before us is roughly this: to give the weaker cadets enough of the right kind of Mathematics, sound so far as it goes, to enable them to apply it intelligently to their navigation, science, and engineering and also to give future specialists more pure Mathematics generally, ... some sound notions of the Calculus, to equip them for continuing their Mathematics after leaving the College. Of the weaker ones particularly, the intention is that they shall have a few tools, but that those few shall be always bright and ready for use, and that they shall be able to choose the right tool for a particular purpose.

Much of the traditional course must be cut out to make way for this new matter. At first, one proceeded with some caution, the tendency being to retain things about which there was any doubt, lest, after all, they might turn out to be of use later. Then, the cutting out had to be done more boldly, the guiding principle being rather that if no reason could be assigned for retaining a certain piece of work, it should go. Something so omitted can be introduced when the need for it arises.

Many people find it hard to believe that what occupied so large a place in their own education can be unimportant and capable of being omitted with advantage. Consider, for instance, the wonderful questions in vulgar fractions, totally unlike anything that could possibly occur in any research. It will perhaps be claimed for such questions that they called for patience, orderly arrangement of work, and accuracy. But there is sufficient scope for the exercise of all these qualities in problems which arise in connection with physics, solution of triangles, etc. These problems have the additional advantage of having some obvious purpose, whereas the others were merely dull.

Although we consider the course of instruction under different headings, in actual practice there is considerable overlapping. We do not want Mathematics separated and shut off in watertight compartments. In solving a problem a cadet is encouraged to make use of any knowledge he may possess. If he finds it advantageous to apply algebra or trigonometry to a problem in geometry, by all means let him do so. His attitude should be 'What have I to do? Which of the implements at my disposal shall I use?'

The effect of giving freedom in choice of methods is that the ordinary boy finds more problems within his scope. As an example, suppose that a class meets for the first time the problem: 'A circular arc ADB has its chord AB 10 inches and its height CD 2 inches, find the radius of the circle.' A clever boy would use the rectangle property: another would apply Pythagoras' Theorem. But most boys, endowed with no special degree of cunning, would use trigonometry while others would perhaps draw a figure to scale and get a result by measurement.

Arithmetic

The old type of arithmetic contained too many financial and artificial problems but the number of separate rules and principles is surprisingly small, and the application of the correct principle to a particular problem is largely a matter of common sense. Our teaching aims at accuracy and facility in the elementary operations, a sound knowledge of the metric system, the unitary method, and the use of four-figure tables. Boys learn not to be afraid of a long piece of computation. They find that accuracy in this kind of thing is important in physics and that in Navigation especially, inaccuracy is fatal. Arithmetical manipulation has not been cut down since, in Geometry and Trigonometry, numerical work, often of a complicated nature, is continually being done, while Navigation requires a considerable power of dealing with masses of figures and in Physics and Mechanics a boy is constantly exercised in elementary arithmetical operations.

The question of arithmetical accuracy is an important and difficult one. We can hardly expect the habit of accuracy to be fully developed in boys of 15, but a great deal can be done if every master, whatever his special subject, for whom a boy has occasion to work out numerical results, will refuse to accept careless and inaccurate work. It is not a question of teaching a boy how to add and multiply, but rather of getting him to realise that to know how to do a thing, and yet to do it wrong, is useless ... He has to learn that every step affects all the subsequent work, and to make sure of each step before proceeding to the next ... This continual criticism of his own work must be insisted on as a necessary part of everything he does until he comes to apply it mechanically. Both the mathematical master and the science masters can and indeed must help, by refusing to accept work which is spoilt by careless arithmetical blunders. Otherwise there is a danger that a boy looks upon the demand for accuracy as a fad of the mathematical master, whereas if he finds that everyone insists upon it, he may be led to believe that it is really important.

In introducing a new idea in any branch of Mathematics, the exercises should be easy and often capable of being done mentally, so that the principle is not obscured by a mass of computation ... Boys should acquire the habit of estimating the approximate value of an arithmetical expression, and where possible, external probability should be used as a check; e.g. if the H.P. of a locomotive is calculated to be .213, it is probably wrong.

To boys who make so many calculations founded on approximate data, the question of degree of approximation is important, and good exercises can be made by making them calculate from measurements made by themselves to estimate the possible errors in their results, due to the inevitable measuring inaccuracies. If the sides of a rectangular sheet of paper are found by measurement to be 9.87 and 7.46 in., with a possible error of .005 in. in each, it is dishonest to state the area as 73.6302 sq. in. without further comment. All that can be stated with certainty is that the area is $73.63 \pm .09$ sq. in.

Algebra

Reasonable skill in straightforward manipulation is expected, but we have no time for tedious examples in simplification of fractions, or with hosts of 'elegant devices' for solving specially constructed problems. With a few simple rules and principles in constant use, and backed up by some common sense, quite difficult problems can be solved by very simple means. The process will then be laborious, and a clever boy will see and use many short cuts. The ordinary boy will see the advantage of these short cuts, and should be encouraged to look for possible simplifications of his work, but he should not be discouraged from proceeding because the method he proposes to himself promises to be tedious.

The work on fractions has been much reduced. Questions on addition and subtraction of fractions are useful as giving practice in multiplication, the rule of signs, and the meaning of brackets. Some drudgery is necessary at the outset ... but we lose none of the necessary drill by limiting ourselves to fractions with monomial and binomial denominators.

The idea of a formula as a compact general statement of the operations to be performed in all questions of a certain type is the starting-point in algebra, and a boy is practised in deriving formulae inductively after consideration of a multitude of particular cases. The evaluation of quantities by the substitution of numerical values for the letters occurring in a formula is a matter of arithmetic, and cadets have constant practice in this kind of work throughout the course.

The simple equation, and the problem leading to it, occur at the very outset. Here the important thing is to be able to express the conditions of the problem in algebraical language, and a few more or less artificial problems may be most useful to this end as involving ideas well within a boy's comprehension. Great pains are taken to inculcate the habit of checking . . ., of verifying that the solution obtained does actually satisfy the conditions of the problem. No problem is considered completely solved without this check.

The idea of functionality is constantly brought out and illustrated by graphical representation. Variation provides a multitude of concrete examples, as do most of the ordinary formulae in use in mensuration, mechanics, etc. Simple numerical questions in geometry can be modified so as to bring out some functional relation . . . The idea of a continuous change in $f(x)$ due to one in x grows on a boy when he has met it in so many different forms, in mathematical and science lecture rooms and laboratories, though it would probably be a mistake to sum it up for him in these words. He learns to look upon a graph . . . as a picture story which he can read. A few examples of exercises in graph-reading are given.

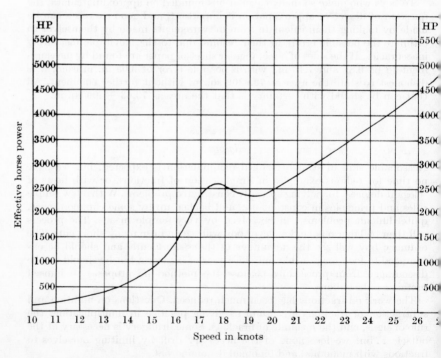

Fig. 1. Relation between EHP and speed of a certain destroyer in 20 ft of water.

1 What is the HP for speeds 13, 16, 17, 19, 24 knots?

2 What increase in HP is required to increase the speed (i) from 11 to 15 knots;   (ii) from 15 to 16 knots;   (iii) from 16 to 17 knots;   (iv) from 21 to 22 knots;   (v) from 22 to 23 knots?

3 The HP increases with the speed as the speed increases from 11 to 17 knots, also as the speed increases from 21 to 27 knots. What difference do you notice in the two cases?

4 What striking feature is shown by the curve?

The application of graphs to the solution of all kinds of equations and the determination of maximum and minimum values of all kinds of functions follow naturally.

A boy is expected to be able to solve simple and quadratic equations in one unknown, linear simultaneous equations in two or more unknowns, and simultaneous quadratics in two unknowns without particular artifices for various types, only those being dealt with in which one unknown can be expressed explicitly in terms of the other and the solution effected by substitution.

Every quadratic is written in the form $f(x) = 0$, and $f(x)$ is resolved into linear factors. If factors can be written down at sight, so much the better. If not, the method of completing the square is used as a device for factorising the left-hand side by writing it as the difference of two squares. This will often save boys from omitting roots.

The resolution of a quadratic function into factors at sight (where possible) is a good exercise in an inverse process. To multiply $3x + 2$ by $4x + 5$ is one thing, to see that $12x^2 + 23x + 10 = (3x + 2)(4x + 5)$ is another. This is one of the first instances a boy comes across of systematic guessing, and at any rate for those who will subsequently do integral calculus, it is important to develop this kind of faculty. A few equations with 'literal' coefficients are solved, the main point being the discrimination between known and unknown. The process is an imitation of that which has previously been employed in many cases in which the coefficients are numerical, and provided that the examples are not made needlessly complicated they provide excellent practice in passing from the particular to the general. The general quadratic $ax^2 + bx + c = 0$ should be set as a 'literal' equation to be solved, and the result may be learnt as a formula for solving any quadratic, but in many cases the method of completing the square is easier.

The teaching of surds is limited to a few fundamental and useful ideas . . . If complicated expressions ever occur, they can be dealt with quite adequately by straightforward application of tables and present no difficulty.

In indices the graph of 2^x or 3^x derived from the values corresponding to $x = 1, 2, 3, 4, 5$, etc., suggests values for $2^{3/2}$, 2^0, etc. Interpretations found in the usual way from consideration of the fundamental multiplication law proved for positive integral indices are seen to agree with these. The formal proof of the general truth of the index laws is not given, but a few particular cases are useful as an exercise in going back to the meanings of symbols. Artificial examples are avoided. The graph of $y = 10^x$ constructed by means of a square root table will now enable a boy to appreciate the meaning of his logarithm table.

Exercise†

Carry out this construction.

The actual use of logarithms precedes a thorough understanding of their meaning. A lesson spent on the graph of 2^x will convince a boy of the reasonableness of taking $3 = 2^{1.6}$, $10 = 2^{3.3}$, $30 = 2^{4.9}$, and thus of saying that $2^{1.6} \times 2^{3.3} = 2^{4.9}$, and so on. The table of logarithms can then be presented as a table by means of which any number can be written as a power of 10, so that 2×3 or 6 is $10^{.3010+.4711}$ or $10^{.7721}$, or in other words log $6 = .7721$, which is the case . . . He will soon learn the use of logarithms, and the more detailed study of the meaning of indices and logarithms as outlined above can be deferred.

Exercise

Is it wise to allow the equals sign to be used in the above approximations? How would you suggest dealing with the errors involved?

The chapter on Variation in the ordinary text-book is, as a rule, somewhat scanty, and boys are apt to look upon the word 'varies' as a signal for the performance of certain tricks with 'k'. As cases of variation frequently occur, and provide excellent examples of functionality, the subject deserves more detailed consideration, and a treatment on the following lines seems to lead to more intelligent work. . . . [*Several examples are now given in Mercer's original.*] Boyle's law is a familiar concrete case of inverse variation.

In all these cases it is important to see that as x gradually changes, so does y; e.g. if the shape of a triangle is fixed, and the length of a side gradually changes, the area will gradually change. If one of the triangles has a side 17 cm, and area 200 cm^2, we can calculate the area corresponding to a side of any other length, and a graph can be drawn showing how the area depends on the length of the side.

Exercise

Draw such a graph.

The forms of the graphs corresponding to $y \propto x$, x^2, x^3, $x^{1/2}$, $x^{1/3}$, $1/x$, $1/x^2$, $1/x^3$, $1/x^{1/3}$, etc. should be known.

Special attention should be given to the graphical representation of the relation 'y varies directly as x'. It is a straight line through the origin, its equation being $y = kx$, k being the value of y when $x = 1$; and conversely . . . It seems a natural step to the case in which the graph is a straight line not through the origin. If a table of values of x and y is given, and we find on plotting y against x that we get a straight line, the relation between y and x is of the form $y = kx + l$, and k and l can be found from the coordinates of any two convenient points on the line.

If on plotting y against x we get a curve like one of the various types of '$y \propto x^n$' graphs, the difficulty will be to guess the probable value of n. The fact that if $y \propto x^n$, we get a straight line by plotting y against x^n is not of

† The exercises have been interpolated by the present authors.

much use if our guess at the value of n is wrong. But if $y = kx^n$, $\log y = \log k + n \log x$, or $Y = nX + K$, where X, Y, K, stand for $\log x$, $\log y$, $\log k$, therefore if we plot Y against X we shall get a straight line, and n can be found as above.

Cases in which y is a function of several variables are treated similarly, e.g. consider the statement that for curves on railways the difference in level of the rails varies directly as the gauge, and the square of the highest speed, and inversely as the radius of curvature.

Exercise

Write down the formula Mercer is describing. Construct a model leading to this formula.

Geometry

The first term at Osborne is devoted largely to practical work. Cadets are supposed to have done some work of this kind at their Preparatory Schools so that when they come to their theoretical work they have picked up a considerable number of geometrical notions and facts, and are in a position to appreciate what they are going to reason about. The object of teaching Geometry is, I suppose, threefold:

1. That a boy may acquire a certain number of geometrical facts.
2. That he may be able to apply this geometrical knowledge to the solution of problems.
3. That he may get some training in sound reasoning and precision of statement.

There is not much in Euclid's first and third books that a boy who has done some rational work with instruments will not accept as a matter of course. For instance, a boy who has constructed triangles to given data sees that a triangle is uniquely determined if a, b, c are given. The fact that the base angles of an isosceles triangle are equal is sufficiently obvious. To present formal proofs of such propositions on Euclidean lines only fogs a boy's mind. These fundamental facts must be properly stated and systematically set down, so as to serve as the basis of subsequent work. The number of propositions a cadet is required to write out formally is not very great, and as much as possible he is led on to discover the truths for himself, and build up the steps of the proof. Often a truth to be proved is suggested by experiment. In whatever way the truth and proofs of the standard propositions are arrived at, a cadet is required to be able to write out a proof more or less on Euclidean lines. He must state clearly what is given, and what he is going to prove, and give full reasons for every step in the proof, quoting only facts which have been previously established. He is not tied down to any particular form of words, but must make it perfectly clear that he understands what he is doing. These formal proofs, if properly done, are very valuable in teaching a boy how to marshal his facts, how to select from his stock of knowledge the particular thing wanted for the purpose in hand, and in showing him the nature of a proof. The propositions naturally fall into a few small groups, and where the proof of one depends on another it will of course be necessary to remember the order. As propositions are proved they are added to the stock of quotable material available for solving problems and riders.

Exercise

Mercer assumed that there were certain 'fundamental facts' that all students should know, and that these were so universally accepted there was no need to list them. What do you think these 'facts' are?

With similar triangles, much of what used to be proved at great length is more or less a matter of intuition to a boy who has been accustomed to draw to scale, and to deal with maps and models. We recall a few facts about congruent figures, such as, a triangle is determined if we know the lengths of the three sides. Thus, if we wish to copy the scheme of points A, B, C, we can do so if we know the three lengths AB, BC, CA. Another way of putting this is to say that if a framework be made of three strips of wood freely jointed at the corners, it is rigid. To copy four points it will not be sufficient to know four lengths. We want three lengths to copy three of the points, and two more to copy the fourth. A framework of four freely jointed rods is not rigid.

These frameworks are rigid.

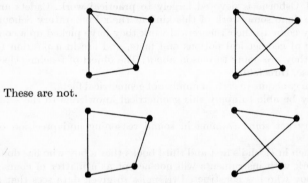

These are not.

Similarly, to copy five points we want $3 + 2 + 2$ or seven lengths, and so on.

Exercise

Prove this last result and obtain a more general one.

The step from congruent to similar figures is easily made, if it be granted that, given one figure, it is possible to have another such that every length in the second is k times the corresponding length in the first where k is any positive number. Plans of fields, cycling maps, models of ships, pictures thrown on a lantern screen, etc., furnish illustrations of this.

There are thus numerous methods of copying a figure, but all are included in this: After two points have been obtained, every other is got by completing a triangle, one of whose sides is already known; and two figures are similar if all pairs of triangles used in their construction are similar.

Exercise

Prove this using induction.

Exercises on constructions of similar figures are done here to drive these principles home, the process being greatly simplified by the use of the theorem that if XY is parallel to BC, a side of the triangle ABC, then the triangle AXY is similar to the triangle ABC.

A similar method of treatment will make it clear that the ratio of the volumes of two similar solids is the same as the ratio of the cubes on any corresponding pairs of lengths. Many numerical problems can be set on these two important principles; in fact, boys will probably have met some before this as exercises in variation.

Exercise

Devise suitable problems. (See Thompson [1].)

Constructions depending on areas of similar figures can, for weaker boys, be helped out by measurement and arithmetic. For instance, consider the problem of constructing a figure similar to a given figure, but with one-third of its area. If AB is any length in the first figure, and $A'B'$ the corresponding length in the second,

$$A'B' = AB \times \sqrt{\tfrac{1}{3}} = AB \times \sqrt{.3333} = AB \times .5774$$

and a very good solution can be got by measuring AB and multiplying by .577. This gives $A'B'$, and the construction can be completed. A clever boy will devise a geometrical construction for finding $A'B'$.

A certain amount of Elementary Solid Geometry, dealing mainly with angles between planes, and angles between lines and planes is done, but informally. Most of the matter of Euclid's Eleventh Book is easily grasped with the help of concrete illustrations, the plentiful use of models, and appeals to familiar ideas such as vertical and horizontal. The main intention of the course in solid geometry is to train boys to 'think in space', and it is found that even the weakest boys can become quite good at visualising three-dimensional figures.

Exercise

Suggest examples of how Mercer's ideas could be put into practice.

A little is done in Descriptive Geometry – plan and elevation – enough to give a clear idea of the principles involved. This kind of work is an excellent medium for further developing the 'space sense'. Practice in the more elaborate technical part of the subject is supplied in the drawing office connected with the engineering shops.

As an introduction to Spherical Trigonometry and Nautical Astronomy, a preliminary course of Spherical Geometry is given in the same term in which Solid Geometry is begun. The cadets draw figures and measure sides and angles of spherical triangles. At the same time wire skeleton models are used so that the connection between arcs of great circles on the surface of a sphere and angles at the centre is constantly in their minds. In other words, they learn to realise figures on the surface of the sphere both from the external and internal points of view. The notions acquired of angles between planes, lines perpendicular to planes, etc., are seen to have their use in Spherical Geometry.

Mercer then gives examples of typical practical and theoretical problems done by pupils.

169

Exercises

1 Write a 'unit' on Spherical Geometry.

2 Discuss whether Mercer's curriculum appears likely to succeed in his three-fold object of teaching geometry.

Trigonometry

Many boys who will never make much headway with Deductive Geometry, soon become quite reasonably good at Numerical Trigonometry. It used to be considered necessary to have done Euclid's Sixth Book before beginning Trigonometry, as otherwise the constancy of the ratio a/b in a triangle with given angles could not be established. As pointed out under Geometry, the truth of this is obvious to the boy who has been in the habit of drawing to scale. The trigonometrical ratios are introduced gradually. The tangent is taken first as being connected with the simplest problems on heights of towers, etc., and it is in connection with the solution of some such problem by scale drawing that the idea of tangent of an angle is introduced. Tables of natural tangents are used straight away, and a boy is interested to find that he has at his command a quicker and more accurate method of solving certain problems than his old one of drawing to scale. The meanings of the tangent, and of the other ratios as they occur, are driven home by easy problems which seem to be much more helpful for this purpose than identities.

The sine and cosine are introduced in connection with problems in which the hypotenuse is involved. After a time the convenience of having the reciprocals of these three ratios tabulated is seen, and the cotangent, secant and cosecant, are defined. At first the arithmetic is made fairly easy; the use of log-sines, etc., will be appreciated when the arithmetic is made more difficult.

In connection with the solution of right-angled triangles, cadets make acquaintance with the Traverse Table, which is very useful for obtaining approximate results in Trigonometry. It is, of course, specially important for cadets to be familiar with this table, as they must be able to use it quite freely when they come to Navigation.

The general triangle is then solved by treating it as the sum or difference of two right-angled triangles. Up to this stage nothing but a knowledge of the trigonometrical ratios of acute angles is required and with no further knowledge a boy has a powerful weapon in his hands.

This method suggests a search for general formulae, and by following out in the general case the method used in the particular cases the sine and cosine formulae are easily obtained. At this stage the definitions of sine and cosine are extended so as to be applicable to obtuse angles.

The problems up to this point are as varied as possible, on heights and distances, and also Navigation, Statics, and Geometry. Even the weakest boys feel that they have something which they can use in all sorts of situations, and they use it quite freely where clever boys would perhaps use algebraical or geometrical methods. Thus, as previously mentioned, they find themselves able to solve many problems, which, under the old system of water-tight compartments and restrictions as to methods, would have been beyond their powers.

The measurement of angles in radians and the simple relation connecting the radius of a circle, the length of an arc, and the circular measure of the

170

THE TRADITIONAL APPROACH

angle subtended by the arc at the centre, must be known, as they occur in Mechanics, e.g., in questions of angular velocity. The approximate equality of sin θ, θ, tan θ when θ is small is also an important fact. For those cadets who will later go on to the calculus, it is important to learn to think in radians.

Exercise

Explain how to convert from degrees to radians (see Whitney [1]) and suggest possible ways of introducing radians in class.

The remainder of the Trigonometry is of a more abstract nature, and therefore more peculiarly adapted to the boys with a mathematical bent. It is, in fact, practically a different subject. It includes the trigonometrical ratios of $(A + B)$, $(A - B)$, $2A$, and the solution of trigonometrical equations, some by graphical methods. This last section includes the extension of the definitions of the trigonometrical ratios to angles of any magnitude, and the use of an auxiliary angle to transform expressions of the form $a \cos \theta + b \sin \theta$ into the shape $r \sin (\theta + a)$. In this part of the subject not much is expected from the weaker cadets. For instance, their acquaintance will amount to little more than the formulae which they use in Spherical Trigonometry.

The better cadets, on the other hand, must have some skill in manipulation of this kind, and a good deal of practice and facility in using the formulae. The forms of the graphs of the ratios, particularly of the sine, should be familiar, but should appear in a restricted form at the outset. The calculation of tangents of various angles from his own figures will make it quite clear to a boy that a gradual change in the angle produces a gradual change in the tangent, and that equal increments in the angle do not produce equal increments in the tangent; and the graph will show distinctly how the tangent increases more and more rapidly as the acute angle increases, and emphasis should be laid on the non-straightness of the graph. It is very difficult to prevent boys from making such mis-statements as tan 40° + tan 30° = tan 70°, and tan 80° = 2 tan 40°, and frequent appeals to the shape of the graph should be helpful. Later, when the definitions are extended to angles of any size, the graphs will be done in their complete form. In connection with the auxiliary angle, it should be noticed that, by suitable choice of scales and origin, the graph $y = \sin x$ can be made to serve as the graph of any of the family of curves $y = a \cos x + b \sin x$.

Exercises

1 Prepare a trigonometry 'unit' based on Mercer's suggestions. What modifications would you think desirable?
2 Find a binary operation □ on the domain of tan, such that tan $(x \square y) =$ tan x + tan y, for those boys mentioned by Mercer, who wish to make tan additive. (See Baker *et al.* [1]).

Calculus

About half the cadets proceed in their last year at Dartmouth to the study of the Calculus. Much more importance is attached to a proper understanding of the meaning and application of the operations than to skill in manipulation, which can be acquired later, if necessary. Before the notation of the Calculus

171

appears, the notions of 'rate of change', 'gradient', etc., are discussed at some length, the illustrations used being such as are already familiar to the boy through his acquaintance with easy kinematics, graphs, and geometry.

Every boy who has done any Dynamics has some idea of the meaning of 'speed at an instant'. Thus the speed at the instant is a real thing in his mind, but for purposes of reasoning another conception is necessary. The first problem he deals with is the calculation of the speed at a given instant from a space–time formula. Suppose, for example, that the formula $s = 2t^3$ is given, and the speed at the end of 4 seconds is required. If the average speed during an interval immediately following the end of the fourth second is calculated, its value will gradually change as the duration of the interval is gradually diminished. Moreover, this average speed will clearly come nearer and nearer to the speed at the instant, and can be made as near to it as we please, if we take the interval small enough. Thus, if a certain speed can be found from which the average speed during an interval immediately following the instant in question can be made to differ by as little as we please by making the interval small enough, that speed is the speed of the instant.

In the particular case which we are considering as an example, the average speeds for intervals of 1 sec., .5, .1, .01, .001 sec., are found to be 122, 108.5, 98.42, 96.2402, 96.024002 ft./sec., and it seems pretty clear that by taking the interval small enough we can get an average speed differing from 96 ft/sec. by as little as we please. To make quite sure, the average speed for an interval of h secs. is calculated. This is $(96 + 24h + 2h^2)$ ft/sec., a result which is seen to include all the numerical results previously obtained, and from which it is quite certain that the speed at the end of 4 seconds is 96 ft/sec. The question of gradient at a point of a curve is treated in a similar way. Everyone has a conception of the meaning of 'tangent to a curve'. Euclid's definition of a tangent to a circle probably expresses the usual conception of a tangent. For working purposes a modification is necessary. Suppose the gradient of the curve $y = 2x^3$ is required at the point P where $x = 4$. If the gradient of a straight line through P, which cuts the curve in a neighbouring point, be calculated, its value will gradually change as the line swings nearer to P. Moreover, the gradient will come nearer and nearer to that of the tangent at P, and can be made as near to it as we like if we take the neighbouring point near enough to P. Thus, if a line through P can be found whose gradient is such that the gradient of a line through P and a neighbouring point on the curve can be made to differ from it by as little as we please by bringing the neighbouring point close enough to P, that line is the tangent at P, and its gradient is the gradient of the curve at P.

In the particular case under consideration, the gradients of chords joining P (where $x = 4$) to the points on the curve, where $x = 5, 4.5, 4.1, 4.01, 4.001$, are 122, 108.5, 98.42, 96.2402, 96.024002. As before, all these are included in the statement that the gradient of the line joining P to the point on the curve where $x = 4 + h$ is $96 + 24h + 2h^2$, and the gradient of the curve at P is 96.

Other cases of instantaneous rate of change are dealt with on the same plan, e.g., if $pv = 1200$, the rate of change of p per unit increase of v, when $v = 20$, is derived from a consideration of the average rate of increase of p per unit increase of v, as v changes from 20 to $(20 + h)$.

This introductory work is necessarily somewhat laborious and takes time, but it is felt to be important that a boy should have a thorough comprehension of what he is doing in a number of particular cases before proceeding to a study of the calculus on general lines.

(It might be mentioned here that those cadets who are not taken on to the calculus proper do much of this preliminary work leading to the general notion of the representation of instantaneous rate of change by the gradient of a curve at a point, which is of frequent occurrence in their science work; for instance, the speed of a body at a given instant is given by the gradient of the tangent at the corresponding point on the space–time graph; the rate at which p is changing per unit increase of v for a given value of v if $pv = 1200$ is given by the gradient of the tangent at the corresponding point of the hyperbola $pv = 1200$.)

This preliminary course, in which each particular case is dealt with on its own merits, leads a boy to look for some general mathematical process which shall replace this separate detailed consideration of each case. For instance, finding the speed of a body at various instants from the formula $s = 2t^3$ requires that the same set of operations shall be performed on different numbers and a general formula for the speed at any instant is naturally looked for. [*Mercer then calculates* $\Delta s/\Delta t$ *for increments* Δs, Δt *and deduces that the instantaneous speed is* $6t^2$.]

Generally if y is a function of x, then in all cases with which we deal, a small change, Δx in x, produces in y a small change, which we call Δy, and as $\Delta x \to 0$, $\Delta y/\Delta x$ tends continually to some limiting value, which we call dy/dx. $\Delta y/\Delta x$ gives the average rate of increase of y per unit increase of x, as x increases by a finite amount Δx; dy/dx gives the *instantaneous* rate of increase of y per unit increase of x.

dy/dx is now found from first principles in a few special cases, such as $y = x^3$, $y = kx^3$, $y = kx^3 + c$, $y = 1/x$, $y = 1/x^2$, etc., and the results applied to such questions as the following, so that the significance of the results may be grasped: [*There then follows a list of problems about finding velocities, equations of tangents, pressure of a gas, and coefficient of cubical expansion of water.*]

A general formula is now obtained when $y = x^n$. The particular cases already investigated have probably suggested this formula. A proof independent of the binomial theorem can be obtained if the theorem

$$\underset{z \to 1}{\text{Lt.}} \frac{z^n - 1}{z - 1} = n$$

be first established.

Exercises

1 Do this!

2 Estimate the error in taking $\Delta y/\Delta x$ for dy/dx when y is a polynomial of degree n.

The cases $y = kx^n$, $y = kx^n + c$, $y = a + bx^m + cx^n$, etc., should be considered and with no further outfit a great deal can be done.

Problems on maxima and minima, geometrical properties of curves, questions on speed and acceleration, and the rate of increase of one quantity with respect to another in general, approximations (depending on the approximate equality $\Delta y = dy/dx \cdot \Delta x$), and relative error, etc., are dealt with, the functional relations involved being always of the type $y = a + bx^m + cx^n + \cdots$, or reducible to this type.

Then the inverse problem – given dy/dx in terms of x, find y in terms of x – is discussed, and this prepares the way for the integral calculus, which is introduced by an attempt to find the area under a curve with a given equation, by a process of summation. The problem is to find the area bounded by the curve $y = f(x)$, the x-axis and the ordinates $x = a$, $x = b$. This is shown to be equivalent to finding

$$\underset{\Delta x \to 0}{\text{Lt.}} \sum_{x=a}^{x=b} f(x) \cdot \Delta x,$$

and cases are first dealt with in which the summation can actually be effected, e.g. $f(x) = 3x$, $f(x) = x^2$. It is very easy to see that generally the summation would be tedious if not actually impossible and a different method must be sought. The formula $dA/dx = f(x)$ is established, and the problem of finding A is thus reduced to that of finding a function of x which when differentiated gives $f(x)$, that is the problem last discussed in the differential calculus.

Exercise

Establish the formula $dA/dx = f(x)$ by a method appropriate to this level.

The notation $\int_a^b f(x) \cdot dx$ is now introduced, the definition being

$$\int_a^b f(x) \cdot dx = \underset{\Delta x \to 0}{\text{Lt.}} \sum_{x=a}^{x=b} f(x) \cdot \Delta x,$$

and it is shown that this is the same as $\phi(b) - \phi(a)$ or $[\phi(x)]_a^b$ where $\phi(x)$ is such that

$$\frac{d\phi(x)}{dx} = f(x).$$

Other problems are similarly treated. For instance, the distance described between the end of a seconds and the end of b seconds by a body moving according to the speed–time formula $v = f(t)$ is $\underset{\Delta \to 0}{\text{Lt.}} \sum_{t=a}^{t=b} f(t) \cdot \Delta t$, which is written $\int_a^b f(t) \cdot dt$. Since $ds/dt = v = f(t)$, s can be expressed in terms of t if we find a function of t whose differential coefficient is $f(t)$. In fact we shall have $\int_a^b f(t) \cdot dt = \left[\phi(t)\right]_a^b$ where $\frac{d\phi(t)}{dt} = f(t)$.

Applications of the integral calculus, to areas, mean values, volumes of revolution, work, centre of gravity, moments of inertia, etc., follow, these being seen to depend on one general principle if the meaning of the integral calculus is properly understood. The graphical interpretation of any definite integral $\int_a^b f(x) \cdot dx$ as the area bounded by $y = f(x)$, the x-axis and the ordinates $x = a$, $x = b$ is important, as it makes it possible to get a good approximation to the value of the definite integral when it cannot be evaluated by ordinary methods.

174

Exercise

Find estimates for the error in using a sum instead of an integral.

For example, to evaluate $\int_0^a \sqrt{(a^2 - x^2)} \cdot dx$ is the same as to find the area of a quadrant of a circle, radius a; to find the work done by a gas in expanding according to Boyle's law from a volume of a c.ft. to a volume of b c.ft. is

$$\int_a^b \frac{k}{v} \cdot dv$$

and the problem of evaluating this integral is the same as that of finding the area bounded by $y = k/x$, the x-axis and the ordinates $x = a$, $x = b$. This volume can be approximately found by applying one of the standard rules, such as Simpson's rule or the mid-ordinate rule.

Exercise

Compare the merits of Simpson's rule with those of the Trapezium Rule and the Monte Carlo method.

(Here again it should be said that cadets who do not take up formal calculus do a certain amount of what may be called graphic integration, without the notation of the subject. They reduce a problem of finding distance from a speed–time formula to that of finding a certain area under the speed–time graph, a problem of finding work done from a force–displacement formula to that of finding a certain area under the force–displacement graph, etc., the approximation to the area being usually made by application of the mid-ordinate rule. Of course the speed–time formula may in some cases be replaced by a table giving the speeds at certain times from which the graph can be plotted.) Up to this stage there is little scope for skill in the manipulation of the symbols of the calculus. The test† that a boy understands his subject is that if he is confronted with a problem he shall be able to make the solution depend on his ability to differentiate or integrate some function of x. The variety of the problems will of course depend on the amount of external knowledge, scientific, geometrical, etc., at the boy's command. The better cadets find time to investigate rules for the differentiation of other functions besides x^n, e.g., $\sin x$, etc., e^x, $\log_e x$, products, quotients, function of a function, etc., and to do problems, of the same nature as before, requiring for their solution this extended knowledge and some greater skill in manipulation such, for instance, as the determination of the meridional parts for a given latitude. Later, they will easily pick up tricks for integrating various functions which they may come across in their work, but such tricks are better learnt when the necessity for them arises. As stated at the outset, cadets learn the calculus in order that they may be able to use it, and the object of an elementary course should be to give a boy some general idea of the various kinds of problem to which he will be able to apply it. The best way of attaining this end seems to be to drive home the meaning of the processes by a great variety of examples demanding a very limited knowledge of the manipulative side of

† Into which of Bloom's categories (see pp. 47 and 327) would this 'test' fall?

the subject. In fact, if a boy can differentiate x^n and perhaps $\sin x$ and $\cos x$, he has, as pointed out above, weapons quite adequate for a host of problems. When he can use these weapons properly, he may be confronted with problems which involve the same fundamental principles, but require the differentiation and integration of other functions, and he will then see the necessity of adding to his stock of standard results and methods of attack. Thus he will gradually increase his power in the manipulation of symbols without losing sight of the main underlying principles.

Algebraical Geometry

About the same time that cadets begin the study of the calculus they do some elementary work in algebraical geometry. No attempt is made to take them through a course of 'Analytical Conics'. The idea is rather that they shall get hold of a few fundamental principles which can be applied to the case of any curve whose Cartesian equation is given. The Cartesian co-ordinates of a point are looked upon as directed steps, and it is convenient first to consider directed steps in one straight line and see that if the step from O to A is x_1 and that from O to B is x_2 the step from A to B is $x_2 - x_1$ $(OA + AB = OB)$. Of course, this conclusion will be arrived at after considering many particular cases. In the same way if the co-ordinates of a point P are (x_1, y_1), and those of Q are (x_2, y_2), we can pass from P to Q by taking an x-ward step $(x_2 - x_1)$ and a y-ward step $(y_2 - y_1) \ldots$

The distance PQ is easily found, also the co-ordinates of a point R which divides PQ in a given ratio.

The next thing is the translation of geometrical statements into algebra with numerous exercises. They need not be difficult, but should be properly understood; e.g.:

(1) Find the equation of the locus of a point equidistant from (3, 4) and $(-1, 2)$;

[*other examples are given.*]

Later some harder examples are done which require the elimination of θ from two equations $x = f(\theta)$, $y = \phi(\theta)$.

Exercise

Compose some such examples.

It is useful, occasionally at least, to plot the locus from the given geometrical conditions and from the algebraical equation independently.

In dealing with these questions on loci a boy incidentally makes the acquaintance of the parabola, the ellipse, and other interesting curves.

The straight line is dealt with in some detail.

1. The facts that $y = mx$ represents a straight line through the origin and $y = mx + n$ a straight line parallel to it are established. Here, as in all cases, the effect of changes in the x and y scales is considered.

The gradient of a straight line can be defined here as the ratio of the y-ward step to the x-ward step from P to Q, where P and Q are any two points on the line.

2. Lines are drawn from given equations, and equations found for given lines.

3. The equation of a line through a given point (3, 4) with given gradient 3 is found.

(a) The equation is $y = 2x + n$. Choose n so that the line shall go through (3, 4).

(b) The general equation of a line through (x_1, y_1) with gradient m is $y - y_1 = m(x - x_1)$.

If (x_1, y_1) and m are given numerical values (a) is easier, and the work can usually be done mentally, and the equation written down straight away.

4. The equation of a line through two given points is written down by means of 3, for the gradient is $(y_2 - y_1)/(x_2 - x_1)$.

5. $y = mx + n$ is parallel to $y = mx + n'$, $ax + by + c = 0$ is parallel to $ax + by + c' = 0$. Hence the equation of a line parallel to a given line and passing through a given point can be written down.

6. It is proved from a figure that $y = mx$ and $y = m'x$ are perpendicular if $mm' = -1$. Hence $y = mx + n$ is perpendicular to $y = -(1/m) x + n'$, $ax + by + c = 0$ is perpendicular to $bx - ay + c' = 0$, and the equation of a line through a given point perpendicular to a given line can be written down.

7. The point of intersection of two lines with given equations and the angle between them are found, but no general formulae learnt.

8. The length of the perpendicular from (h, k) to $ax + by + c = 0$ is

$$\frac{ah + bk + c}{\sqrt{(a^2 + b^2)}}.$$

This last formula is only arrived at after several particular cases of the problem have been dealt with. The length of the perpendicular from (3, 4) to $2x + 3y + 7 = 0$ would in the first place be found as follows:

(1) Write down the equation of the line through (3, 4) perpendicular to $2x + 3y + 7 = 0$; it is

$$3x - 2y - 1 = 0.$$

(2) Find the intersection of $2x + 3y + 7 = 0$ and $3x - 2y - 1 = 0$; it is

$$\left(-\frac{11}{13}, -\frac{23}{13}\right).$$

(3) Find the distance of (3, 4) from $(-\frac{11}{13}, -\frac{23}{13})$; it is

$$\sqrt{\left\{\left(\frac{50}{13}\right)^2 + \left(\frac{75^2}{13}\right)\right\}} = \sqrt{\left\{\left(\frac{25}{13}\right)^2 (2^2 + 3^2)\right\}} = \frac{25}{\sqrt{13}}.$$

After solving a few exercises of this kind and noticing that a simplification like that in the last line is always possible, an investigation of the general case is suggested and the length of the perpendicular from (h, k) to $ax + by + c = 0$ is found by the same method. The simple form of the result suggests the existence of some simpler method of proof, and a geometrical proof is found.

When boys have done enough calculus to know that dy/dx gives the gradient of the tangent at any point of a curve, they can establish interesting geometrical properties of curves whose equations are given, the range of operations being determined by the extent of their skill in differentiating.

It seems worthwhile to devote a little special attention to the circle, as the results obtained can be checked by geometrical knowledge, and this helps to give confidence. The only really essential thing is to see that the general equation of a circle is $x^2 + y^2 + ax + by + c = 0$. This, together with its converse, that $x^2 + y^2 + ax + by + c = 0$, can be written in the form

$$\left(x + \frac{a}{2}\right)^2 + \left(y + \frac{b}{2}\right)^2 = \frac{a^2}{4} + \frac{b^2}{4} - c$$

and hence represents a circle with centre

$$\left(-\frac{a}{2}, -\frac{b}{2}\right) \quad \text{and radius} \quad \sqrt{\left(\frac{a^2}{4} + \frac{b^2}{4} - c\right)},$$

is all that need to be known about the circle, and it is interesting to deduce some of the elementary properties of the circle by purely algebraical methods.

Exercises

1 Write out such a deduction, including, for example, the rectangle property of the circle.

2 In (8) above, pupils are often confused by the statement that 'the perpendicular from (x, y) to $ax + by + c = 0$ is $(ax + by + c)/\sqrt{(a^2 + b^2)}$'. Attempt to resolve this confusion by constructing the 'levels' $f(x, y) = 0$ (perhaps by using the notation $\{(x, y) \mid f(x, y) = 0\}$).

3 Does the definition of gradient on p. 176 agree with the definition that uses dy/dx?

4 Attempt to formulate statements of the purpose, content, method and means of assessment for each section of Mercer's course.

5 Try to find out what the 'traditional course' was to which Mercer refers.

6 Compare Mercer's criteria for 'cutting out' material (p. 162) with those of MacLane [1] (a description of curriculum design at Chicago University).

7 If Mercer were planning a course for naval cadets today, what changes in purpose, content and method do you think he would be likely to make? Which modern notations do you think he would adopt, taking into account what he says about the futures of his pupils?

8 Discuss Mercer's observation that the simple form of a result suggests that there should be a simple proof, giving examples outside coordinate geometry.

9 Does Mercer's curriculum satisfy the criteria of the Dainton Committee listed on p. 22?

14

Some Modern Approaches

We now look at some curricula proposed during the 1960s following the spate of reforming activity described in Chapter 12. Two of these are intended for the British 'grammar school' pupil. It will be noted that both were constructed to satisfy the constraints imposed by the traditional examination system, although each demanded its own examination syllabus. A third is a more theoretical study and is a curriculum proposed for American primary schools. In Case Study 4 we see an attempt to bring mathematics to a hitherto neglected part of the school population, while Study 5 concerns the construction of a curriculum in a university.

1. CASE STUDY 2

The SMP O-level course

The following paper is a shortened version of one which appeared in the 1962–3 Director's Report.† As we stressed earlier and as the paper itself emphasises, the curriculum is constantly changing and there are now, therefore, changes of emphasis in the SMP approach. We shall refer to these later (p. 184). The examination syllabus (see p. 183) was one of the first to be drawn up on modern lines at the GCE O-level.

The course was originally for pupils from 13 to 15 (Books T, T4) but it was later extended to begin at 11 (Books 1–5). Pupils following the course were drawn from the top 25 per cent of the ability range and the 'eight schools' to which the paper refers included some of the country's leading schools.

ON THE O LEVEL COURSE

One might well ask what the difference is between the SMP syllabus and others; after all, quite a number of experimental syllabuses of one kind or another have been tentatively proposed. This is not, in fact, a particularly profitable question since we reject its underlying assumption that there is a best or

† The paper is reprinted in full in Thwaites [3].

definitive syllabus for school mathematics; there is far too much mathematics which can usefully be taught at school for that to be true. However, the SMP syllabus is particularly distinguished by the fact that it is the joint work of the staffs of the eight central schools who are putting it into immediate use; thus it is what these schools prefer at the moment. Views nevertheless alter with the years, and it is no part of the SMP policy to define a syllabus for a long-term.

One of the aims is to investigate the feasibility of more or less continuous syllabus change. As the new approach becomes more familiar we may well find that some topics can be introduced at an earlier stage than is now contemplated . . .

The syllabus has been constructed with the pupil who will do no mathematics beyond O level primarily in mind. Nevertheless, we believe that this syllabus is also wholly suitable for those who will study the subject at a more advanced level. The foundations of many topics which are studied in depth at A level will have already been laid in the more elementary treatment at O level. For this reason, any reform of GCE syllabuses, in our opinion, must begin in the lower school and it would not be sound policy to attempt to initiate the first changes at A level.

A major aim of the syllabus is to make school mathematics more exciting and enjoyable, and to impart a knowledge of the nature of mathematics and its uses in the modern world. In this way it is hoped to encourage more pupils to pursue further the study of mathematics, to bridge the gulf which at the moment separates university from school mathematics – both in content and in outlook – and also to reflect the changes brought about in the world by increased automation and the introduction of electronic computers.

As with most new syllabuses, one of the main changes lies in the increased emphasis on algebraic structure. Sets, matrices and vectors are all mentioned in the examination syllabus and such concepts as associativity, commutativity, relations and groups are introduced in the text. We attempt to give a treatment of modern algebra similar to that which has been evolved over the last fifty years for teaching the differential and integral calculus. In this the pupil learns at a fairly early stage how to draw graphs of functions and how to find rates of change and areas under graphs by drawing. At the age of 16 or so the more formal language of the calculus is introduced and Stage B of the pupil's calculus education begins. Finally at university the student is given a Stage C treatment, that is a completely formal and rigorous account. It is our aim to establish a similar three-stage treatment for some of the topics of modern algebra. At present these concepts are first met by the student at the university where they are presented in a Stage C manner, and the lack of suitable preparation for this causes much of the difficulty encountered by university students. We hope that one of the results of our work will be an easing of this particular difficulty. In addition to this, we feel that such an introduction to modern algebra is an important part of the mathematical education of those who do not go on to university. It cannot be too strongly emphasised, however, that the members of the Project have no desire whatsoever to introduce work into the schools which is more properly done in universities.

At O level, then, we seek to convey something of the nature of various algebraic concepts rather than to impart a definite body of knowledge. It is our general belief that definitions and axioms must be allowed to grow out of concrete illustrations with which the pupil is familiar and that a deductive

180

treatment is better deferred. We do not want to see GCE questions of the type 'Define a group'. Nevertheless the pupil must be helped to realise the distinction between assumptions and definitions and consequences, and between intuitive feeling and proof . . .

Time for the introduction of new topics, here and in other parts of the syllabus, comes from the reduction of the complication of the examples to which ideas are applied rather than by the elimination of large parts of the existing syllabus. Thus although we would expect a pupil to have a reasonable facility in solving simple and simultaneous linear equations, factorising, operating on simple algebraic fractions and using indices, the emphasis would be on his understanding the processes involved rather than on his ability to cope with complicated applications.

Our proposals in geometry are a natural consequence of the rejection of axiomatic teaching in algebra. We do not dispute that the training in deductive reasoning offered by the formal geometry of the conventional syllabus has been of great value to some pupils (though such pupils would in all likelihood benefit equally from an axiomatic treatment of algebra). But for the majority of pupils, formal geometry offers little training in logical reasoning and emphasises, instead, practice in the memorising of theorems and proofs of no particular worth. In place, therefore, of a 'watered-down Euclid' approach we have substituted the study of Euclidean space by means of the geometrical transformations of rotation, reflection, translation and enlargement, and we hope that in this way the child will come to have a feeling for such spatial relationships. Thus a perfectly acceptable answer to a question†️ such as:

In the figure, AB and PQ are parallel chords of two circles with a common centre O. Prove that

(a) *the angle AOP is equal to the angle BOQ,*
(b) $AP = BQ$.

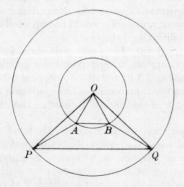

would refer to the symmetry (duly and correctly established) of the figure about a line through O and perpendicular to AB.

Coordinates, vectors and displacements are all introduced much earlier than is now usual and throughout the course the interplay between algebra

† Summer 1963 O level Pure Mathematics (Syllabus B) Paper I: reproduced with the kind permission of The Senate of the University of London.

and geometry is emphasised. For example, it is shown how the geometrical transformations can be expressed in matrix form and yield examples of groups.

It is our intention to give more attention than is usual to the study of three-dimensional geometry. Work on the construction of polyhedra and, later, on plans and elevations can help a pupil to accustom himself to the nature of three-dimensional space. The replacement of the conventional geometry, obsessed as it is with two dimensions, should also help to make him less fearful of working with problems in three dimensions.

Our approach to trigonometry is based on the study of components. Besides being excellent preparation for later work on vectors, this has the advantage of introducing the trigonometric functions as functions of angles and not as ratios connected with right-angled triangles. With such a starting-point, the extension to angles of any magnitude is immediate and natural.

The study of inequalities, both algebraically and graphically, is high-lighted; it leads to many interesting yet simple examples on linear programming and helps to direct attention to the reasons underlying some of the routine algebraic processes. Indeed, inequality is perhaps an easier concept to grasp than that of equality and there seems little justification for restricting elementary algebra to the single relation of equality.

In a similar manner, much light can be thrown upon the four basic operations of arithmetic by the binary and other number scales. It has proved very easy to diagnose and remedy misunderstandings of these operations when working in a scale as simple as the binary. Here the multiplication table is so trivial that the only problems to be solved by the pupil concern place value and an understanding of the operations. The binary scale also provides a simple introduction to the mathematical formulation of logical problems.

Elementary statistical ideas are introduced, with special attention to the presentation, tabulation and grouping of data. The mean and median of a distribution are defined and the idea of dispersion introduced. The emphasis here is not on the technique of calculating statistical measures but on the organisation and interpretation of simple data. Some work is also included on elementary probability, and pupils would be expected to be able to calculate easy compound probabilities by deduction.

It is thought important that boys and girls should consciously judge the degree of accuracy appropriate to any particular problem. The use of slide rules will be encouraged and 3-figure trigonometrical tables will be used. It is thus hoped to save time spent on numerical computation and also to encourage pupils to form the habit of estimating the size of the answer they expect to obtain.

In this syllabus, therefore, we have constantly tried to shift the emphasis towards mathematical ideas and away from manipulative techniques. Considerable facility in manipulation is, of course, required by pupils who are hoping to become mathematicians, physicists or engineers, but it is the opinion of those in the Project that the acquisition of techniques is best left until the post-O level stage; then, technique will come more rapidly, since the need for it will by then have become apparent. This will also free the pupil who stops at O level from much unnecessary learning. Thus altogether, we are confident that this O level course will provide the student with a better mathematical education and that, in addition, the teacher will find that much of the new material we propose is far more suitable for classroom use than certain topics still to be found in established syllabuses.

SOME 'MODERN' APPROACHES

THE O LEVEL SYLLABUS

What follows is the syllabus on which the O level examination will be set in July 1964. It is not a teaching syllabus: individual schools (and the texts) may well go beyond what is here, in their fifth-form work. It is also provisional in the sense that examinations in succeeding years may follow a syllabus duly modified by experience.

General

The emphasis of the examination will be on the understanding of simple basic mathematical concepts and their applications.

Importance will be attached to clear expression and careful reasoning; candidates will be expected to understand the correct use of the signs $\Rightarrow$, $\Leftrightarrow$.

Questions requiring lengthy manipulation will not be set.

Candidates will be expected to be able to express physical situations in mathematical symbols, and to use their judgement as to the degree of accuracy appropriate to any particular problem.

Slide rules with A, B, C, and D scales, the usual geometrical instruments, and an approved set of 3-figure trigonometrical tables with a list of formulae will be required.

Knowledge will not be required of the rectangle properties of the circle; angle bisector theorem; extension of Pythagoras; secant, cosecant and co-tangent ratios. Questions will not be set explicitly on proofs of theorems and 'ruler and compass' constructions. Questions will not be set involving the solution of quadratics by formula or by the completion of the square; nor on the '$\frac{1}{2}ab \sin C$' or 's' formulae for triangles.

Syllabus

The important units of weights, measures and money, including metric units (Quantities will not be expressed in more than two units with the exception of £. s. d.)

Fractions, decimals, ratios, percentage.

Approximations and estimates, significant figures, decimal places, limits of accuracy and the use of inequality signs.

The idea of scales of notation other than the denary.

The expression of numbers in the form $a \times 10^n$ where n is a positive or negative whole number.

The use of the slide rule.

Length, area and volume: mensuration of common plane and solid figures – the rectangle, triangle, circle, cylinder, cone and sphere.

The use of Pythagoras' theorem. Sine, cosine and tangent ratios of acute angles. Solution of triangles in cases reducible to right-angled triangles. Simple applications to three-dimensional problems.

The notation and idea of a set; union, intersection, complement, subset; empty and universal sets; Venn diagram; the number of elements in sets and the unions and intersections of sets. (Approved symbols: $\in$, $\cup$, $\cap$, $'$, $\subset$, $\varnothing$, $\mathscr{E}$.)

Locus.

The use of symbols to represent numbers, sets, transformations.

Conditional and identical equations: rearrangement of formulae.

Factorisation of $ax + bx$, $a^2 - b^2$, $a^2 \pm 2ab + b^2$, simple manipulation of fractions, $x \times y = 0 \Leftrightarrow x = 0$ or $y = 0$.

183

Inequalities and their manipulation. Simple and simultaneous linear equations and inequalities in not more than two unknowns. Applications of inequalities, especially to linear relationships and graphs.

Rectangular cartesian coordinates. 2×2 matrices. Vectors as matrices. Matrix multiplication, the unit matrix, the formation of the inverse of a non-singular matrix and applications to simultaneous equations and linear transformations.

Relationships, especially linear, square, and reciprocal, and their graphs. The exponential law of growth. Proportion of variables related by simple power laws.

The gradients of graphs by drawing and the estimation of areas under graphs. Applications to easy linear kinematics involving the distance–time and speed–time curves and other rates of increase.

The use of graphs in linear programming.

Similarity and congruence. The geometry of euclidean space based on the operations of reflection, rotation, translation and enlargement. Symmetry about planes, lines and points. Combination of transformations.

The circle, including the constant-angle property and tangents.

Applications of similarity including the areas and volumes of similar figures, scales and simple map problems.

Simple plans and elevations.

The earth considered as a sphere: latitude and longitude, great and small circles, nautical miles, distances along parallels of latitude and along meridians.

Simple probability. (Specific knowledge of the sum and product laws will not be required but problems on the combination of probabilities may be set.)

Graphical representation of numerical data; calculation of the mean, median and quartiles.

Exercise

Prepare an examination syllabus for the Mercer curriculum. What, apart from the inclusion of the calculus, are the main differences between Mercer and SMP? Do they display a similar spirit?

2. SOME NOTES ON THE EVOLUTION OF THE SMP O-LEVEL COURSE

(A) First thoughts

The papers published in 1964, although correctly describing the aims (*purposes*) of the SMP founders, do not describe the course (*content*) as it was originally envisaged. In the two years' work which preceded the publication of these documents many original ideas had been modified and new topics had crept into the course. In retrospect, the biggest change was the realisation that matrices had an important role in the O-level course. It was only at the last moment that matrices took their place in the syllabus, and at first little attention was paid to them, as we see by reference to the extracts reprinted. At that time, those involved found it difficult to conceive a treatment of matrices fundamentally different from what they had encountered at university. It was feared also that examiners would look on the multiplication of matrices as suitable material to fill the gap caused by the omission of the solution of quadratic equations. Perhaps more striking, though, was the original attempt to introduce non-Euclidean geometry to the O-level course. Classroom material was prepared which was intended to provide pupils with an introduction to hyperbolic and elliptic geometries, as well as the more usual projective geometry. The material was rarely taught – perhaps because of the teachers' lack of background knowledge – and by 1964 the experiment had been dropped.

Other minor changes involved the abandonment of an attempt to teach only the trigonometry of acute angles, in favour of an approach through components; and this opened the way to consideration of the general angle (it will be observed, though, that the 1964 syllabus includes trigonometric functions of acute angles only).

(B) *Later thoughts*

If one compares the 1964 syllabus with that for 1969 (Thwaites [3]) and the original O-level books (*T*, *T*4) with the later ones (*1–5*) then several swings in emphasis can be discerned.

Perhaps the most striking omission in the 1964 papers is any specific reference to relations, functions and their composition. This oversight had been rectified by 1969, when the examination syllabus specifically referred to them and to relation diagrams and matrices; for, by then, the need to devote a considerable time to these fundamental concepts had become clear. The use of matrices to describe relations was but one of their applications to have found a place in the 1969 course. Experiments had shown that matrix descriptions of routes and networks were suitable O-level material (SMP *Book 2*) and the way in which they were introduced made it possible to consider matrices of all shapes.

The additional time available for teaching new material also affected the course. The work on vectors, in particular, was greatly extended and some elementary topology included (SMP *Book 2*). Pupils were now explicitly required to know about rational, irrational and prime numbers and to find the solution sets of equations in various domains. In trigonometry, the general angle was now considered. Greater coverage was given to statistics, probability and other applications of mathematics.

On the other hand – and significantly for mathematical educators – some of the more abstract work contained in the earlier course had disappeared. Although the idea of structure and of a group was retained (SMP *Book 5*), the more abstract concept of isomorphic structures (SMP *Book T4*) was omitted. Again, the frontal attack on 'proof' (SMP *Book T*) was dropped in favour of a more restricted attempt to show how certain conclusions in geometry can be reached from given data (SMP *Book 4*). This does not reflect any change in the overall aims of the SMP, but rather is a reframing of specific objectives in an attempt to bring them into line with what experience has shown to be feasible. Thus we have examples of feedback being used to determine changes not only in approach (*method*) but in specific objectives (*content*).

Exercises

1 Establish the symmetry of the figure in the Director's report (p. 181). What needs to be assumed first?

2 Such an object as a smoker's pipe has mirror symmetry about one plane, yet it can be inserted into a suitable close-fitting case in only one way. What is the relationship between this kind of symmetry and that of the child's 'posting-box' toy (see SMP *Book 1*, p. 223)?

3 There is a reference on p. 182 to the manner in which arithmetical misunderstandings can be diagnosed through the use of the binary system. Design tests for some common misunderstandings.

4 The SMP now provides an alternative route to O-level through its *A–H* and *X, Y, Z* books. Discuss changes of emphasis, of both content and method, apparent in this latest O-level course.

5 Write a unit on non-Euclidean geometry.

3. CASE STUDY 3

The Cambridge Conference and Contemporary School Mathematics Project Syllabuses

The two flow charts refer to mathematics courses for 5–12-year-olds (in the USA) and 11–16-year-olds (in England). The former (K–6) is discussed in detail in Chapters 15 and 18. The second (CSM) is a course designed for pupils of above-average ability studying for the GCE O-level mathematics examination. As with the SMP syllabus on p. 183, this flow chart represents 'second' thoughts – and there have since been others.

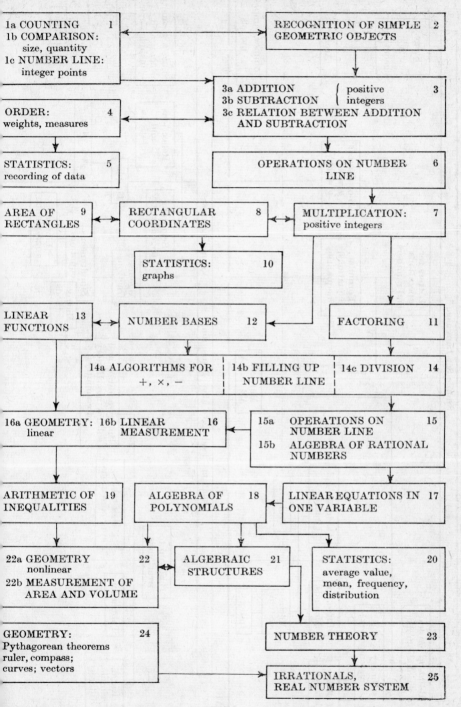

1a COUNTING   **1**
1b COMPARISON:
    size, quantity
1c NUMBER LINE:
    integer points
	RECOGNITION OF SIMPLE **2**
	GEOMETRIC OBJECTS

3a ADDITION   { positive   **3**
3b SUBTRACTION   { integers
3c RELATION BETWEEN ADDITION
    AND SUBTRACTION

ORDER: **4**
weights, measures

STATISTICS: **5**
recording of data

OPERATIONS ON NUMBER   **6**
LINE

AREA OF   **9**
RECTANGLES

RECTANGULAR   **8**
COORDINATES

MULTIPLICATION: **7**
positive integers

STATISTICS: **10**
graphs

LINEAR   **13**
FUNCTIONS

NUMBER BASES   **12**

FACTORING   **11**

14a ALGORITHMS FOR   14b FILLING UP   14c DIVISION   **14**
+, ×, −   NUMBER LINE

16a GEOMETRY: 16b LINEAR   **16**
    linear   MEASUREMENT

15a   OPERATIONS ON   **15**
       NUMBER LINE
15b   ALGEBRA OF RATIONAL
       NUMBERS

ARITHMETIC OF   **19**
INEQUALITIES

ALGEBRA OF   **18**
POLYNOMIALS

LINEAR EQUATIONS IN   **17**
ONE VARIABLE

22a GEOMETRY   **22**
    nonlinear
22b MEASUREMENT OF
    AREA AND VOLUME

ALGEBRAIC   **21**
STRUCTURES

STATISTICS: **20**
average value,
mean, frequency,
distribution

GEOMETRY: **24**
Pythagorean theorems
ruler, compass;
curves; vectors

NUMBER THEORY   **23**

IRRATIONALS,   **25**
REAL NUMBER SYSTEM

187

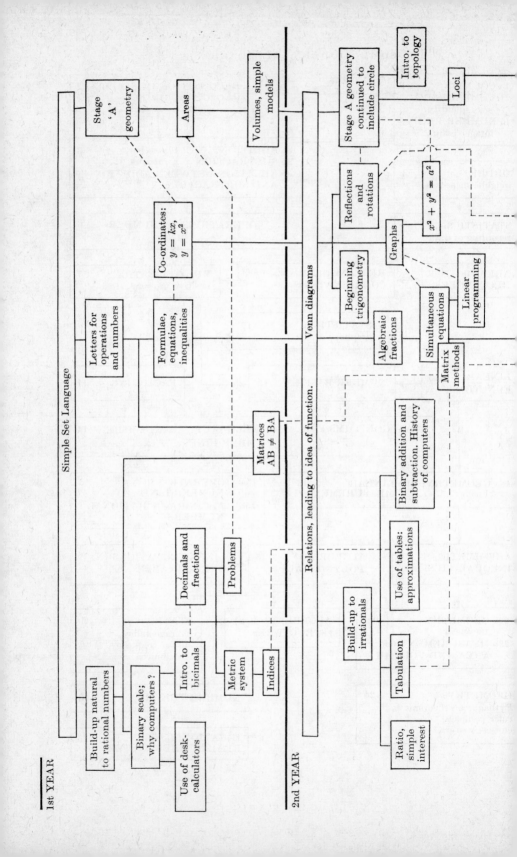

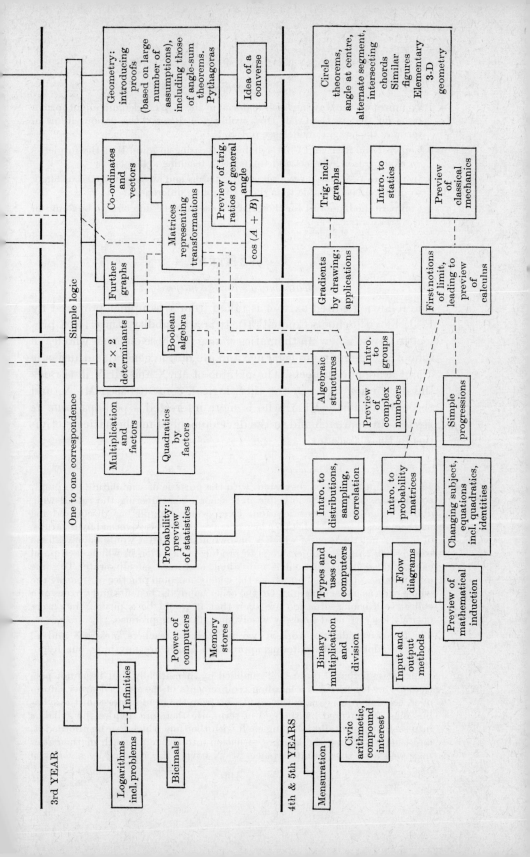

3rd YEAR

4th & 5th YEARS

Logarithms incl. problems

Infinities

Bicimals

One to one correspondence

Simple logic

Mensuration

Civic arithmetic, compound interest

Power of computers

Memory stores

Binary multiplication and division

Input and output methods

Types and uses of computers

Flow diagrams

Preview of mathematical induction

Probability: preview of statistics

Intro. to distributions, sampling, correlation

Intro. to probability matrices

Changing subject, equations incl. quadratics, identities

Simple progressions

Multiplication and factors

Quadratics by factors

Algebraic structures

Intro. to groups

Preview of complex numbers

2 × 2 determinants

Boolean algebra

First notions of limit, leading to preview of calculus

Further graphs

Gradients by drawing; applications

Co-ordinates and vectors

Matrices representing transformations

Preview of trig. ratios of general angle

$\cos (A + B)$

Trig. incl. graphs

Intro. to statics

Preview of classical mechanics

Geometry: introducing proofs (based on large number of assumptions), including those of angle-sum theorems, Pythagoras

Idea of a converse

Circle theorems, angle at centre, alternate segment, intersecting chords Similar figures Elementary 3-D geometry

Exercises

1 Comment on the way in which material in the K–6 curriculum might form a basis for that in the CSM. How would you describe the various items in terms of the 'stages' model (see p. 158)?

2 Contrast the SMP and CSM syllabuses (bearing in mind that the former is an examination syllabus and the latter a teaching syllabus).

3 In what respects are an examination syllabus and a teaching syllabus likely to differ? Is it desirable that there should be differences?

4 Compare the blueprint for the CSM course with the outcome to be found in the group's texts.

4. CASE STUDY 4

Mathematics for the majority

The following edited extract is taken from a pamphlet prepared by P. J. Floyd (Schools Council [3]). It is a feasibility study for a project to devise a new mathematics course for those children who leave school at the minimum legal age and are not, in general, enthusiastic about academic subjects. The parable of Mr X and Mr Y illustrates the distinction between the tidy, logical, content-based Mr X, and the individualistic Mr Y who is more interested – perhaps *faute de mieux* – in approach and in the development of mathematical activities in the classroom.

42. The approach to be suggested, with the purpose of stimulating the pupil, could prove perfectly practicable in a school, remembering the rather wide ability range under consideration. Even when a pupil has discovered his mathematical road, the distance he can travel along it is governed by a variety of factors, not the least of which is his own ability, and it would be unrealistic to overlook this. The objective is to produce a situation in which each pupil can progress at his particular, individual, and most effective rate. To some extent at least this must cut across the widely accepted practice of treating the whole class as a teaching unit. On the other hand, the benefits and virtues of a collective learning situation are such that it would be a pity if they were entirely lost; we need to apply wisely the gift of compromise.

43. Let us consider the work of two hypothetical teachers, Messrs X and Y, as they adopt widely differing approaches which are anything but hypothetical.

Mr X is, perhaps, quite well qualified mathematically, but he is not particularly in tune with the learning requirements of the slower learner. Moreover, echoing in his ears, is the injunction from the headmaster that the syllabus must be covered by the end of term. He therefore divides the syllabus into sections, mentally labelling each with the time which can be afforded it, and thus converting the learning situation into a race in which mathematics may well take the hindmost place. Mr X covers the syllabus by a series of lessons delivered to the class as a whole. In any particular lesson, his exposition is clear, his recapitulation also and he uses the blackboard artistically. And so the stage of the class working some examples has been reached. Here he becomes conscious of some difficulties and he would like to subdivide the class into groups. But such a practice does not find favour with his superiors in the school, so he drops the idea. In the terminal test, the results are disappointing, both to Mr X and to the pupils. His zealous and continuous hard work does not seem to be rewarded by a satisfactory outcome, and this he attributes to the lack of mathematical ability on the part of the particular set of pupils with which he has had to deal.

Mr Y is a man whose mathematical qualifications are rather less than those of Mr X, but he is very much *au fait* with the individual characters and foibles of the members of his class of fourth year leavers. At first sight, his classroom appears untidy . . . unlike Mr X's classroom, with the pupils working around the edges of the large table space formed by grouping the desks. The pupils themselves move freely about, to a side bench or a store room to gather materials, or to a shelf in the corner to consult a book of reference. They discuss their work freely with Mr Y and with one another. This leads to a certain amount of working noise. The observant Mr Y, however, never misses a move, and knows full well when and where to intervene, quickly and to no uncertain effect. He knows that there is a form syllabus in mathematics, but this he regards as a general framework within which he can operate and he does not attempt to complete the syllabus in any given time, or even at all in some individual cases. He is careful and meticulous in recording the work and progress of each individual pupil. Mr Y (and those who think with him) claims that this type of approach, combined with a measure of class teaching, solves to a great extent the problem of a wide ability range and widely varying learning rates. In general, he prefers the pupils to work in pairs of broadly equal ability; but he finds that some of the abler of these pupils are sufficiently self-reliant and competent to work on their own on a topic of their own choosing, after discussion with him. At present about half the weekly time is devoted to this procedure and the other half to a more traditional procedure with the class working together as a group. Much of this work, so presented on assignment cards, is often geared to the syllabus topic under general consideration at that time – it may be perimeter, area, volume or some other aspect of shape; but together with this, some provision of time is made for a pupil to pursue his individual bent. In addition, Mr Y has acquired a certain skill in making mathematical issues out of situations currently arising; with his initiative and vision he could make even more of this were he of heavier mathematical calibre. One example will suffice.

As part of the school's social service programme, the form had undertaken the purchase of a quantity of railway sleepers, and after reducing them to bundles of firewood, they were to be distributed free to deserving cases in the neighbourhood. After studying the essential geometrical and physical qualities of a railway sleeper, discussion, measurement and estimation determined the best length of the blocks to be sawn off to avoid waste of wood and unnecessary sawcuts. How many blocks came from one sleeper? Then after the sawing of the sleeper came the chopping of the blocks into sticks. What was a fair figure for the number of sticks produced from one block? When the sticks were assembled into bound bundles, how many sticks to a bundle? How many bundles to a block? How many bundles to a sleeper? How many bundles from all the sleepers? Was this expectation in fact realised? If not, why not? How many sticks would be produced in all? In this realm of relatively large numbers the computational ability of some members of the class was severely tested, to say the least. The intelligent use of a desk calculator would have eased the labour. This being a class encouraged to discuss mathematical problems verbally, and accustomed to the responsibility of a fair degree of freedom, they then embarked on what was for them, a matter of high finance. Had they sold the bundles of firewood at the retail price prevailing in the neighbourhood, what profit would they have made?

44. The point to be noted is that Mr Y was prepared to shelve his prepared plan and syllabus content and to devote some of the time allotted to mathematics for such a study. It demanded from him not only vision and initiative but organisational and professional powers. The project was a source of some justified satisfaction to him, and above all it was a course of considerable satisfaction to the pupils engaged on it. Indeed such was true of most of the mathematics they encountered in this permissive but controlled and orderly atmosphere.

45. It must be emphasised that Messrs X and Y are hypothetical teachers. If they indeed exist in the flesh, they have yet to be met as whole persons. The elements of the composite pictures presented are drawn, however, from experiences of discussing such problems with teachers, and also from case book notes.

46. The parable of the two teachers illustrates not only differences in approach but also a reorientation of outlook. 'Logical' teaching, in the first case, is replaced by a degree of psychological learning in the second case. It is suggested therefore that serious consideration be given to a more widespread adoption of the second approach; an approach based on the conviction, supported by psychological findings, that all children, but particularly the slower learners, really learn mathematics through close personal involvement with physical models or manifestations . . .

48. Many of the mathematics syllabuses of the CSE examining boards proffer offerings to the average pupil far in advance of anything hitherto met on a large scale. With an increase in the number of schools thinking of the Mode II form of examination (internal syllabus externally examined) and the Mode III form of examination (internal syllabus internally examined and externally moderated) an even wider variety of suitable examination content should result. Syllabuses under Mode I examination (external syllabus externally examined) are published by the various examining boards, and are readily available for study to anyone interested. The list of topics for an individual study by a pupil, as given below, might well prove more fruitful in one or all of these circumstances:

(a) where the Mode I CSE examination permits the submission of individual studies as part of the examination work; or

(b) where such work is acceptable to an examining board under Mode II or Mode III arrangements; or

(c) in any school, and to all pupils, including those of below average ability where it is recognised that the natural practice of reading and writing form an integral part of the mathematical education of a pupil.

The list is not intended to be exhaustive by any means and the subjects are listed entirely at random.

(1) A *historical* study of some particular field of mathematics with which the pupil is already fairly mathematically familiar, e.g.: English weights and measures, metric measures, time and timekeeping, the mathematics of ancient Egypt, aids to computation.

(2) Construction of, and some mathematics arising from, some geometrical models – polyhedra perhaps.

(3) The mechanics of flight, or of a bicycle, or of the human body.

(4) A study of a computer – its parts and their functions – the signals to which it responds (probably best limited to binary code at this level), punched cards and punched tape.

(5) Construction of simple logical circuits, again involving a two-state system study and the accompanying binary mathematics.

(6) Computational aids. A wide variety possible here ranging from conversion graphs to slide rules, nomograms and desk calculators.

(7) Mathematics in nature. Natural forms related to their mathematical models; growth rates; food conversion rates.

(8) An environmental study demanding the collection and presentation of some statistical data.

(9) Where appropriate to a given area, a study of tides, or of navigation or the mathematics of a swimming pool, or of a school farm unit.

(10) Aspects of geometrical design associated perhaps with needlecraft, or the layout of a formal garden, or church architecture.

(11) Some interesting loci – the harmonograph or spirograph – cycloids, epicycloids – hypocycloids.

(12) Trains of gear wheels and some applications commonly found.

(13) A study of the 'Golden Section' – its properties and manifestation.

(14) A study of a particular regular polygon, e.g. hexagon. Various constructions possible – properties – symmetries (including rotational), tessellation possibilities. Linked possibly with arithmetic (modulo 6).

Exercises

1 Name some 'benefits and virtues of a collective learning situation', which the author of the extract does not wish to lose (Para. 42).

2 Frame a definition of 'mathematical ability' which would justify Mr X's conclusion in Para. 43. Frame a definition which would not.

3 Devise some questions and problems relative to the handling of the firewood by Mr Y's boys.

4 What kind of discussion would you expect to ensue if the pupils were asked to say what might be done with any profits that could have been made by selling the wood on the open market?

5 Could Mr Y have worked as he did if his pupils needed to pass an examination?

6 After the extract, the author writes 'The pupil then approaches the mathematical road with a desire and a will to travel along it and to discover relationships between the ideas previously found. He is now a genuine mathematician, even if he never becomes a very good one in the accepted

193

sense of the term.' In what sense is a genuine mathematician *not* a good one ? Must a good one be clever ?

7 'There is no defence of a demand, made in a locally recognised examination for 15-year-old school-leavers, for pupils to be involved in such expressions as

    (i) $1\frac{13}{15} + \frac{5}{7} \div 2\frac{1}{7} - \frac{8}{9}$ of $1\frac{4}{5}$,

    (ii) $\sqrt{3.3489}$ without using tables,

    (iii) £15-16s-7d × 31.' (*Mathematics for the Majority*, p. 24.)

What defence do you think the examiners might adduce ? Is there any other defence ? (See Chap. 18, Section 9.)

The author goes on to systematise the work he is suggesting by presenting his 'Chart A', which 'presents streams of development stemming from the three themes, number, relationships and shapes . . . they are inter-related, and this linkage is shown on the chart'. He goes on to emphasise that a 'Stage A' approach is intended. Four other charts are given by the author, of which Chart B is a sample: like the others it expands part of Chart A in greater detail. To be able to devise such expansions is a very important activity of the curriculum designer.

Exercises

1 Write similar expansions of your own for other parts of Chart A.

2 'I hate "teaching", and have had to do very little . . .; I love "lecturing" . . .' (Hardy [1], p. 149). Discuss this statement and its relevance to the story of Mr X and Mr Y.

3 Into which of the Beeby categories would you place (*a*) Mr X, (*b*) Mr Y, (*c*) G. H. Hardy ?

5. CASE STUDY 5

Mathematical curriculum studies

The course described below was designed, as an option, for Honours mathematics students in their final year at Southampton University. Both authors of this book have participated in this course which, indeed, led to the writing of the book.†

The course is provided so that students, whether or not they intend to become school-teachers, can learn more about the problems of designing mathematical curricula and of the work being carried out in different parts of the world. Such a course has to satisfy various constraints arising from the structure of the usual British degree course as a whole. It has to be offered as an option in

† An earlier version of this paper was presented by one of the authors at a Colloquium held in Bucharest in 1968 (see UNESCO [4]).

Mathematics for the Majority : Chart A

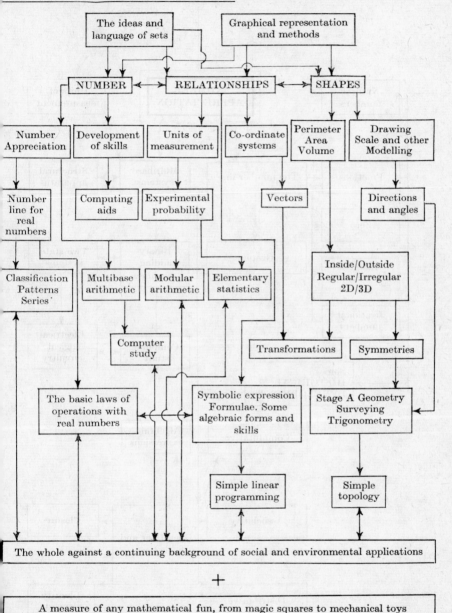

The whole against a continuing background of social and environmental applications

+

A measure of any mathematical fun, from magic squares to mechanical toys

Mathematics for the Majority : Chart B

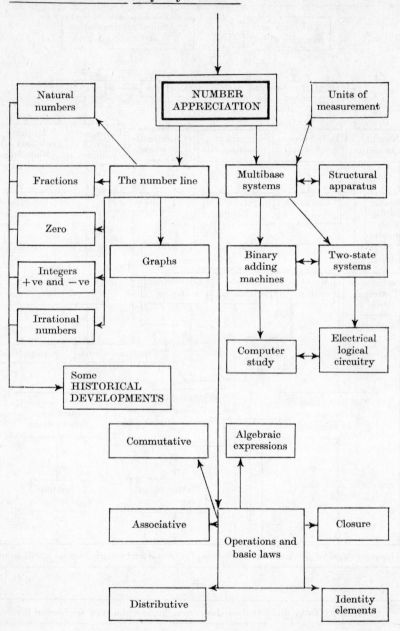

competition with about thirty other mathematics courses, of which each student chooses eight. During the year, he must take an examination in each of the options, and in 1966, when the course was first given, all examinations were of the conventional timed type. The course had, like the others, to consist of 36 contact hours followed by an examination at the usual English Honours level.

The examination paper had, therefore, to be comparable in difficulty with the papers in, say, topology or fluid dynamics. (Because of the financial differentials that the award of an Honours rather than a Pass degree can carry in British life, this constraint – which readers from other cultures may find surprising – must be considered when planning any new course.) The students taking the course could all be expected to have reasonable backgrounds of real and complex analysis, algebra and groups, some geometry and (depending on their earlier choices) such specialities as statistics, quantum theory or fluid dynamics. Some intended to teach, but others would go into industry as specialists or managers.

It seemed unsuitable to give them a purely historical account of the famous projects like SMSG or SMP – even though the latter had originated in that very Mathematics Department – because the form of examination would then lead to mere memory work. Detailed references to changing classroom techniques were also not suitable, since this was the sphere of the School of Education. Nevertheless, something of these matters would have to be included.

Similarly, some of the sociological questions we have discussed earlier in this book would need to be included, but a wide coverage could not be given, owing to shortage of time and of a suitable text.

To fix ideas, therefore, the first courses were based on duplicated notes and took as a starting point the flow diagram on p. 187 of this book. This diagram indicates a course in primary school mathematics suitable for children aged between 5 and 12. It begins with discrete mathematics such as counting and cardinal numbers, and with continuous mathematics such as length, area and volume, developing and intertwining the two. The diagram had been used as the first step in designing a course for primary teachers (see p. 209); and during the many hours of discussion about its construction it became clear that a considerable mathematical competence was necessary for those who took part. Starting with this, the students were challenged to see if they could talk about the underlying mathematics which at first sight seemed to them to be trivial. They were asked what we meant by counting, cardinal number, length and area. The students could not usually explain and then they realised that if *they* as specialists couldn't, they could not expect primary teachers either – and even less could the primary children themselves understand such concepts. It followed that the students needed formal knowledge of such topics as cardinal number and area, at a rigorous level; and two objects had to be kept in mind:

(*a*) To show that mathematics specialists must be able to communicate mathematical ideas with the highest precision to other mathematicians.

(*b*) To show that once they can do this, then they can contemplate the harder problems of watering down the presentation to make it suitable for those less experienced than themselves. Here they almost certainly need co-operation from those acquainted with the intellectual attitudes of their target audience.

With (*a*) the lecturer was in a thoroughly orthodox situation because one normally attempts in a university to teach topics like transfinite arithmetic, number theory, Jordan content, etc., 'with the highest precision' and at a professional level. Moreover, this material could be set to students in examinations of the traditional kind. However, the real purpose was to develop (b) and test it in examinations. To this end, one can illustrate within a student's experience by showing how some results in, say, calculus are simple deductions from theorems (e.g. the mean-value theorem) which are obvious and intuitively acceptable but hard to prove. On the other hand, some proofs are not suitable for such treatment and an analogue approach may be necessary – for example, to impart conviction that $(-1) \times (-1) = 1$ as in the K–6 primary course discussed in Chapter 18. Clearly, such examples give plenty of scope for discussion with the class about the finer details of technical proofs and about the general concept of proof in mathematics. It is precisely such discussion that students miss in their orthodox courses of lectures on analysis, algebra, etc.

Each student did one sustained piece of work over the vacation. These were of the form: devise short units on geometry (or statistics or computation) suitable for schools; write a review of certain of the newer school texts from the point of view of advising a school whether or not to adopt them; find a school involved in a new project and describe its difficulties; try out an idea of your own on a group of children and write an account of what happened. On all these topics excellent work came back as well as essays on broader topics like 'What is the point of doing mathematics?', 'A survey of material on modern mathematics available for parents' or 'An account of some of Piaget's experiments'. Since some of them gathered a great deal of factual information the students were encouraged to circulate their work among the class, but there was unfortunately not time for the whole class to discuss it in detail.

Finally the students took a written examination, and we have included (see pp. 199–202) some of the questions set. The open-ended† ones tended to be avoided or not well done, but this may be due to the time factor (full marks are obtained for doing four questions in two hours). On the other hand British university students of science are traditionally trained intensively to answer very closed-ended questions, and patience will be needed to reverse the tradition. This tradition may also have a bearing on the way in which the students select themselves for the course. Comparing them very subjectively with those taking another course (such as Geometric Topology) we infer that this course attracts the less introverted students – the ones more interested in people than in mathematics for its own sake; with one or two exceptions the potential research-students have avoided it. On the other hand, the type of student in the course has been frequently the sort to engage in the mathematical dialogue we wanted to generate. British students, unlike American ones, are difficult to get talking in a formal classroom, but there is a trend in our department to try to teach mathematics in such a way that students can talk about it (to each other and their instructors) with a feeling of confidence, just as professional mathematicians do. In the past they have relied solely on doing written mathematics. Of its nature, the course has to be treated in a non-authoritarian way, since unique 'best' solutions to the problems do not exist. Here also a great deal was learnt by discussing with the students the nature of examinations and some reasons for the failure rate. Perhaps they also learned something of our problems as setters of examinations, and of our general difficulties as administrators of an educational system in which all kinds of

---

† By definition, an 'open-ended' question is one without a unique solution, which usually involves an opinion, or choice of direction to follow. A 'closed-ended' question on the other hand is of the kind 'Prove that $X = Y$'.

compromises have to be made, in view of such external constraints as shortage of time, money and manpower. Any dialogue of this kind is surely also worth-while in view of the present world-wide trend among students against formal examinations and towards having a greater say in the running of universities; we want their grumbling to be constructive! For several students the course was reinforced by an option on the History of Mathematics given from a modern point of view.

While the course has been enjoyed by students it still has defects (though these are being improved through a liberalisation of the examinations by placing greater emphasis on project work) and by moving some of the mathematics (such as problem-solving) into a separate course. It has been noticeable that many students take advantage of the open-ended nature of much of the work to move away from mathematics. Though this may provide therapy for weaker students and excellent preparation for later educational studies, it runs counter to the aim of the course that the student should learn to *relate* his mathematics to educational problems.

Exercises

(The following questions were set in the University of Southampton for the degree of BSc in Mathematics.)

1 Define cardinal numbers with their addition, multiplication, and ordering. Verify that the ordering is compatible with addition and multiplication.
  Explain how the usual geometrical model of the whole numbers $\mathbb{N}$ in the number-line $\mathbb{R}$ can be reconciled with these definitions, and discuss the operation $a - b$ in the two cases $a < b$, $a > b$ (in $\mathbb{N}$).

2 How can $\mathbb{R}^2$ be used as a model for plane Euclidean geometry? Investigate those isometries of $\mathbb{R}^2$ which leave 0 fixed, and indicate how they may be used to introduce the notion of angle into $\mathbb{R}^2$.

3 Give geometrical constructions which show how to add and multiply real numbers on the number-line $\mathbb{R}$. Using analytic geometry, show how the operations you define satisfy the CAD laws. Give visual proofs that $(-1)^2 = 1$ and that the operations agree with those of cardinal arithmetic when we identify $\mathbb{N}$ with the usual lattice-points in $\mathbb{R}$.

4 Define 'length' of a curve in $\mathbb{R}^2$, and prove that the unit circle has a length. Obtain simple approximations to this length, with estimates of error. What steps in the argument would you omit from a 'school' treatment, as compared with one in a University course?

5 Is it a definition or a theorem that the area of a plane rectangle is length times breadth? Define an area for plane polygons, showing it to be translation-invariant and additive in a suitable sense. Reconcile your definition with the standard area of a triangle.
  If $T:\mathbb{R}^2 \to \mathbb{R}^2$ is a linear transformation and P is a polygon in $\mathbb{R}^2$, show that $T(P)$ is a polygon, and by direct calculation display a relationship between the areas of P, $T(P)$ and det T.

6 Prove the theorem that to each bijection $f: A \to B$ there is a function $g: B \to A$ such that $f \circ g = id_B$, $g \circ f = id_A$, and g is also a bijection,
  Show how to modify your proof, using Papygrams, to make a convincing visual demonstration of the above theorem, at an elementary level. Use

199

this technique also to demonstrate that a composition of two bijections is a bijection.

7 Construct a flow-diagram for setting out a course at sixth-form level on one of the topics: Statistics, Logic, Approximation, Mechanics, Differential Equations.

Explain whether or not your diagram can be simplified to make a more elementary treatment of your topic.

8 The author of a certain text wishes to estimate the error in calculating a root $x_0 = 1 + d$ of the equation

$$(1 + a)x^2 - (3 + b)x + 2 + c = 0 \ (a, b, c \in \mathbb{R})$$

when $|a|, |b|, |c|$ are all $\leqslant 1/100$. Substituting for x_0 he gets

$$-d + (a - b + c) + d^2 + 2da - db + ad = 0. \quad (*)$$

He states 'We make the assumption that d, a, b, c are so small that any products of them can be neglected', and hence reduces equation (*) to $d = a - b + c$. He goes on to check the above assumption by showing that $|d| \leqslant 3/100, |2da| \leqslant 0.0006$ etc., and therefore that the neglected term in (*) is $\leqslant 0.0019$. Because this is $< |d|, |a|, |b|, |c|$, he says 'our assumption is valid'.

Criticise the author's logic, and use the usual quadratic formula to give a correct argument. How would you deal with the author's reply that the reader of his book was not expected to know about the Binomial expansion of $\sqrt{(1 + x)}$?

9 Suppose that you are an External Examiner for a Technical College, and that you are asked for comments on a paper which includes the question 'Differentiate $(1 - x^2)^{-1}$, $\sin^{-1} (1 + x^2)$, $\sqrt{x}$.'

Reword (or otherwise modify) the question to make it sensible

  (a) in professional language, and

  (b) in the case that the candidates could not be taught to understand (a) in time for the exam.

Indicate the form of solution you would regard as acceptable in the two cases.

If the College retorted that 'the students know what *we* mean, and won't understand what *you* mean', what would you reply?

10 Explain what is meant by a convex set in $\mathbb{R}^2$ and show that sets of the form $\{P \mid f(P) \leqslant \text{const.}\}$ are convex, when $f(x, y)$ is of the form

$$ax + by, \quad (a, b \in \mathbb{R}); \quad x^2 + y^2.$$

What if $f(x, y) = x^2 - y^2$?

Show that a linear function g on a convex polygon P achieves its bounds at some of the vertices. If g represents the profit for a 2 variable problem (such as the 'nut-mix') discuss what you would do if g took a maximum all along one edge of P.

11 A skeleton outline of a course for teachers gives a list of topics of which the first two are:

  (i) Define factor, prime and composite by looking through a (partial) table for multiplication of positive integers. Discover the uniqueness of prime factorisation for small numbers. Pose the general problem.

  (ii) Demonstrate Wilson's Theorem, that $(p - 1)! \equiv -1 \bmod p$ for each prime p.

Give a more detailed account of what you think would be an adequate exposition for the teachers to follow. What background knowledge would you consider to be essential?

12 For each positive integer $n > 0$, let $p(n)$ denote the highest power of 2 dividing n. Show that $p(nm) = p(n) + p(m)$, and hence prove that $\sqrt{2}$ is not rational.

Show how the function p might be used to introduce logarithms, and write a flow diagram to show what material you would need to derive the principal properties of the log function in a School Calculus course.

13 Explain the terms *Physical Induction*, *Mathematical Induction*, and the *Principle of Descent*. Show how the latter Principle follows from a property concerning the ordering relation in $\mathbb{N}$.

In what sense does the following statement hold 'by induction': If n resistances of $R_1, \ldots, R_n$ ohms are connected in parallel across two terminals, then they are equivalent to a resistance of R ohms where

$$\frac{1}{R} = \frac{1}{R_1} + \cdots + \frac{1}{R_n}.$$

Use Mathematical Induction to prove that a finite set with n elements contains exactly 2^n subsets.

14 Let $\mathbb{Z}_p$ denote the field of residues mod p where p is a prime, and let A denote the set of all matrices $X = \begin{pmatrix} a & b \\ -b & a \end{pmatrix}$, where $a, b \in \mathbb{Z}_p$. Show that A is a commutative ring in which $-I$ has a square root, where I is the identity matrix. (You may assume the CAD laws for matrices.)

If -1 is not a square in $\mathbb{Z}_p$ show that A is a field, and conversely.

Discuss a way of using A to throw light on various features of the system $\mathbb{C}$ of complex numbers. Can you introduce 'polar co-ordinates' in A? (Consider the case $p = 3$.)

15 A pack of twelve cards, numbered 0, 1, 2, ..., 11, is shuffled in the following way: first the pack is broken into two parts each containing six cards, and then the two 'sub-packs' are interleaved to form a single pack.

Thus, if the pack is originally in the order 0, 1, 2, ..., 11, then it will be broken down into the two sub-packs

0, 1, 2, ..., 5 and 6, 7, ..., 11

and interleaving will produce a single pack in the order

0, 6, 1, 7, 2, 8, 3, 9, 4, 10, 5, 11.

The process of shuffling is then repeated.

Obtain a formula which gives the position after the nth shuffle of the card numbered x.

After how many shuffles will the cards appear in their original order?

This experiment has been carried out by many classes of schoolchildren who have usually solved the above question by making use of the number patterns which appeared after the results of three or four shuffles were tabulated. Suggest topics in mathematics to which this and similar experiments could provide insight.

16 In a paper presented in 1964 to the International Commission on Mathematical Instruction, Alexander Wittenberg wrote 'if the aims (of teaching mathematics) include the mastery of *ideas* by the student (for instance, an

201

understanding of mathematical structure), then the examinations must include tests for mastery of ideas, that is, essay-type questions such as "*Explain and discuss the idea of mathematical structure*".'

Answer the question suggested by Wittenberg as one would hope to see it answered by an A-level candidate who has followed a 'modern' course.

17 (*a*) Make out a case for the inclusion of the differential calculus in the school mathematics curriculum.

(*b*) Set a question on the definition of a derivative and on the techniques of differentiation which is suitable for inclusion in an A-level paper. List the precise points which the question is designed to test.

(The two questions which follow were set under different conditions from those above. Although the answers had to be written under examination conditions – thus putting emphasis on content rather than presentation – candidates were told the questions some three weeks prior to the examination and were allowed to bring skeleton notes to the examination. These notes were handed in together with the scripts. Exercise 1 on p. 217 and Exercise 2 on p. 329 were also set under similar conditions.)

18 'The fundamental difference between school geometry and university geometry is that the former is a branch of physics and the latter a branch of mathematics.' (C. V. Durell, *Math. Gazette*, 1948)

Give an explanation of this statement with examples.

Do you think that Durell's statement is still valid, and, if so, do you think that this state of affairs is necessary or desirable? Give your reasons.

19 'The foremost goal of mathematics on a scientific level is the study of structures.

The most important means for the attainment of this goal is the axiomatic method.

If this is accepted, what are the consequences for school teaching?' (B. Christiansen, *Proc. 1st ICME*, 1969)

Give a brief explanation, with examples, of the view of mathematics as expressed by Christiansen. If you do not accept this viewpoint, give your reasons for rejecting it. If you do, then answer the question posed by Christiansen.

6. SOME GENERAL POINTS

To summarise, let us record what seem to be the general demands made on a potential curriculum designer. These were suggested to us in designing our course in 'Mathematical Curriculum Studies' – although we had more freedom than would be typical because we were concerned with relatively small classes in one institution, in which we were both lecturer and examiner, albeit in an existing framework. Nevertheless, the same principles apply to all the curricula in the Case Studies. In roughly logical order the general demands are:

(1) What topics is it hoped to teach?

(2) Why? What are the arguments for replacing old material by

new? These arguments must have a managerial content to take into account the fact that those who taught the old material may now become technologically obsolescent.

(3) What examination system is to be used and why? (Often the 'why' question here cannot affect the issue immediately, since the form of examinations is often imposed externally. Nevertheless if we keep asking the question we may be able to civilise the climate of thought.) What might be the effects on the students if the new curriculum itself leads them to fail the examinations?

(4) Can new course material be written? How? By whom? Can plenty of good problems be devised, or borrowed from mathematicians in industry or elsewhere?

(5) Can new course material be taught? How? By whom? How is the technologically obsolescent person in (2) to be retrained? What curriculum will be needed for this? One then asks (2) to (5) again.

(6) How expensive will all this be? Who will finance it? What would be the potential freezing effect of a heavy investment in new books and materials?

Observe that the view throughout (1)–(6) is taken, that a rather rigidly formal curriculum must first be designed even if there are gifted teachers available who can develop their own classroom methods of treating the material. Some such teachers refuse adamantly to use printed textbooks but nevertheless they surely need some kind of blueprint before they can actually go into action. Our aim in Case Study 5 was to help train mathematics graduates who could write such blueprints if required, either as advisers or as teachers themselves; and we hope the readers of this book will be able to act similarly.

15
Some Anti-Garbling Philosophies

However good a curriculum may look on paper, it is worthless if, by the time it is translated into real live lessons, its whole spirit is lost, and the pupils are turned into unthinking automatons. Thus, it must always be kept in mind that the teaching may cause it to degenerate; its message may be garbled (cf. p. 143). A teacher may expound the mathematics with perfect correctness, without communicating it to his pupils, as did Mr X in the parable of Case Study 4 in Chapter 14. On the other hand, Mr Y's enthusiasm and persuasiveness will have little worth if what he is telling his pupils is mathematical nonsense. In terms of the Beeby model (p. 62) we must allow for the Mr Xs who may be trained but not educated, and the Mr Ys who may be educated but not trained.†

In this chapter, we deal with certain philosophies about mathematical activity, which might, if they pervade the teaching of a curriculum, make teachers more aware of hidden dangers and lead them to watch out for garbling tendencies in their work. Self-awareness combined with a deep commitment to professional standards, are probably the only protection against degeneration. This must be set against the need for curricula that can be followed by poorly-trained teachers and those children who must frequently change school.

1. THE CONFERENCES AT CAMBRIDGE (MASS.): THE 'GOALS' PAMPHLETS

We begin by considering two theoretical works which were produced by groups of mathematicians at two conferences near Cambridge, Massachusetts (USA). They were organised by Educational Services Inc., as it then was (see p. 145), the first in 1963, the second in 1966. The job of the first was to look ahead, accepting the existence of the first wave of reform work being carried out by such projects as

† Some readers might not agree with the authors' classification.

SMSG, UICSM, etc. (see p. 139), and to consider 'what might be accomplished in a second wave of reform after the new courses and fresh attitudes to mathematics had been established in the schools and accepted by the public at large'. That conference lasted for several weeks, and resulted in the publication of a pamphlet *Goals for School Mathematics* (CCSM [1]) in which proposals (rather than practical recommendations) were put forward for mathematics curricula suitable for grades K through 12 (in American terminology), i.e. kindergarten to the 18-year-old age-group. As well as delineating the topics to be included in this education of the future (say twenty years from that time), they emphasised a philosophical attitude to mathematics, by asserting the importance of free and honest discussion of mathematical ideas by pupils and teachers: drill, authoritarian techniques – and their consequence, dishonest proofs – were condemned. It is impossible effectively to summarise the pamphlet, since it contains so much of value, reflecting the talents and character of its contributors. We can only urge the reader to go to the pamphlet itself and enjoy it.†

But of course, the authors knew there was a snag. This was the problem of getting the material taught by either existing or future teachers – 'how do we teach tomorrow's teacher to teach the mathematics of the day after tomorrow?' They set aside this problem, however, and in a characteristically American way, supposed first that there were to be no constraints at all: what was best *mathematically*? Even this simplification of the model did not produce just one answer, as we shall see. They realised too that there might be such constraints in practice as the ability or otherwise of the pupils to handle the ideas, but they felt that knowledge of this limitation is scanty and many experiments need to be undertaken to see what is possible – a refreshing contrast to the approach of certain traditionalists who assert dogmatically that since nobody has ever previously taught topic X to age-group Y, nobody ever will be able to. There was agreement that the 'Spiral Approach' would be best; that is, pupils should make a guided tour of mathematics and return to important ideas after encountering further material and gaining experiences that would increase their understanding of those ideas. In this way the basic unity of mathematics would be automatically stressed, in contrast to the splitting up into year-long compartments which was traditional in the USA (cf. p. 18). Here again we see the importance

† He should also read Stone [3], a review of the pamphlet.

of crossing the frontier from one subject to another. This reflects, within mathematics particularly, the tendency for mathematics (as a growing organism) to slough off detail by subsuming it under some new (*and simpler*) general theory – which is the only way to keep it comprehensible as it grows (see Zeeman [1]). The pamphlet lays great stress, too, on the motivation for mathematical ideas – that each occurred first as someone's solution to a problem. It followed that 'the design of imaginative problem sequences involving combinations of routine techniques and "discovery" procedures was a matter of the greatest importance in curricular development'.

But how were these 'ideal' speculations to be realised in practice, taking into account the real-life constraints of teacher training, pupil psychology, etc.? The first priority was given to the *mathematical* training of teachers, since it is useless to talk about teaching unless they have something to teach; not that this platitude has been accepted in the past by certain Education Departments, especially in some American Colleges, where students have spent all their time learning about Education, with only the background technical knowledge they acquired at High School!

This problem was considered by the Second Cambridge Conference (1966) on Teacher Training (CCTT), which had a different membership from that of the first. Its report is a pamphlet entitled *Goals for Mathematical Education of Elementary School Teachers* (CCSM [2]), for it restricted itself to the already formidable problem of training teachers for grades K–6, just half of the full period envisaged by the first conference. The new report has a deliberately sketchy style, so that no publisher could advertise a range of textbooks as being exactly what the CCTT recommended. No such conference should be naïve about commercial pressures!

The conference lasted for four weeks, and it is interesting to describe how it worked. First, there were preliminary discussions between the full membership of twenty-five or so. Most conferences on education, unlike those on traditional scientific topics, must still provide time for participants to describe their models to each other and to establish a common language. Once essential points of agreement are located it becomes clearer where energy must be directed. Then two groups were formed to work out the details of differing approaches that had become apparent in the preliminaries. From time to time the two groups met to compare notes, and sometimes members changed from one to the other. Since no fusion between the approaches was possible, two possible curricula (each with variations) were finally designed and these formed the basis of the final report. The reader should be aware of the hard work put in by participants: this was no mere speechifying kind of conference, and indicates the modern trend in educational planning to make people think professionally in written terms, using working papers, rather than the older approach, using off-the-cuff thoughts. Any idea, however far-out, was allowable provided its proponent could write out the details to give his colleagues an opportunity for considered criticism: 'put up, or shut up' was the motto.

1.1 *Two philosophies*

The two kinds of proposal given in the report reflect two important differences in the philosophy of mathematics, arising from the general question, 'What is a significant piece of research in mathematics?'. One reply is that of the great British eccentrics: one follows one's whims and tastes, and if a man is talented, his work will be good and hence significant. Another reply is that which has influenced some great Germans: clearly there are in mathematics some profound problems which can be easily stated but may take centuries to work out in full, and the techniques necessary for the working out will influence other parts of mathematics (for example, the problem of Fermat's Last Theorem, the Riemann Hypothesis, the problems of Hilbert (see p. 231), the three-body problem in mechanics, etc.). In this 'German' approach, then, significance is always relative to some larger goal. Neither approach is black and white, of course, and mathematicians frequently adopt both approaches.

Exercise

Find out what the problems mentioned above are, and try to state their significance, if you can. Find examples of the 'British' approach by looking up Boole, Heaviside, Turing, etc.

1.2. *A 'British' approach*

In the conference, the 'British' approach was adopted by one of the groups (which had only one British member), and its proposals embody the idea that future teachers should be presented with good mathematics – chosen in a fairly arbitrary way – and its inherent quality would prepare them to adapt to any future curriculum they were required to teach. A teacher might later need to learn more mathematics, but his initial exposure to good mathematics would give him the curiosity and confidence necessary for reading on his own, and the attitude to mathematics that CCSM [1] wished to foster. The consequences of this approach form 'Proposal Two' of the conference report, and a sequence of 'units' is outlined, each unit containing a list of its mathematical contents followed by a commentary on the spirit in which to teach it (mentioning motivation, number of lecture-hours, place in the sequence, etc.). We list just the titles of the units, and refer the reader to the report for details:

Unit IA, Number theory; IB, Vectors in line and plane, IC, Functions and transformations.

IIA, The real number system; IIB, Counting problems and probability.

IIIA, Intuitive differential calculus; IIIB, Linear transformations and matrices.

IVA, Isometries and symmetry groups; IVB, Quadratic forms and conics; IVC, Intuitive integral calculus.

A selection from these topics, depending on the time available for mathematics in the future teacher's training, would form his mathematical curriculum. They are therefore deliberately chosen with only loose interconnections, and they are intended to develop properties of systems assumed to be well known (in an intuitive way), rather than containing long developments of definitions, theorems and proofs. An alternative collection of units is also supplied including topics such as circular functions and complex numbers, induction, rings and unique factorisation, probability, statistics, and 'impossibility' proofs in elementary geometry.

One possible criticism of this proposal is that there is no built-in guarantee that those who learn in this way will not go back into the schools and teach traditional material by the traditional approach they themselves were given as children. 'Transference' of ideas and approaches, from one course of training to another, is often hoped for, but rarely occurs satisfactorily in practice. It would seem that if a curriculum designer *hopes* his curriculum will induce later specific behaviour in pupils, then *he must provide* for it in the curriculum. Experience shows, depressingly, that conventional methods of following curricula do *not* of themselves produce people who think broadly, seeing systems as a whole.

1.3. *An alternative approach*

The second group produced 'Proposal One', which is much more tightly structured. They began by asking first what mathematics was envisaged by CCSM [1] in the K–6 curriculum, and a flow diagram (see p. 187) was constructed to make clear how the various topics inter-related – all to see just what the future teachers were to be expected to cope with. It was then decided that it would not be right to teach the trainee teachers that material directly, but that they should be taught a syllabus containing material that was deeper and more advanced. The hope was that this would enable the trainees eventually to understand existing teaching 'units', and to write their own 'units' on the various items in the flow diagram. A few of these units were also written by members of the group, who had found both exacting and stimulating the discussions about how best to approach certain topics with children. An elaboration of one such unit, with the deeper underlying theory, forms the substance of Chapter 18 of the present book. As explained there, the 'unit' *material* is what would be taught to the children as part of the K–6 curriculum, while the *theory* is what would be taught to the trainee teachers. Similar considerations were to apply, so the group thought, to other parts of the K–6 flow diagram.

Now, that diagram reveals an essential framework of four topics: the arithmetic of the real numbers, introductory algebra, informal geometry, and simple applications such as statistics, weights and measures; with algebra providing a central bridge to tie together numerical arithmetic and geometry. This important role of algebra led the group to choose matrices as the topic that the trainee teachers would first meet at their training colleges, partly also because the topic would be new to many of them, it would give them practice in elementary arithmetic at which many of them were known to be bad, and it would apply to geometry. Further, well-tried texts existed from existing school projects. This then formed the starting-point of Course A devised by the group, to be taken by the trainees in their first year. Two more courses, 'B' and 'C', were then devised, intended as half-year courses for the second year. Course B was to bring out the algebraic aspects of the number system, by comparing it with the system of polynomials and other integral domains. The idea behind Course C – 'Mathematics, Science, and Society' – was to introduce the trainees to applied mathematics, through the techniques of algebra, calculus, probability and statistics, and logic. We shall enlarge on Course C in Chapter 19.

The first two courses turned out to be not greatly different in plan from those relevant topics chosen by the other group (which ignored applied mathematics except for probability theory). But if the reasons for selection of topics were known to the students, and especially if their mathematics was constantly illuminated by references to the K–6 flow diagram, then the effect on the trainees might well be quite different from that of Proposal Two. This illustrates the vast potential difference between a curriculum as a list of topics and as a realised, communicated discourse between expositor and student.

1.4. *A method of working*

Besides having a different philosophy about the selection of suitable topics for the proposed course, the two groups worked in totally different ways. Group I began by formulating a distinctive language as a common mode of discussion for its problems. Because it had begun by asking general questions about the nature of mathematical *activity* (in order to foster such activity by any proposed course), it was forced to invent a language for discussion, and with it some crude models. Briefly, the group agreed that all mathematicians are concerned with certain† important 'categories' of

OBJECTS AND MANOEUVRES

with the objects of essentially four kinds. DEDUCTIONS are then made about these, motivated by curiosity. The four kinds of object and manoeuvre are:

I. *Physical*. The objects are those of the real world, its solids, collections, etc., while the manoeuvres consist of making new objects from old, either by putting things together or by splitting them up. We sometimes also manoeuvre manoeuvres, by intelligent planning and organisation.

II. *Naming*. The objects are names for those in I, together with abstract names like the numerals. The manoeuvres are: forming

---

† The informed reader will observe the influence of the mathematical theory of categories and functors on the following account. See, for example, Griffiths [1], Ch. 37.

combinations of symbols according to rules, such as $3 + 6 = 9$, $(0, 2, 4)$, $\begin{pmatrix} 5 & 7 \\ 0 & 8 \end{pmatrix}$, combination of short algorithms to work out sums, etc.

III. *Construction of systems.* From particular integers like 3, 5, in II, we pass to such new objects as the ring $\mathbb{Z}$ of all integers, the field $\mathbb{Q}$ of rationals, the plane $\mathbb{R}^2$, space $\mathbb{R}^3$, the algebra of matrices, linear graphs, etc. Manoeuvres are such combinations as direct sum, forming quotient structure, etc.

IV. *Mappings between systems.* The objects are 'functors' (i.e. mappings) between sub-categories of the first three, and the manoeuvres are compositions of such functors, inducing new ones, etc.

At each stage, there is also a study of the objects and manoeuvres themselves; in this study, one manoeuvres thoughts about the objects and manoeuvres of that stage. Note that Stage I can enrich the others by forcing the construction of new models in one of the later stages.

Experience with these stages leads to the construction of strategies for solving problems, and to show that a strategy will work, one needs to devise proofs for convincing people that the claims made for a strategy are justified. Thus another category arises, whose objects are strategies and whose manoeuvres include their combination and assembly.

This, then, with admitted drawbacks, was a conceptual framework within which Group I argued out the curricula it devised. When the group explained this framework to the other members of the conference, the 'outsiders' rejected it fiercely, either on the grounds that (i) many more categories were necessary for an adequate description of mathematical activity, or (ii) that there are only two relevant ones – that of familiar mathematics and that of the unfamiliar mathematics one happens to be wrestling with at any time – or (iii) that some mathematicians do not analyse their activities in that way at all. In short, the conference would not endorse these ideas, but nevertheless Group I persisted with the framework as a scaffold to help its work, removing all trace of it at the end. An interesting observation was made, however, by those who could observe each group at work: Group II, without its framework, kept having increasingly heated discussions, whereas Group I was far more calm! One sees here how difficult it is to have discussions, designed to improve each participant's initial statements, unless there is agreement about what form an improvement should take. We have seen this before in mathematics (see p. 106), with regard to answering the question 'What is a number?'. Of course, the initial agreement about the form of an answer may be naïve and even wrong-headed relative to the inquiry, but the method at least shows all participants in the inquiry that all are then to 'blame' – since we must take account of such psychological factors as fear of being shown to be wrong, and the aggressive wish to seem superior by being correct. It was assumed that in such discussions, the object is not for one side simply to *impose* its initial ideas on the other side, thinking that no improvement is possible. Criticism is essential for the improvement of arguments and solutions, as the histories of science and mathematics show.

We hope to have shown the reader that the CCTT was a valuable experience for its participants, that the way in which it was conducted holds lessons for all designers of curricula, and that its report will repay careful reading.

Exercises

1 We have spoken of the need for participants at conferences 'to describe their models to each other' and the Proceedings of the Exeter International Congress (ICMI [1]) claim that 'although mathematics possesses a basic vernacular which has international validity, the words used in mathematical education have to be interpreted afresh by each congress member in the light of his educational environment'. How significant is this problem of language and, if it does exist, how is it to be ameliorated?

2 Do you consider the model of mathematical activity described in Section 1.4 helpful? What deficiencies has it? Can you suggest a better model?

3 'On all sides misgivings were expressed as to the propriety and the value of any suggestions university mathematicians might offer about the teaching of mathematics in an environment as remote from their own professional experience as the elementary school classroom necessarily must be.'
(CCSM [2], p. 11)

Is it proper that university mathematicians should comment on primary school teaching? Would such comments be 'professional' (see Exercise 2 on p. 152)? What value would one expect such comments to have?

4 'The older medieval philosophers like Anselm had said "I must believe in order to understand". Abelard took the opposite course: "I must understand in order that I may believe".'
(Clark [1])

Interpret these two points of view in terms of mathematical education in general and of teacher training in particular.

5 What could be done for those children, for example of servicemen, who must frequently change schools and, hence, mathematical courses?

6 Exercise 11 on p. 200 is based on an extract from the second *Goals* pamphlet.
Take other extracts from these pamphlets and expand them similarly.

2. MATHEMATICAL INSIGHT AND MATHEMATICS CURRICULA†

We have now seen many examples of mathematics curricula. It is
now necessary to discuss the question of mathematical *insight*, which
is something no curriculum can incorporate on paper, but whose
cultivation is a test of that curriculum.

We explained in Chapter 9 how the new mathematics curricula
relate to the changes that have occurred within university courses in
the last twenty years. These university changes have not been so
explicit, or carefully planned, as those for more elementary courses –
perhaps because of the greater freedom from central direction that
university teachers enjoy. But regardless of the level, the impetus
for change has come from two directions; and the relative emphasis
in any planned course has depended on the mathematical level of the
course.

2.1. *Two tendencies*

First, then, there is the tendency to change which comes from the
growth, internal logic, and reorganisation of mathematics itself;
this tends to be teacher-centred, because the need for change is seen
by the experienced worker in mathematics, and he explains the
subject accordingly. We shall refer to this tendency as the 'Renewal'
tendency; for details see Chapters 9 and 10.

Second, there is the tendency to change which springs from the
spread of humanitarian, liberal, anti-authoritarian attitudes in
schools and society at large; this tends to be pupil-centred, in that
mathematics (as any other subject) is developed with constant
reference to, and respect for, the pupil's reactions. We shall refer to
this tendency as the 'Open' tendency. (See Chapters 4 and 7.)

Both tendencies are present to some extent in the professional
development of mathematics, where youthful minds are constantly
creating new mathematics, and scepticism rather than authority is a
guiding force. Within the educational system proper, however, the
Open tendency has been associated above all with the teaching of
very young children, through the names of Froebel, Montessori,
Gattegno, etc. (See Chapter 11.) By contrast the first, Renewal,

† Earlier thoughts on this topic can be found in Griffiths [2].

tendency manifested itself most importantly at university level, especially through the work of Bourbaki.

2.2. *The influence of Bourbaki*

We mentioned in Chapter 9 that Bourbaki developed his own course as recorded in his book *Eléments de Mathématique*, a record still incomplete and under revision, still growing because it reflects the same trends in mathematics. To begin writing the book, Bourbaki had to devise mathematical goals which his readers were expected to reach, and then he had to devise appropriate mathematical paths to these goals. From existing mathematical treatments he had to devise suitable language, especially by selecting suitable definitions, so that he could then deduce all the mathematics he wanted, exposing it for his readers in a logical and beautiful way. To do all this involved immense toil and presumably some false starts, but these difficulties can hardly be inferred from the text itself, where Bourbaki tells his students the mathematical story selecting for them the definitions and the main logical development; the students do as *he* says. They may hear about his *human* story through gossip, or by noticing changes in successive editions, but the only official hint about the construction of the work is tucked away in the 'Notes Historiques' at the ends of chapters. (See, however, the 'inside story' as told by Dieudonné [3].)

Bourbaki's approach has had immense influence among younger university teachers, who have (often unconsciously) modelled their style on his, giving lectures that unfold before their listeners with inevitable logic, and often with a clarity that is fine for a certain kind of listener. Unfortunately, most young teachers do not read historical notes on mathematics, and are pathetically ignorant of the history of its growth and motivation – even of their speciality (see also Wilder [1]). Consequently their exposition does not convey the mathematical *insight* that it should, and which Bourbaki has, but fails to spell out. The result is that these teachers fail to teach: they clarify their own minds in the process and may even get some students to pass examinations in the material, but their educational success is dubious because insight is lacking.

2.3. *Insight*

But what do we mean by 'insight'? We obviously cannot define it

214

formally, and must convey the notion ostensively, by pointing to examples. Also, we shall be distinguishing between 'insight' and 'intuition'. Let us then look at examples of insight (or lack of it).

(1) If a student takes a course in cardinal arithmetic, and cannot even begin to see how the theory might help him to explain why $3 + 5 = 8$, then his course has given him no real insight.

(2) A student who (as is common) is taught to grind out derivatives of functions, and is content to write down the derivative of arc $\sin (1 + x^2)$ without further comment, has no insight into the matter.

(3) It is quite usual to find mathematics students who are well versed in the theorems about solving n linear equations in m unknowns; and yet such students often cannot relate the theorems to their knowledge of intersections of planes in three-dimensional space. Here is a very common lack of insight.

(4) When Kelvin said that a mathematician was 'anyone who could see that it was as obvious that $\int_{-\infty}^{\infty} e^{-x^2}\, dx = \sqrt{\pi}$, as that $2 + 2$ makes 4', he was talking about insight. (See E. T. Bell [3].)

(5) A course on groups that simply plays around with the axioms and makes no reference to groups of geometrical transformations, nor gives many examples and practice in their recognition, is unlikely to yield insight.

One could give many other examples. However, a common feature is that 'insight' into a mathematical theory seems to be related to the realisation that the theory has a model in physics, geometry, or some more familiar or accessible part of mathematics. A computer cannot show insight in this sense when it checks the steps of a formal proof for correctness, and this merely manipulative checking is unfortunately all that many students seem to get – regardless of whether or not the subject matter is 'modern'. In case (2) above, we have the difficulty of teaching the algebraic manipulations of the calculus, without forgetting the motivating model of functions and their graphs (drawn on paper). In case (3) we see the typical 'modern mathematics' student's failing of not building up a repertoire of examples against which to check the abstract theory (and the ignorance of the way in which the general theory sprang from those very examples). In (4), either Kelvin was perhaps trying to be funny, or else he had some physical model in mind, with whose analysis he was thoroughly familiar. The insights in (5) are related to situations of the following kind, where *intuition* also plays a part.

Suppose a person finds a book on analytic geometry which is purely algebraic in treatment, with no sketches.† He will probably find the text easier to follow if he is given the insight that the algebra models geometry in the usual Cartesian way. The insight then makes it legitimate to say, for example:

'It is intuitively obvious that two equations $f(x, y) = 0 = g(x, y)$, of the second degree, have no more than four common roots'

but *only when the model is known to be a good one*. Thus, in beginning courses of analysis, the definition of a continuous function is chosen to model certain properties of curves drawn on paper, and we often hear declarations that 'it is intuitively obvious that the sum of two continuous functions is continuous'. On the contrary, such non-intuitive basic theorems of continuity give confidence that Weierstrass, in his development of analysis, was showing the ability to model a subjective feeling (about graphs) in an objective language – here in the language of mathematics. The skill he used for this modelling was that of a poet, who pins down an elusive concept in a few words. Of course, once he had become accustomed to working with this model, building up an experience against which to contrast results and conjectures, his insight became enriched by 'internal skill'.

Exercise

State whether or not you agree with the assertions about lack of insight or otherwise, made in Examples 1–5 above. Give reasons, where you can.

Supply examples of your own.

2.4. *Formulation of definitions*

Consider again the important question of mathematical definitions. Bourbaki simply writes them out, and young enthusiasts hand them down to their classes. Consequently, few students ever formulate a definition themselves, so that their 'poetic skill' is never developed in a creative way. As it is, the students are using other people's formulations, without real understanding, and they frequently write nonsense as a result. Thus, the definition handed to them may

† Compare, for example, the books on linear geometry by Dieudonné [1] and Gruenberg and Weir [1]. See also p. 264 of Griffiths [1].

damage their primitive thought, instead of improving it as intended! The need to formulate one's own definitions is not one that applies only to advanced students. Indeed, this need was recognised many years ago by Benchara Blandford [1] who in 1908 wrote:

Definitions are the working hypotheses of the child; they develop gradually with the growth of his knowledge.

To me it appears a radically vicious method, certainly in geometry, if not in other subjects, to supply a child with ready-made definitions, to be subsequently memorized after being more or less carefully explained. To do this is surely to throw away deliberately one of the most valuable agents of intellectual discipline. The evolving of a workable definition by the child's own activity, stimulated by appropriate questions, is both interesting and highly educational. Let us try to discover the kind of conception already existing in the child-mind – vague and crude it generally is, of course, otherwise what need for education ? – let us note carefully its defects, and then help the child himself to re-fashion the conception more in harmony with the truth. This newer and correcter conception, sprung from the old, will itself subsequently be replaced by a truer, but it has thereby played its essentially useful function as a link whereby the vague becomes slowly transformed into the more accurate and true. Only thus can we make sure that the child assimilates knowledge and it is really prepared for the digestion of more complex mental food. Contrast this procedure with that of forcibly thrusting into the mind a fullborn definition of which the child neither perceived the need nor understands the beauty and the truth.

This lack of practice in 'poetic skill' is best countered by cooperative class discussion, and the skill is hard to test by means of timed examinations. Similarly with the formulation of problems: even if the question is clear, discussion may be necessary as to an acceptable form of answer. For example consider what has to be done to answer the question 'Does $\sqrt{2}$ exist ? '. Such questioning is hardly avoidable whenever the Open tendency is strong in a curriculum as with Mr Y in Chapter 14.

Exercises

1 Discuss the statement printed above in italics and illustrate by means of examples how the 'definition' of a concept can develop at school and at university.

2 Why do you think Blandford's arguments have been ignored for so many years ?

2.5. *Intuition*

Now it is often easy to ask important *questions* in a naïve language, whereas a thoroughgoing *answer* may require within mathematics the creation of a strict formal language. This is the root of the constant arguments between curriculum designers about rigour versus intuition (see, for example, Thom [1, 2]). To give an example, a child's exercise book contained a sequence of equations '$\frac{1}{2}$ of $2 = 1$', '$\frac{1}{3}$ of $3 = 1$', and so on, but then it caught his imagination and he worked out several more than required, jumping exponentially to '$1/90,000$ of $90,000 = 1$'. Clearly, he realised the equation $(1/y) \times y = 1$, but he was far too young to have that language imposed upon him: it might well have destroyed his insight, and at that stage he was best left to his intuitive beliefs. This seems to be what mathematicians mean when they say that things are 'intuitively obvious' – that they are convinced but lack the formal language of a proof (and their internal skill gives them confidence that a formal proof can eventually be given).

Again, consider the teacher who asks a class of children, 'Who can tell me what parallel lines are?' No reply. 'Who knows what parallel lines are?' Everybody knows. A rigorist might say that none of the children really did know, because they could not convey their knowledge in an objective language; but an 'intuitive' would argue that they could be tested for it behaviouristically. This latter is included in the philosophy of many primary school projects with their emphasis on calculations associated with practical measurements; and in many technical institutions in Britain, 'mathematics' now means the derivation and application of algorithms, often quite advanced but expressed in non-repellent 'common-sense' terms.

Exercise

A definition is sometimes chosen so as to facilitate the answering of particular questions, rather than to reveal the essential properties of the concept. Thus IMU (p. 98) uses the definition 'two lines are parallel if and only if a third line can be drawn which is perpendicular to them both'. Show how this may be a 'useful' definition for certain purposes. However, is it one which is likely to foster understanding and insight? Is it a 'parachute postulate' (p. 239)?

It might nevertheless be maintained (because of the Renewal tendency) that students ought to be trained to appreciate the power that comes from being able to make general statements like 'For all x, ...', but then they need also to learn the notion of deductive proof (as distinct from the 'natural' proof that arises in algorithmic manipulation). That is when 'definitions' are of natural importance. Suppose then that it was thought desirable that the schoolchildren we mentioned should have a definition of parallel lines. The Open tendency would suggest that it might be better to get the children to formulate a definition themselves, cultivating their 'poetic skill', than perhaps to repel them by imposing a definition. Such an imposed definition is likely to be given by a teacher who had 'the' definition imposed on him as a student; and if he himself got no insight at that stage of his own training, his pupils may be doubly deprived. Their reaction is likely to be mere degenerate parroting, especially if they have not appreciated the point of proving anything at all. It should therefore not be taken for granted that pupils will immediately understand why definitions are made, yet this groundwork *is* frequently assumed, and is perhaps a strong reason for poor understanding by students. Even when a teacher attends to such points, there is *still* a grave danger that preoccupation with formal language and proof will drive insight out of the window; for the motivating models are forgotten in the effort of acquiring the language.

Sometimes it is possible to use a model whose description is isomorphic to the formal one, but where the language is more natural, as with that of the number system described in Chapter 18. For example, the student of cardinal arithmetic in Example 1 on p. 215 might do better to use a calibrated number-line to explain why $3 + 5 = 8$, rather than talk of bijections; but it needs a good insight on his part to give a non-confusing explanation.

2.6. *An unnecessary clash between the Renewal tendency and the Open tendency*

An important example of the difficulty with formal treatments arose in connection with the BEd course, which began developing in Britain around 1965 and which may well develop elsewhere (but perhaps under other names). At that time† teachers of younger children in England usually trained, not at universities, but at 'colleges of education'. There, they took courses in Education and other subjects, and were sent into schools at certain times for teaching practice. A student in such a college would often take extra courses in some speciality like Mathematics or Art, and after a three-year course he was awarded a certificate allowing him to teach in schools

---

† The following discussion refers to the past, because of changing patterns of training and organisation. However, it is unlikely that the basic faults here depicted will be eliminated from the system by mere changes in name.

but at a lower salary than a university graduate. For various reasons, the Government arranged that a good college student could be selected to stay for a fourth year of supervised study, planned in conjunction with a nearby university; and if he were successful in the course, he would be awarded a university degree, entitled Bachelor of Education (abbreviated to 'BEd'). He would then be presumed to be expert in *both* Education *and* his main technical subject, so as eventually to teach at senior high school level, or become an educational administrator. In particular it was hoped by this means to increase the supply of mathematics teachers, because the universities were not supplying enough to meet the demand. The resulting attempts to plan curricula for the mathematics BEd candidates form an interesting case study in the discipline of mathematical education (or rather in the non-application of some basic principles). In explaining it, we must oversimplify for brevity.

The worst feature of the design of the mathematical education of the BEd candidates was *ignorantly* to subject them to both the tendencies, Renewal and Open, mentioned in Section 15.1, with disastrous results. In their three years of training for the Teacher's Certificate, the Open tendency was always uppermost; perhaps because the colleges also train teachers for primary schools. The students were taught in small classes with plenty of time for discussion, because timed examinations, though used, never had the supreme importance that a conventional university treatment would have given. Thus continuous assessment and project work were usual, with an intuitive approach to practical or appealing problems without a great emphasis on formal language or deductive proof. The notions of logic and proof were implicit, arising unobtrusively through the use of statistical techniques, algorithmic methods, and intuitive calculus. This way of doing things has gradually evolved, because the students were selected from the slightly less academic ones in high schools; if they had been good at formal examinations they would probably have gone to a university rather than to a college of education. Such a student, then, generated much of the actual work that he submitted, and if he were to be selected for the fourth, BEd year, this was because he gave promise of being a good teacher of mathematics. In the fourth year, however, the course was usually planned by the mathematics department of the local university, not by his college teachers. Thus he was handed out a formal treatment of some branches of mathematics, often very up-to-date because of the effect of the Renewal tendency in the local university,

220

and tested only by timed examinations. But the student, almost by definition, was bad at timed examinations; and he had spent the previous three years doing a very different kind of mathematics, usually left quite unrelated to the rigorous material that now was forced upon him. Consequently such a student lost interest; 'modern' does not always mean 'refreshing'! Worse still, he lost confidence and ran the risk of being a *poorer* teacher of mathematics as a result of his extra year of training!

On the other hand, just as he was taught mathematics in a watertight compartment, he had to spend half his time in another compartment called 'Education' where he scored well, being basically very intelligent. The (possibly weighted) average of his low mathematics mark and his high education mark gave him his final, quite satisfactory, 'career' mark; and that mark would get him a teaching job with special pay allowances and seniority as an expert mathematician! Mathematicians cannot quickly change the 'Education' compartment of these courses, but they can improve matters a lot by being more sophisticated about the Renewal tendency. We should add that corrective action soon began, because the university mathematicians eventually were forced, by the bureaucratic processes for planning the BEd degree, to consult with the mathematicians in colleges. Some mutual sympathy and understanding arose, and the academic mathematicians began to lose arrogance when challenged by such questions as those in the Exercises of this book. These very questions are also not suitable for testing by timed examinations or for authoritarian lecturing techniques, so that the Open tendency can be presented to the universities as something for improving their own educational methods. In return, the college teachers (who have traditionally not concentrated on mathematical research) can be introduced to the Renewal tendency. Also moves began to integrate the 'Education' requirement with the mathematics, taking into account such matters as history and philosophy of mathematics, and the growth of concept-formation. The leaders in these moves could see the BEd course as a way of producing a better mathematics teacher than the conventional mathematics graduate, whose education as a potential teacher has been unsatisfactory for many years.

Exercise

Students in a college of education take two mathematically based courses of study: one is called 'Mathematics', the other 'Computation'. Three tutors,

A, B, C, prepared 100 candidates for examination in these two subjects, with the following results:

	A	B	C	Total
Passed both subjects	18	12	6	36
Passed 'Mathematics' only	7	11	6	24
Passed 'Computation' only	5	6	7	18
Failed both subjects	5	6	11	22
Total	35	35	30	100

Examine the hypotheses that:

(i) there is no essential difference between the tutors' abilities to teach 'Mathematics';

(ii) there is no association between passing in 'Mathematics' and passing in 'Computation', taking all the candidates together.

Discuss the implications of any statistically significant results you find.

[Hull BEd, 1971]

2.7. *Mathematical education*

The two questions mentioned in the previous paragraph are typical of one aspect of the discipline of mathematical education. They have the characteristic form of asking for explanations *suitable for specified levels of understanding or experience*, by contrast with conventional mathematics in which explanation is given solely in a language suitable for mature professionals. In mathematical education, then, we have to take account of sociological factors outside mathematics, as explained in Case Study 5 (Chapter 14): but of course we need the mathematics first in order to process it appropriately. For this reason, insight is essential; one cannot modify an accurate, 'official' mathematical treatment into an intelligible account with appropriate gaps and plausible jumps, if one has only a parrot's knowledge of the 'official' version. This is why it is useless to teach mathematics, to future mathematics teachers especially, without paying great attention to insight so that they may be able to prepare good 'watered down' treatments of 'official' mathematics. Such insight is likely to be lacking so long as the notion of a 'curriculum' is retained, in the sense of nothing more than a planned syllabus or list of topics to be taught by the teacher and learned by the pupil. Consequently a striking change to be noticed in current work is the departure from the 'global' standpoint of earlier schemes, and a move to 'local' emphasis on special topics. For example, many of the papers in the journal *Educational Studies in Mathematics* are about ways of introducing one particular topic into a course, sometimes through a disguised model. In *Mathematics Teaching* much of the material is about creating mathematical 'situations' which are explored by classes in detail; this is very much an example of the Open tendency, but often the material is left in an 'intuitive' state without a formal, summarised theory. Its primary aim is to stimulate interest in both teacher and pupil, and it makes great demands on the teacher particularly, by placing him in positions where he must admit to his pupils that he may be more ignorant than they are. He must also be quick to respond appropriately to unforeseen reactions by his pupils as they develop the mathematical 'situation' in class. Eventually he must ensure that the pupils' work is not always left in the 'intuitive' state, but that they learn to cope with the formal, summarised mathematics, and know why such mathematics has arisen.

Such ideas are clearly going to be of great importance in the future training of mathematics teachers (and hopefully of their pupils).

PART 6
THE CURRICULUM IN THE SMALL

In Part 5 we considered some of the problems presented by curriculum planning in the large – that is, the drawing up of a global blueprint for content and method. The difficulties do not end, however, when one possesses a flow diagram of the type shown on p. 187, for the details have still to be supplied, and indeed it is in their approach to this task that the attitudes and achievements of the various projects differ most. Clearly in a book of this nature it is only possible to deal with two or three examples of curriculum planning in detail, and we shall, therefore, confine our attention to the following important topics: the axiomatic method, the number systems, and ways in which pupils can learn about the applications of mathematics. By means of these examples we hope to show how an approach to a topic at school level can be based not only on an understanding of how a professional mathematician would nowadays view it, but also on a knowledge of the topic's ontogenesis – its history and growth.

16

The Axiomatic Method and the Teaching
of Geometry

1. AXIOMS

Historically, axioms are associated with the notion of mathematical proof, as devised by the Greeks. This notion of proof is of vital cultural importance, and is discussed in detail in, for example, Griffiths ([1], Ch. 16).† Briefly, however, if to 'prove' a statement is to convince someone of its validity, then axioms can be picked out as statements beyond dispute (at least for the time being), from which a potentially disputable statement, S, is then deduced. If the axioms and the steps of the deduction are agreed, then S must be granted also, and S becomes a 'theorem' provable from these axioms. Thus, axioms are an important tool for the activity of proof and as such are vital in mathematics: one establishes mathematical facts by *proving* them, even if they are located by guesswork.

We shall assume in this introductory section that the reader has met the idea of the axioms of a geometry or of groups, etc. We hope to clarify and reinforce these notions.

First, we again stress that we regard axioms as statements *picked out* as being *beyond dispute* for the time being. To explain this sentence, let us consider some contexts in which the word 'axiom' is used. To take the Greeks first, Euclid's axioms were of two kinds: (*a*) axioms stating what rules of procedure to allow (e.g. 'If equals be added to equals, the wholes are equal'), and (*b*) axioms stating the relationships between basic objects (e.g. '(that one can) draw a straight line from any point to any point'). These axioms were thought to be 'beyond dispute' because no sane man could possibly object to them and hence 'for the time being' was taken to be 'for all eternity'. Now Euclid himself realised that these two types of axiom were essentially different and, indeed, referred to them by different names – he called the former 'common notions' and the latter

† A more light-hearted (but still deep) discussion can be found in P. J. Davis [1].

226

'postulates'. For those of type (a) refer to mathematics generally, not just geometry. They modelled the reasoning processes used, and were supplemented by deductions made by Aristotle about syllogisms, together with other processes not made explicit until the nineteenth century. Such axioms are called *logical axioms*. The axioms of type (b) model the Greek notion of the geometry of the observed plane, and are *geometrical axioms*. They were picked out because they appeared to describe accurately the system of geometry it was intended to reason about.†

In a similar manner, after various systems now called 'groups' had been observed within mathematics and seen to have common features, the simplest such features which served to characterise these systems were written down as the *axioms of group theory*, and the theorems deduced from the axioms applied then to all the various observed groups. Here, however, the emphasis has altered: it is recognised that the group axioms cannot be 'eternal truths', since some objects do not satisfy them – there are systems which, although very like groups, are not groups. The Greeks, on the other hand, knew of nothing 'resembling' plane geometry which was not *exactly* plane geometry.

The notion has grown, then, that in mathematics there are several different kinds of 'system', groups, vector spaces, geometries, topologies, algebras, . . ., and an object belonged to a particular system provided it could pass certain tests – the *system axioms*. One can deduce theorems from such axioms and these theorems apply to *all* objects that pass the test for that particular system. The proofs of such theorems are often easier to handle than the proofs for particular cases, since by concentrating on the *system* properties one is not confused by possibly irrelevant properties of individual objects. (In an analogous manner one can deduce from the 'system' for early Wild West films, that 'good' and the cowboys will finally prevail – to that extent, the detailed stories of individual films are irrelevant.)

When a mathematician considers a set of axioms for a system, A, he is likely to do several things.

(1) He may check to see if any one of them can be deduced from

---

† We note that they contain certain 'terms' such as 'line', 'point' and 'draw'. Euclid attempted to 'clarify' matters by giving definitions of these terms, thus, for example, 'a *point* is that which has no part', 'a *line* is breadthless length'. His attempt was, of course, bound to fail since his definitions could only be framed using other undefined terms. As we shall see later, those who design axiom systems nowadays realise that in their system they will have to include certain 'undefined terms', such as 'set', 'point', 'line' and 'plane'.

the rest and so be 'redundant'. He may then seek the most economical set of axioms with no redundancies at all (*independent* sets), and in his search he may find different sets which will suffice (*logically equivalent* sets).

(2) He may find that he can deduce something contradictory about A, that his system is not *consistent*. This may be because he has drawn up his system axioms incorrectly or because his arguments depend upon a mathematical system containing errors. This latter situation can never be completely resolved – all mathematicians proceed in the belief that the system in which they are working is consistent, yet there is no general way of establishing this (see Griffiths [1], Ch. 38).

(3) He may find that an important theorem can be deduced using only a subset of his axioms (cf. absolute geometry, p. 256). That theorem will then apply to a more general system than the original A. (Bourbaki startled some of his contemporaries by taking such theorems as axioms for a new kind of system.)

(4) He may attempt to vary the axioms slightly and to see how this affects the validity of his theorems.

When engaging in such considerations the mathematician will often be transferring his interest from the objects and problems which gave rise to the system to the form of the axioms themselves.

We note that to show that an object is a system of a particular kind, i.e. that it satisfies the system axioms, it is often necessary to prove theorems in another part of mathematics. Thus, for example, to show that the distance function

$$d(\mathbf{x}, \mathbf{y}) = \left[\sum_{i=1}^{n} (x_i - y_i)^2 \right]^{1/2},$$

$$\mathbf{x} = (x_1, \ldots, x_n), \qquad \mathbf{y} = (y_1, \ldots, y_n)$$

on $\mathbb{R}^n$ satisfies the triangle axiom for a metric space, one must call on a lot of algebra (see Griffiths [1] and Rhodes [1]).

So far we have considered axioms in mathematics. Now let us consider their use in mathematical education – as 'pedagogical axioms'.

It may happen that within a mathematical system specified by axioms, we can deduce from some statement S other statements easily and interestingly. For example, S might be the statement P: 'All (correct) ways of counting a finite set give the same answer for the number of elements in the set.' To deduce the validity of P

from the axioms of set theory is long and difficult, yet one accepts P as true, and uses it as the starting point for the development of elementary arithmetic in primary school; technically P is a *theorem* of set theory, but, as we shall term it, a 'pedagogical axiom' within schools. Again, in sixth forms, it is customary to take the Fundamental Theorem of Algebra (that every polynomial $p(x)$ with complex coefficients has a root in the complex plane $\mathbb{C}$) as a pedagogical axiom. Even earlier, if $p(x)$ is a real polynomial such that for some $a < b$ in $\mathbb{R}$, $f(a)$ and $f(b)$ are of opposite signs, then it is taken as a pedagogical axiom that $p(x)$ is zero for some x between a and b. At a more advanced level, in courses of mathematics for engineers, Cauchy's Integral Theorem (Ahlfors [2]) can be taken as a pedagogical axiom because it is plausible and has many consequences of interest to engineers, with proofs of greater palatability to them than the rigorous proof of Cauchy's Theorem within the theory of functions.

In the same spirit, the original aims of the AIGT (see p. 15) included the right to use pedagogical axioms, and asked for the removal of limitations on 'the restriction . . . of axioms to those only which admit of no proof'.

In these examples, the pedagogical axioms correspond to 'cut-off' points in a *mathematical* development, thought of as an exposition flowing linearly in time. Each begins a portion of a theory which follows a different portion of the same theory that began with more primitive axioms. This situation is comparable to case 3 above. We may, however, choose as a pedagogical axiom the statement E: 'Every statement provable in coordinate geometry is valid in Euclidean geometry.' This is because it can be shown (with much effort) that the axioms of Euclidean geometry are all represented by theorems of coordinate geometry in the following sense. If we allow the words 'point', 'line', 'on', etc., of Euclidean geometry to correspond to 'number pair (x, y)', 'equation $ax + by = c$', 'satisfies equation', etc., then the axiom 'there is exactly one line passing through two given points' corresponds to the statement 'there is exactly one equation $ax + by = c$† which is satisfied by two given number pairs (x_1, y_1), (x_2, y_2)', and the latter is *provable* in coordinate geometry, using our knowledge of solving equations in algebra. Now coordinate geometry is easier to describe than a satisfactory set of

† Up to multiplication by a non-zero constant; strictly there is an *equivalence class* of equations corresponding to each line.

axioms for Euclid; and, as we remarked above, it contains all the axioms (and hence the results) of Euclid. Therefore, we may gain a great deal pedagogically, and lose nothing mathematically, by working with coordinate geometry rather than with its isomorph, Euclidean geometry. (Of course we also gain such advantages as the ability to deal analytically with various loci that would be difficult, if not impossible, using only the Euclidean vocabulary.)

In this spirit, it should not be surprising that a pedagogical axiom *system* may well include redundant axioms, as in (1) above; economy of axiom systems tends to be an aesthetic criterion of mathematics rather than a practical virtue of mathematical education.

Thus, for example, when setting out the group axioms it is usual to demand the existence of both a right and a left neutral element, and a right and a left inverse, whereas the existence of a right neutral element and a right inverse will suffice. In doing this we select axioms which emphasise symmetry, and simplify technical work, i.e. are pedagogically desirable; rather than those which are most economical and therefore logically desirable.

Observe that to designate a statement as a 'pedagogical axiom' involves a judgement within the discipline of mathematical education and is irrelevant to mathematics itself. Nevertheless a great deal of mathematics may need to be done to uncover the most suitable pedagogical axioms, as we shall see later with the related notion of 'parachute postulate' (which is usually lacking in both plausibility and fruitfulness).

To summarise then, axioms play an important role in the 'aesthetic' side of mathematics, but they also belong to the 'practical' side. Complicated mathematical systems can be described by sets of axioms. Sometimes such an axiomatic description is the only mathematically satisfying one known, as with the real numbers (see Chapter 17); sometimes it is a way of introducing pupils to work within a branch of mathematics, long before they are mature enough for a detailed proof that the axioms are satisfied. In any case, the consequences of an interesting set of axioms are frequently of great utility.

Before proceeding further, however, we give two quotations about axiom systems, to try to dispel the widespread misunderstanding that 'pure' mathematicians simply play around with axioms as perhaps suggested by case (4) above, varying them at will to avoid the hard realities which confront their 'applied' brethren. The first is from Klein's *Elementary Mathematics from an Advanced Viewpoint*

THE AXIOMATIC METHOD AND GEOMETRY TEACHING

(Vol. 2), p. 187 (see p. 107), and his remarks have been amply confirmed by the evidence of the seventy years of mathematical activity that have elapsed since he wrote:

In (modern theory) . . . we may set up axioms arbitrarily, and without limit, provided only that the laws of logic are satisfied and, above all, that no contradictions appear in the completed structure of statements. For one, I cannot share this point of view. I regard it, rather, as the death of science. The axioms of geometry are – according to my way of thinking – not arbitrary, but sensible, statements, which are, in general, induced by space perception and are determined as to their precise content by expediency.

The second quotation is from Courant and Robbins's book *What is Mathematics?* and reflects something of the attitude that led Courant to oppose the direction he feared American mathematics might take, by founding the Institute in New York (see p. 121).

A serious threat to the very life of science is implied in the assertion that mathematics is nothing but a system of conclusions drawn from definitions and postulates that must be consistent but otherwise may be created by the free will of the mathematician. If this description were accurate, mathematics could not attract any intelligent person. It would be a game with definitions, rules, and syllogisms, without motive or goal . . .

By Courant's own argument, *serious* creative mathematics could not have taken the direction Courant feared. The danger, however, still exists, as A. K. Austin's satirical 'Modern research in mathematics' (*Math. Gazette* **51**, 1967) shows, and we cannot underline it too heavily for the reader.

In practice, many professional mathematicians do not work with explicitly stated sets of axioms at all, but instead concentrate on narrowly defined problems. Sometimes they can be blind to similarities between apparently different problems, and yet a new problem sometimes yields to a mathematician who sees that it is 'like' an old one – in some sense they both belong to the same axiomatic system. This view was taken by E. H. Moore in Chicago, who said 'When two mathematical theories are seen to yield the same theorems, it is a mathematician's duty to uncover the common underlying structure.' The 'duty' expresses itself in using the axiomatic method to express the 'underlying structure': and skill here can be learnt from making mistakes of the sort (2) above.

The role of the problem is well exemplified by Hilbert who, in a famous address to the 1900 ICM Congress, gave a list of unsolved problems that have since occupied the energies of many mathematicians. Some are now solved, and a particularly famous one was the

231

fifth, on Lie Groups. Hilbert and his contemporaries had not the technique even to express it properly, and the axiomatic method was needed (to express the notion of a Lie Group) before the real attack could be planned. Hilbert's attitude to problems was this (in the same address):

The great significance of specific problems for the advancement of mathematics in general, and the substantial role that such problems play in the work of the individual mathematician are undeniable. As long as a branch of science has an abundance of problems, it is full of life; the lack of problems indicates atrophy or the cessation of independent development. As with every human enterprise, so mathematical research needs problems. Through the solution of problems, the ability of the researcher is strengthened. He finds new methods and new points of view; he discovers wider and clearer horizons.

In the remainder of this chapter, and in the following one, we consider the effects of the axiomatic method in geometry and on our understanding of the number systems. We choose these topics because mathematicians, pure or applied, are deeply concerned to find *pattern* (or 'structure') in what they contemplate, to look at what seems at first to be chaos, and there to find order. This order is usually expressed in terms either of geometry or of some object constructed from a number system.

Exercises

1 (*An exercise in 'economy'*)
The common logical connectives are defined in terms of truth tables as follows:

X	Y	Negation $(\sim X)$	Conjunction $(X \wedge Y)$	Disjunction $(X \vee Y)$	Implication $(X \Rightarrow Y)$	Equivalence $(X \Leftrightarrow Y)$
T	T	F	T	T	T	T
T	F	F	F	T	F	F
F	T	T	F	T	T	F
F	F	T	F	F	T	T

(*a*) Show that all the other logical connectives above can be defined in terms of $\sim$ and $\vee$ alone (for example, $(X \Rightarrow Y)$ is equivalent to $((\sim X) \vee Y)$, in the sense that they have the same truth table).
(*b*) Show that all the logical connectives above can be defined in terms of the Sheffer stroke, $|$, defined by

X	Y	$(X \mid Y)$
T	T	F
T	F	T
F	T	T
F	F	T

2 Show that if S is a *finite* set of elements together with a binary operation $*$:

$S \times S \to S$ which is associative, then $(S,*)$ is a group if and only if right and left cancelling is permitted, i.e. if each of the equations

$$A*X = B*X \qquad \text{and} \qquad Y*A = Y*B$$

implies that $A = B$.

(Thus in the particular case when S is finite, the axioms of a group concerning the existence of an identity element and of inverses, can be replaced by a single, simpler axiom.)

3 Show that the following definition of a group is logically equivalent to that more usually given (see, e.g., Griffiths [1], p. 288).

'A group is a non-empty set G together with an operation $*: G \times G \to G$ which is associative and which satisfies

$$\forall a, b \in G \quad \exists x, y \in G \quad \text{such that} \quad a*x = b = y*a.'$$

(Papy [2])

Which definition do you consider to be more suitable for (a) pedagogical and (b) mathematical purposes? Give your reasons.

2. AXIOMS AND THE DEVELOPMENT OF GEOMETRY

Geometry, as we know it, had its beginnings in Egypt and Babylonia where it was a scientific study in which facts were established by empirical procedures. It was the Greeks who transformed the subject into one in which deductive reasoning took the place of trial-and-error experimentation. They were the first to introduce chains of propositions each one of which could be deduced from those stated earlier. Hippocrates, Leon and Theudius are all credited with attempts to give a logical presentation of geometry in the form of a chain of propositions stemming from a few initial definitions and assumptions, but their efforts were overshadowed by those of Euclid who, about 300 BC, produced his famed *Elements*.

This work – divided into 'Books' corresponding to what we should term 'Chapters' or 'Parts' (cf. p. 13) – had an enormous impact on mathematical thought and method, and marked the beginning of what can be termed the 'axiomatic approach'.

Euclid set down a few 'Postulates' and 'Common Notions' – what we now think of as axioms – which were in fact statements thought to be undisputed facts about the world, and from them he deduced the body of results that we know as Euclidean geometry. One of his postulates is the 'parallel axiom' which can be written (in an equivalent form due to the nineteenth-century geometer Playfair, whence its† name 'Playfair's Axiom' (cf. p. 130)): 'Through a

---

† The statement here is Playfair's postulate as it is used by Hilbert in his *Foundations of Geometry*.

given point A not on a given line m there passes at most one line (in the plane determined by A and m), which does not intersect m.'

This axiom was eventually regarded in a different way to the others; it seemed less 'obvious' and mathematicians were tempted to try to 'prove' it from the remaining axioms. This corresponds to the search for economical axiom systems as in (1) on p. 227. Some mathematicians obtained spurious proofs (see, for example, that by Posidonius reprinted in Tuller [1]); others, such as Saccheri (1667–1733), generated much interesting mathematics in an attempt to find a proof, but it was not until the beginning of the nineteenth century that the great breakthrough occurred. For, in an attempt at proof by 'Reductio ad absurdum', Bolyai (1832), Lobachewsky (1829) and Gauss (1813) (all independently) supposed that the Parallel Axiom was false, to see whether they could deduce something that contradicted the remaining axioms. But, although they could work out a body of interesting results, they were unable to obtain a contradiction. Thus, they had worked out new kinds of geometries (now called non-Euclidean) that might be equally valid descriptions of the geometric space we live in (see, for example, experiments on vision reported in *Science*, 9 March 1962). The search for a proof of the Parallel Axiom ended when it was shown by Beltrami (1868) and others that one could model non-Euclidean geometry within a certain kind of coordinate geometry, just as Descartes had modelled Euclidean geometry, so that each kind was as consistent as coordinate geometry: if the latter had no contradictions, neither had non-Euclidean geometry, so Bolyai, Lobachewsky and Gauss could not have found the 'Reductio' proof they sought, and their new geometry was valid.

By the end of the nineteenth century, other kinds of geometry were known, and Klein had stressed that what really distinguished them was their different symmetry groups. The group of a geometry, however, can only be described if one can describe the geometry first in some other way (like Euclidean, Projective, Affine or non-Euclidean geometry). In thinking about the logical interconnections between their axioms, Hilbert realised that many mathematical systems such as geometry could be described solely by the axioms one used to build them up. That is to say, if one knows the axioms one needs no further description of the system: any other facts can be *deduced*, given sufficient effort, In particular, it is unnecessary to know what the elements of a system (for example, points, lines and planes) *are*, just as one does not need to know what a pawn looks like in chess: its functions are explained by the rules of chess, and any shape would do. In the words of Hilbert (Reid [1]), 'one must be able to say at all times – instead of points, straight lines and planes – tables, chairs, and beer mugs'. See also the Exercises on p. 243.

3. TWO ASPECTS OF GEOMETRY

Much of the discussion about the teaching of geometry reflects the psychological attitudes of different mathematicians. There are those who get great aesthetic pleasure from the contemplation of geometric pictures, of shapes and visual patterns; others prefer the contemplation of algebraic formulae or elegant verbal arguments. Cutting across these are the 'problem-solvers' and the 'general theorists', mentioned earlier. The former delight in particular situations, analysing them with detailed and penetrating argument. The latter prefer to look at general situations in the hope of fitting the particular problems into the theory as special cases; they are frequently not good at particular problems because they are reluctant to use the special information in case they lose sight of a powerful general method. The 'problem solvers' often miss applications of arguments because they are reluctant to generalise in case they lose a tiny but vital feature inherent in their special problem.

In terms of school geometry, these attitudes are represented by those who wish to stress the axiomatic structure of geometry, and those who wish to proceed as rapidly as possible to the more complicated parts, such as the analysis of triangles (Simson's Line, Morley's Theorem, Desargues' Theorem) and such geometrical constructions as are useful in perspective drawing. Both sides are represented in such a book as Coxeter ([1]), where it is made clear that much interesting geometry can be done without going through the long axiomatic theory that occupies Part III of his book.

Apart from aesthetic reasons, the axiomatic part of geometry has been put forward as something by which to train pupils in the notion of proof (or to 'train the mind'). Certainly this was always a prime argument for introducing Euclid into secondary schools. It would seem, however, that such a 'global' approach to proof is less suitable for most pupils than a 'local' approach via special problems. For example, granted certain basic facts (pedagogical axioms) about the incidence of planes, one can teach an attractive deductive proof of Desargues' Theorem of perspective triangles; but to establish those basic facts rigorously is hard work, and might seem – to bored pupils – to be proving the obvious. Similarly, given certain basic facts, it is a good exercise in deduction (and powers of spatial intuition) to complete an accurate perspective drawing of an object.

Words like 'boredom' and 'aesthetic appeal' crop up frequently in discussions about geometry teaching. They are inseparable from the relationship of geometry to our visual sense, and of course are important constraints on the planning of a geometry syllabus. Such constraints are not part of the subject *mathematics*, but they are very much part of the subject *mathematical education*; a fact frequently overlooked by Dieudonné ([1–2]) and others.

The attitudes we have mentioned can be seen in the articles of the two issues of *Educational Studies in Mathematics* devoted to geometry (**3**, Nos 3/4, **4**, No. 1). Several of the contributors are clearly unconscious of the possible differing psychological attitudes. One (Villa) tells of the severe educational consequences in Italy that followed the laying down, half a century ago, of a strict axiomatic treatment of geometry in schools; the then Ministry of Education had followed the advice of several eminent Italian geometers, who knew little of mathematical education. Unfortunately, the substitute course suggested by Villa offers little of interest to the 'problem solver'. An interesting attitude is shown by the article of Marshall Stone – who combines both the local and the global points of view, but lays down some formidable practical challenges to schoolteachers. He also makes a point that we cannot ignore: that the axiomatic approach in geometry is such an enormous achievement in human culture, that any education should include a treatment of it. This is the difficulty: practice in 'local' deduction may produce a good 'problem-solver', but it is unlikely to give insight into the axiomatic approach *to mathematics*. On the other hand, most pupils are drawn to mathematics by their delight in solving problems (most notable was Gauss!). Certainly, we must be cautious of the approach that treats the training of mathematicians like that of 'cellists: to be a good 'cellist, perhaps the child should emulate the man and bow and finger correctly from the beginning. Nevertheless that child might well give up if at the same time as he is acquiring technique, he must play only the most mature Bach.

Keeping these differences in mind, we therefore now look in more detail at various arguments concerning the teaching of geometry.

4. AXIOMS IN SCHOOL GEOMETRY

Although Euclid's system is always taken as a paradigm of argument that all should follow, the critical approach of the late nineteenth century showed that Euclid's exposition was not logically perfect (see, for example, Russell, *Math. Gazette* 2 (1902), reprinted 55 (1971)). For example, Euclid attempts to define points and lines, but the definitions are unsatisfactory; and he assumes without proof such 'obvious' facts as that the diagonals of a parallelogram meet inside it. He does not even say what 'inside' means and there are geometries in which no satisfactory definition of 'inside' is possible. Hilbert's book (1899) was written to supply a correct version of Euclid, but the resulting theory is unsuitable for all but mature students, since the arguments are long. Nevertheless, an attempt was quickly made (1904) to produce a version of Hilbert's book for schools (Halsted [1]), for it was becoming widely recognised that although Euclid was placed in the curriculum to teach pupils a perfect example of the proof method, his material was unsuitable for that purpose. Two questions then arise for curriculum designers:

(1) How should geometry be taught?
(2) Is there a simpler part of mathematics that would teach the notion of proof?

The most usual British reply to (1) has been to teach geometry as a branch of deductive physics until the pupils are mature enough to understand that coordinate (analytic) geometry is a satisfactory model of Euclidean space. (Compare the methods by which Mercer allowed his pupils to solve the problems on p. 162 above.) As to (2), within the traditional syllabus the response has been to lay less and less stress on the notion of proof at all, on the 'liberal' ground that the only suitable proofs are geometrical, and these become mere exercises in memory work (for the question is, of course, affected by examinations). Alternatively, they pose too many problems for those pupils who are not mathematically gifted. This further widens the gulf between school and university mathematics, mentioned in Chapter 9, for pupils are often plunged into the Definition – Theorem – Proof form of mathematics as soon as they begin a university course, whereas their school course has contained only proof of manipulation of equalities, and rules of substitution. Curiously, this poor equipment has been thought to be academically superior to that which might be gained from one answer to question (1), namely, a course in geometrical drawing. The 'reason', if any, seems to be ignorance combined with the traditional snobbish dismissal of Engineering by the British; yet a most interesting course in Engineering Drawing was designed sixty years ago by D. A. Low [1]. More recently, T. J. Fletcher [1] has pressed the claims of perspective drawing as an aid in the teaching, and understanding, of geometry.

The British schools have always treated geometry – and indeed, all other mathematical subjects – in a 'spiral' manner. That is to say, the topic is introduced in the lower forms of the school and as the child progresses through the school he meets the topic time and time again, seeing it on each occasion, hopefully through maturer eyes. This has not been the case in the United States, where the tradition of 'Tenth Grade geometry' (that is, the concentration of geometry teaching into a single year) has established a firm hold. The disadvantages of such a system are obvious, but in the case of the teaching of geometry it does mean that pupils begin the 'formal' study of the topic somewhat later than their British counterparts, at about the same age as the Greeks of Euclid's Academy. It is, perhaps, for that reason that attempts to teach Euclidean geometry in what can be described as a 'Euclidean manner' have persisted longer in the USA than in Britain.

We have already mentioned Halsted's attempt to adapt Hilbert's axiom system for school use. Something similar was to happen later with an axiom system that G. D. Birkhoff [1] devised in 1932. He assumed the fundamental properties of the real numbers and was able to reduce his axiom system to four postulates; basically he assumed that one could use a ruler and a protractor in a 'natural' way – and with these 'strong', pedagogical axioms it proved possible to reach interesting theorems very quickly. This system was later (1940) incorporated into a school text by Birkhoff and Ralph Beatley.

A further attempt to proceed on these lines was made by the SMSG on its establishment in 1958, when it prepared a pedagogical axiom system which combined the ideas of Hilbert and Birkhoff.†

5. SOME RECENT EXPERIMENTS

The approaches outlined in the previous section were all aimed at proving the body of results of Euclidean geometry by traditional, synthetic methods.

† The axiom systems of Euclid, Hilbert, Birkhoff and the SMSG are reprinted as appendices to Tuller [1].

At the Royaumont Seminar of 1959, however, Dieudonné (a member of the Bourbaki collective) launched a vigorous and influential attack on such practices (see OEEC [1]) and declared that the only possible way to teach geometry was as the study of a vector space with a scalar product. Such an approach is essentially an approach through coordinate geometry, and it might obscure the way in which the theory models the geometry of Nature. This did not greatly worry Dieudonné, partly for reasons concerning the Bourbaki philosophy of mathematics, and partly because he taught only university students.

Approaches in the spirit of Dieudonné – or, as it happened (see below), in the spirit of what was thought to be the spirit of Dieudonné – were soon prepared.

Foremost amongst them were those due to Choquet and Papy. Significantly, both came from countries where the requirements of utility could be lightly shrugged off, owing to the local prestige of the 'intellectual'.

Choquet's book [1] was not primarily intended to be used in schools, but was addressed to teachers and those training to be teachers. In his preface, however, Choquet did express the opinion that 'It [the book] can also be profitably used by pupils between 15 and 18 years of age.' In its originality and lucidity of presentation it is outstanding amongst recent mathematics books relevant to mathematical education; and few mathematicians could deny its brilliance. Choquet clearly differentiates between affine and metric ideas, and reveals the structure of Euclidean space with great clarity. Pedagogically, however, one has strong reservations. The axioms used by Choquet (in particular, his Axiom IV) are by no means 'statements beyond dispute' that would appear 'obvious' to pupils. On the contrary, they are examples of what have been called elsewhere (CCSM [1], p. 79) 'parachute postulates', i.e. attacks on the problem from the rear, which have not been framed because they are natural assumptions. They are chosen with hindsight, because one sees that they will get round what in the normal way of things would be points of difficulty. It is on the grounds of the obscurity of his axioms – which must eventually be replaced by those of a vector space – that Dieudonné [1] later chose to criticise Choquet. Dieudonné suggests a development based on the more usual vector space axioms; in fact, an approach to linear algebra and geometry on advanced university lines.

An axiomatic approach which in some ways resembles Choquet's has been developed by Papy [3], who cleverly combines the development of the real number system and the affine plane. Like Choquet's approach, this is one which any reader of this book could study with benefit. Whether it is pedagogically any more worthwhile is doubtful.† Indeed it must be doubted whether any axiomatic development of such a rich structure as the Euclidean plane, could be prepared in terms which would prove acceptable to 11- and 12-year-olds. The very maturity of the idea of axioms and an axiom system militates against its being understood and followed by pupils of this age.

Nevertheless, many mathematicians feel, like Papy, that 'the most fundamental and central topic of the secondary school programme is, without doubt, vector spaces' and couple that belief with a desire to develop the topic in an axiomatic manner. Thus in addition to the systems described above, alternatives have been proposed by Levi [1], Pickert [1] and others.

It is necessary, though, to question the value of such approaches not only on pedagogical grounds but also on mathematical ones. As Hans Freudenthal [1] has pointed out, although linear algebra pervades much of modern mathematics, the geometry to which it lends itself is restricted, and by no means of the type that is likely to interest or inspire young pupils. Heaven help the child brought up on Dieudonné's approach, who seeks to prove that the plane can be tessellated with regular hexagons but not with regular pentagons! Of course, algebra is necessary for many problems of geometry, such as the 'impossibility proofs' (Courant [1]), but it is not *linear* algebra, and its spirit is different.

Some projects have seemingly realised that a full axiomatic treatment of the Euclidean plane entails either an artificially complicated axiom system, or a long slog of dreary reasoning, or both. Instead, they have opted for smaller, less ambitious axiom systems which, although not leading to the results of traditional Euclidean geometry, will still give the student an inkling of what a mathematician understands by the axiomatic method, and even of the abstractness of such systems as revealed by Hilbert. Thus the Secondary School Mathematics Curriculum Improvement Study (SSMCIS) takes as an axiom system:

Axiom 1: (a) Plane Π *is a set of points, and it contains at least two lines.*

† Papy's ability as a teacher is such that he might well be highly successful if he taught the material personally. We are here concerned, however, with the suitability for average teachers.

(b) *Each line in plane* Π *is a set of points containing at least two points.*

Axiom 2: For every two points in plane Π, *there is one and only one line in* Π *containing them both.*

Axiom 3: For every line m and point E in the plane Π *there is one and only one line in* Π *containing E and parallel to m.*

Simple logical consequences are developed from these axioms and both geometrical and non-geometrical models of the axioms are exhibited.

Thus situations are described in which when 'point' is replaced by 'commands' and 'line' by 'team', the axioms and hence the logical deductions are still satisfied. This shows the abstract nature of axiomatic reasoning.

It is possible that, if it is felt necessary to develop any part of geometry in an axiomatic manner carried out at school, this approach will provide a reasonable solution. Certainly, if it were carried out at sixth-form level then it would be possible to deduce the interesting formula for the cardinality of the affine plane, given that of any of its lines. Also the substitution of Axiom 3 – the Parallel Axiom – by

Axiom 3: For every two lines in plane Π, *there is one and only one point in* Π *contained in them both,*

would yield projective geometries. The effect of changing Axiom 3 could now be investigated, the duality of the new axiom system observed, and new results concerning the cardinality of the plane found.† (See, for example, Albert and Sandler [1].)

Such a programme also has the advantage of quickly bringing pupils into contact with an unsolved problem on the existence of finite projective planes. This is still attracting the attention of geometers (see, for example, Room [1]) and, because of the interconnection between finite projective planes and Latin squares, of combinatorial analysts. (See, for example, Ryser [1].)

As we mentioned in Section 3, the traditional British approach to the teaching of geometry came to place less and less emphasis on axioms and a logical development of the subject. Not surprisingly, therefore, modern projects in Britain have fought shy of any attempt

† The introduction of finite affine and projective planes can be compared with the increasing use of finite algebraic structures for teaching purposes; for example, finite rings and fields (clock arithmetic) (e.g. Maxwell [1]) and finite vector spaces (Glaymann [1]).

to teach geometry axiomatically. The SMP were greatly influenced by attempts made in Germany and Switzerland (see Jeger [1], OEEC [2], ATM [1]) to base the study of geometry on that of transformations, in the Klein tradition. This is done up to O-level in an empirical fashion, which would appear to many to be physics rather than mathematics. Only in the specialist sixth-form course (SMP *Further Mathematics*) is any attempt made to treat geometry formally. Transformation geometry is then developed from the assumed properties of a vector space with a scalar product.

The (Scottish) SMG treatment is more formal than that of the SMP and combines vector and transformation approaches. The treatment, which is due to Geoffrey Sillitto, is based on coverings of the plane by parallelograms and, though the axioms are never explicitly stated, the development is a logical one. (Interesting correspondence about this approach is contained in *Math. Gazette* **53**, 1969.)

The MME based its geometry course on a study of vectors and their properties, again giving a treatment which would be seen in some quarters as being essentially physics.

In Section 3 we mentioned the two views of geometry – those of the 'problem solver' and those of the 'general theorist'. Those of the latter are clearly evident in some objectives for teaching geometry set out by Begle [2]:

(1) to give the students some facts about the configurations of Euclidean geometry;
(2) to give them some experience with deductive reasoning;
(3) to give an example of an axiomatic system;
(4) for the sheer attractiveness of the subject;
(5) as a foundation for future mathematics.

There is no mention here of the numerous applications of geometry to science, surveying, engineering, architecture, etc., nor any indication that the principal reason for teaching geometry could well be its utility – the part it plays in the solution of problems. Neither is it emphasised that we teach geometry so that students themselves can solve geometrical problems.

This last objective does, however, give rise to many of the difficulties associated with the teaching of geometry. As all teachers know, problems in pure geometry generally rely on insight and inspiration for their solution, rather than an ability to perform well-rehearsed techniques or an ability to reason logically. (It is for this reason that pupils find analytic geometry so much easier.) This is true whether one is thinking of the riders to be found in traditional texts or the problems to be found in, say, Jeger [1] and Yaglom [1]. Yet far too few pupils acquire the insight to grapple with such problems. What is to be the remedy? One possibility for the less gifted geometrically – and this will include many intelligent pupils – has been mentioned earlier, engineering drawing. Another is to drop any attempt at 'systematic' geometry and to engage them in 'geometrical activities'. Some idea of what this last phrase can mean is given by Engel [1]; other examples can be found in ATM [1], Chs. 8 and 9.

Exercises

1 'If Euclid failed to kindle your youthful enthusiasm, then you were not born to be a scientific thinker.'

(A. Einstein, quoted by Pólya in ICMI [1])

'The early study of Euclid made me a hater of geometry.'

(J. J. Sylvester, 'A plea for the Mathematician', *Nature* 1)

Comment.

2 Siddons (of the Godfrey and Siddons partnership) wrote that at one period 'much time was wasted over accurate drawing without much aim'. Consider the place of geometric drawing in a school syllabus. How does one prevent it from over-emphasising objectives in the psycho-motor domain and utilising only low-level cognitive skills?

3 Are the axioms on pp. 240–1 a pedagogical system?

4 In a 'two-dimensional projective geometry' there are two kinds of undefined element, called 'knots' and 'strings', and a relation described either as 'the knot lies on the string' or 'the string passes through the knot'. The following axioms are satisfied:

A. Through any pair of distinct knots there passes one and only one string.

B. On any pair of distinct strings there lies one and only one knot.

Show that each of the following two systems exemplifies such a geometry:

(i) The knots are lines through a point Q in three-dimensional Euclidean space, the strings are planes through O.

(ii) The knots are pairs of antipodal points on a sphere (i.e. points at the two ends of a diameter), the strings are great circles.

Explain how one of these examples could be derived from the other.

In system (i) the knots can be described by equations of the form $x/a = y/b = z/c$, and the strings by equations of the form $px + qy + rz = 0$. Let (a, b, c) be called the 'coordinates' of the knot, and $[p, q, r]$ the 'coordinates' of the string. Explain why (ka, kb, kc) is the same knot as (a, b, c), and similarly for the strings; and why the coordinates, $(0, 0, 0)$ and $[0, 0, 0]$ are inadmissible. Express the relation 'the knot (a, b, c) lies on the string $[p, q, r]$' algebraically.

Find the coordinates of the string passing through $(1, 1, 1)$ and the knot common to $[0, 1, 2]$ and $[1, 2, 3]$.

(SMP, A-level (*Further Mathematics*))

5 A squash match is played between two teams, each player playing one or more members of the other team.

It is arranged that:

(i) any two members of the same team have exactly one opponent in common,

(ii) no two members of the same team play all the members of the other team between them.

Prove that two players who do not play each other have the same number of opponents and hence deduce that any two players, whether in the same or different teams, have the same number of opponents.

(Southampton BSc, 1971)

6 Pupils normally find difficulty with problems of the type 'Verify that . . . satisfies the axioms for a . . .'. One reason may be that in ordinary language we do not make fine distinctions and a gradual progress to axioms may be necessary. Devise a sequence of procedures, each more refined than the last, for deciding whether an object is an elephant, a woman, a boy, a Welshman, a gas, a symmetry group of a solid, a surface, a metric space.

7 Several situations in real life have a structure which people describe by reference to famous stories – 'a poacher turned gamekeeper', 'new clothes for the emperor', etc. Find other such examples which might be used as an introduction to mathematical structure.

6. AXIOMS OUTSIDE GEOMETRY

All the British projects have emphasised their belief that geometry is not the only training ground for logical thought, nor is it the only branch of mathematics in which an axiomatic approach is apposite. The claims of other structures, such as groups or Boolean algebra, have been advanced. In these it is easy to see exactly what follows from the axioms – our algebraic intuition is not as well developed as our geometric, and for that reason we are less likely to dismiss propositions as 'obvious'. Unfortunately, the accessible theorems are few and not very interesting. In any case, if one studies groups, it is mathematically necessary first to spend time on many examples, in order to recognise different types of group. Now although such a familiarity with many groups is valuable, the common examples teach one little about proof, because one uses the axioms as things to verify, rather than as the foundation of the proved theorems that come at a more advanced stage. And the verifications tend to be too simple; in contrast to more advanced situations such as the verification that the fundamental group in topology is a group. There is an urgent need for examples to be constructed which fall between these two extremes (not necessarily in group theory); since pupils often do not know how to begin the task of verifying axioms.

Other approaches – and alternative answers to Question 2 (p. 237) – are to approach probability in an axiomatic way (e.g. Durran [1]) or to teach the elements of mathematical logic – propositional calculus and the notion of quantifiers (e.g. SMP (*Additional Maths*, Part 1), and the materials of the SSMCIS and the CSMP projects).

Unfortunately, this is difficult to motivate unless the pupils have a rich background of mathematical experience; otherwise they have little to be logical about. Alternatively, one can postpone attempting a solution of the problem until pupils are ready to appreciate mathematical analysis. Then one can use elementary topology as a training ground in proof (as with Papy [4]); or elementary, but rigorous, number theory and analysis (as with Weston [1]).

Finally, one can use the axiomatic method to demonstrate proof by selecting specific applications. One example of this is that of Weil (Kemeny [1]), who axiomatises the rules of a primitive community's marriage system, and gives a valuable discussion in terms of permutation groups. Another example is that of Steiner [1], who axiomatises a notion of voting to analyse the limitations of a system. His example has the advantage that it also suggests to the pupils another aspect of the axiomatic method – the possibility of generalisation by 'weakening the axioms'. Some such generalisations are significant and others are not, and the subjective notion of 'mathematical significance' is well worth discussing with pupils, although such discussion is rare.

An alternative to all these approaches is to teach proof by example, by looking at specific problems and solving them. Examples of such problems can be found in graph theory, elementary number theory and the theory of inequalities. The proofs are not manipulative; and the pupil must know what he is doing. The 'Swansea' scheme of Weston, mentioned above, includes material of this type, as a 'starter' before the real analysis begins. In particular, elementary number theory is included; it is excellent material for imparting the concept of proof in a 'natural' way, partly because one can establish statements which are attractive but incapable of verification by computation (e.g. that $n^{37} - n$ is divisible by 1,919,190 for all $n \in \mathbb{N}$).

SOME FURTHER READING

Dieudonné [4], Freudenthal [1], OISE [1], Shibata [1], Thom [1, 2]

17
Axioms and Number Systems

The variety of approaches to number is far smaller than that of approaches to geometry. Consequently in this chapter our style changes. We show how the axiomatic method can help with the discussion of the number systems in imparting their most important features to pupils.

1. THE NATURAL NUMBERS

If we were asked to explain what real numbers are, to a class of pupils, then our reply would depend upon their maturity. Let us, then, consider the easiest case first and suppose that the pupils are mathematically as mature as we are: thus, how do we explain to *ourselves* what whole numbers are?

We could say 'Whole numbers are the counting numbers 1, 2, 3, and so on', but 'and so on' will hardly suffice for doing many mathematical problems. It will not suffice to answer questions like 'Is 100 a whole number?' 'Is the real root of $x^3 - 4x + 1 = 0$ a whole number?' To have a description that allows us to do mathematics we need to account for the whole numbers, not individually, but as a *system*: we describe not the chess pieces, but the *rules of the game*. The first to do this was Peano, in 1889.

Peano observed that, however $\mathbb{N}$ was to be described,† everyone agrees that the description must allow us to play a game in which, among other things, addition and multiplication could, in principle, be performed by a finite number of operations of the form 'Change n to $n + 1$'. Also, the usual method of proof by induction was essential to the game. Recall that this method must be used to prove such statements as the following, which we label P_n. Thus

P_n: 'For all whole numbers n,
$$1^2 + 2^2 + \cdots + n^2 = \tfrac{1}{6} n(n + 1)(2n + 1)'.$$

---

† We denote the set of natural numbers by $\mathbb{N}$. There is no fixed convention on whether or not 0 is a natural number. In what follows we exclude 0 from $\mathbb{N}$.

(Many such statements do not involve algebraic formulae so directly.) We give a proof of P_n in the following way, because it may be rather different in lay-out from what the reader has met before, and the present form is vital for later discussion.

Let X denote the set of those whole numbers n for which the equality holds. Certainly 1 lies in X. Also, if n lies in X, so does $n + 1$; for, to verify this we have to check that

$$1^2 + 2^2 + \cdots + n^2 + (n + 1)^2$$
$$= \tfrac{1}{6}(n + 1)(n + 1 + 1)[2(n + 1) + 1],$$

knowing that $1^2 + 2^2 + \cdots + n^2 = \tfrac{1}{6}n(n + 1)(2n + 1)$ because n lies in X.

The verification consists of simple algebra and we omit it. But now we have picked out a subset X of whole numbers such that 1 lies in X, and whenever n lies in X so does $n + 1$. The only such subset of whole numbers is the entire set $\mathbb{N}$ itself. Hence, the proposition P_n holds for all n.

These observations led Peano to describe the game of arithmetic with whole numbers in a rather strange way; but its strangeness prevents us from making the mistake of saying that things are 'obvious' when we are dealing with subtle matters where traps abound. Thus Peano showed that we work with the set $\mathbb{N}$ of whole numbers whenever we use a set S of things with the following properties:†

(1) For each x in S, there is another element x' in S; x' is called the *successor* of x, and x has just the one successor, x'.

(2) Different elements of S have different successors (that is, if $x \neq y$ in S, then $x' \neq y'$).

(3) There is just one element (denoted by b) which is *not* the successor of anything in S, but for all $x \neq b$ in S, $x = y'$ for some y in S.

(4) If X is any subset of S such that X contains b, and whenever x is in X then x' lies in X, then X is all of S.

Peano was then able to show that this strange description does allow us to deduce that S satisfies all our requirements for doing the arithmetic of whole numbers. Thus he showed – and a good account of the details can be found in Thurston [1] – how the rules (1) to (4) allow us to add and multiply in S, with b playing the role we expect of 1, and where $x' = x + 1$. Moreover, (4) then becomes the rule that allows us to do proofs by induction, after the manner of the proof given above: it is called the *Axiom of Induction*.

† (1), (2) and (3) can be abbreviated to 'There is an injection, suc: $S \to S$, whose image is $S - \{b\}$.

As a further example of its use and of the way in which we can now establish propositions, let us prove that *every x in S is either even or odd*. (We cannot say that 'every whole number is obviously even or odd, if we do not know what whole numbers are: for example, is the number of sand-grains in the world obviously even or odd?) Our proof will assume the hard work of Peano in showing that we can add in S. We shall show that every x in S is either odd, i.e. is of the form 1, or $y + y + 1$ for some y in S, or is even, i.e. $x = y + y$. We leave it as an exercise for the reader to show that x cannot be both odd *and* even. The proof proceeds as follows.

Let X be the set of all those x in S that have such a form. Clearly 1 is in X. If x is in X we must – to apply the Axiom of Induction – verify that $x + 1$ ($=x'$) is in X. But since x is in X we have either $x = 1$, $x = y + y$, or $x = y + y + 1$ for some y in S (y need not lie in X). If $x = 1$ then $x' = x + 1 = 1 + 1$ which is of the allowed form $y + y$; if $x = z + z$, then $x' = x + 1$ is of the allowed form $z + z + 1$; and if $x = z + z + 1$, then $x + 1$ is of the form $y + y$ with $y = z + 1$.† Always, then, x' lies in X if x does so. Hence by the Axiom of Induction, $X = S$, so *every x in S* has one of these three required forms.

If the reader now understands the role of the Axiom of Induction, then we can use it to satisfy his curiosity as to how addition is defined in S. We set

$$\text{(i) } x + 1 = x', \qquad \text{(ii) } x + y' = (x + y)'$$

for all y in S, thus defining $x + z$ '*inductively*' for a fixed x and all z in S. To see that we have done it for *all z in S*, we use the Axiom of Induction and let X consist of all those z in S for which $x + z$ is defined. Then 1 lies in X by (i) above; and if z lies in X, we know what $x + z'$ means – by (ii). Thus z' lies in X. Hence, by the Axiom of Induction, X is all of S, so we know what $x + z$ means for *all z in S*. It is necessary, of course, to show that addition, thus defined, has all the usual properties, and we leave this as an exercise to the reader. Similarly, multiplication may be defined inductively through the rules

$$x \cdot 1 = x, \qquad x \cdot y' = x \cdot y + x$$

and knowledge of addition as defined above. Again we leave it to

---

† Note how we have here used the commutative and associative properties of addition in S.

the reader to prove that this multiplication works in the usual way.

In fact, Peano proved (for details see Griffiths [1], Ch. 23) that any two sets S and T satisfying the axioms on p. 247 are copies of each other; technically they are *isomorphic*. Thus we may choose one such, and call it $\mathbb{N}$, the set of whole numbers with addition and multiplication given by rules (i)–(iv) and those for $+$ and $\cdot$ above. Each of the operations is associative and commutative, multiplication is distributive over addition, i.e. $x(y + z) = xy + xz$, and for all x in $\mathbb{N}$,

$$1 \cdot x = x, \qquad x' = x + 1.$$

$\mathbb{N}$ is then a model of our vague notions of the system of whole numbers that are induced in us by doing elementary arithmetic; in this model, the familiar numbers 2, 3, and so on, appear as

$$2 = \text{suc } (1),\ 3 = \text{suc } (2) \quad \text{and so on}$$

(where suc $(x) = x'$) and we do *not* need to say what 'and so on' means *as part of the specification of* $\mathbb{N}$. The number names are part of our language *about* $\mathbb{N}$, not *in* $\mathbb{N}$.

Exercise

Give a rule for 'choosing' a copy of a Peano system to be called $\mathbb{N}$.

2. INTEGERS, RATIONALS AND REALS

Once we have an understanding of what we mean by $\mathbb{N}$ then we can approach the set $\mathbb{Z}$ of integers in two possible ways.

The first of these is to describe the set $\mathbb{Z}$ of all integers $0, \pm 1, \pm 2,$ ..., as a commutative ring† R with unit element 1, which

(*a*) contains a copy of $\mathbb{N}$ (we may denote this copy for the moment by R^+) that behaves like $\mathbb{N}$ with respect to the addition and multiplication it gets from R; and such that

(*b*) if x is in R and $x \neq 0$ (0 is not in R^+), then either x or $-x$ (but not both) lies in R^+. (Hence, every element of R is either 0 or $\pm n$, where n lies in R^+. Such an n is called 'positive'.)

Again, this serves to specify $\mathbb{Z}$ to within isomorphism, for if R' is another ring satisfying properties (*a*) and (*b*), it can be shown that

† An algebraic structure in which addition and multiplication are defined and are both commutative and associative; in which multiplication is distributive over addition; and which is a group with respect to addition. For an introduction to such structures, see Griffiths [1] Part III.

there is an isomorphism $R \to R'$ between the rings R, R' which preserves the 'positive' elements (and, hence, zero and the 'negative' elements): R and R' differ only in notation.†

Notice the economy of this definition: we have not needed to say what the elements of R 'are' (no need to be philosophical about, say, -1), nor what addition or multiplication 'means'. Just from the ring axioms, zero is *described by its role* – all we need to know about it is that for all x in R, $0 + x = x$; and for any x, $-x$ is defined by its role that $-x$ is the unique solution in R of the equation $x + y = 0$. From the ring axioms alone we can deduce that $-(-x) = x$ and $(-x)(-y) = xy$, without needing to know what x 'is'. Indeed, this lack of knowledge then allows us to assert these facts for *any* rings.

Exercise

Prove that $-1 \times -1 = 1$ by establishing the following steps from the ring axioms:

(i) $-(-x) = x$ in any group written additively,
(ii) $a \cdot 0 = 0$ (consider $a \cdot (a + 0)$),
(iii) $-(a \cdot b) = a \cdot (-b)$ (consider $0 = a \cdot [b + (-b)]$),
(iv) $-(a \cdot b) = (-a) \cdot b$,
(v) $a \cdot b = (-a) \cdot (-b)$.

For all its simplicity and elegance this method is, of course, highly sophisticated; and considerably more discussion and work will be needed if we want to use $\mathbb{Z}$ as a model for doing real-life arithmetic (see Chapter 18). Our second approach will, in fact, lend itself more readily to models suitable for school use; for we shall use a construction rather than Peano's 'parachute' attack. However, for the moment, we continue this first approach, by considering next the system $\mathbb{Q}$ of rational numbers (fractions).

To define the set $\mathbb{Q}$ of rational fractions, we cannot say that a 'rational fraction is the result of dividing one integer by another', for until we have described $\mathbb{Q}$ we do not know that there need be a 'result' of even a division of 1 by 2. The axiomatic method used for obtaining $\mathbb{Z}$ prompts us to describe $\mathbb{Q}$ by saying that it shall be a field F‡ such that (a) F contains a copy Z of the system $\mathbb{Z}$, and (b) for

---

† The reader with the appropriate technical background will appreciate that R^+ is more rigorously defined as the image of an injective homomorphism $p: \mathbb{N} \to R$ for which $p(1) = 1$. The later isomorphism $R \to R'$ must then commute with p and $p': \mathbb{N} \to R'$.
Similar remarks will also apply to our description of $\mathbb{Q}$ below.
‡ A field is a commutative ring in which to every non-zero element x there is a y in F such that $xy = 1$.

each x in F, there is an integer r in the copy Z of $\mathbb{Z}$ such that rx lies in Z.

The elements in Z are the 'integers' of F. It can be shown that if F' is another field satisfying (a) and (b), then there exists an isomorphism between the fields F and F' under which the integers of F and F' correspond isomorphically. Hence, any two such fields differ only in notation, so we may choose one copy of F and call it $\mathbb{Q}$. From the definition we can also deduce that if x lies in $\mathbb{Q}$, then integers p, q exist in $\mathbb{Z}$ such that $x = pq^{-1}$ (and $q \neq 0$), thus retrieving our intuitive picture that x should be the result of dividing p by q.

The second method by which one can obtain $\mathbb{Z}$ and $\mathbb{Q}$ from $\mathbb{N}$ is by means of a direct construction, which should be understood before 'school' presentations are designed.

To obtain $\mathbb{Z}$ we consider the set of all ordered pairs of natural numbers and partition these into equivalence classes by saying that the two pairs (x, y) and (x', y') – where x, y, x' and y' are natural numbers – are to be equivalent if and only if $x + y' = y + x'$. Thus, for example, $(1, 3)$ will be equivalent to $(7, 9)$, and $(3, 1)$ to $(10, 8)$. We now define $\mathbb{Z}$ to be the set of all such equivalence classes, so that a typical element of $\mathbb{Z}$ will be the class

$$\{(1, 3), (2, 4), (3, 5), \ldots\}.$$

It would, of course, be very difficult to deal mentally with integers having such a complicated form, so we denote these equivalence classes by certain new symbols, namely:

the class $\{(1, 1), (2, 2), (3, 3), \ldots\}$ by 0,
the class $\{(2, 1), (3, 2), (4, 3), \ldots\}$ by $^+1$,
the class $\{(1, 2), (2, 3), (3, 4), \ldots\}$ by $^-1$,
and so on

We must now say what addition and multiplication are to mean in $\mathbb{Z}$, i.e. to define $z_1 + z_2$ and $z_1 - z_2$. This is done by selecting one pair (x_1, y_1) from the equivalence class z_1 and one pair (x_2, y_2) from the class z_2, and by defining $z_1 + z_2$ to be the class containing the pair $(x_1 + x_2, y_1 + y_2)$, and $z_1 z_2$ to be the class containing the pair $(x_1 x_2 + y_1 y_2, x_1 y_2 + x_2 y_1)$. (These definitions are chosen because we want the formulae

$$(x_1 - y_1) + (x_2 - y_2) = (x_1 + x_2) - (y_1 + y_2)$$
and $\quad (x_1 - y_1)(x_2 - y_2) = (x_1 x_2 + y_1 y_2) - (x_1 y_2 + x_2 y_1)$

to hold.)

One notes here that it is necessary to show that our definition is valid in the sense that $z_1 + z_2$ and $z_1 z_2$ will not depend upon the choice of the ordered pairs chosen to represent z_1 and z_2.

Given these definitions, it is easily shown that $\mathbb{Z}$ is a commutative ring, that $\{^+1, {}^+2, {}^+3, \ldots\}$ is a copy of $\mathbb{N}$, and that $^-1, {}^-2, {}^-3, \ldots$ are $-1, -2, -3, \ldots$ in the sense that $^+n + {}^-n = 0$; moreover if z is the equivalence class containing (m, n), then $z = {}^+m + {}^-n$.

Exercises

1 Verify directly that this construction of $\mathbb{Z}$ gives a commutative ring.
2 Verify directly that in this particular construction $^-m \cdot {}^-n = m \cdot n$, thus avoiding the need for a proof along the lines of the Exercise on p. 250.

In a similar way one can construct $\mathbb{Q}$ by considering equivalence classes of ordered pairs of integers (x, y) where y is taken to be non-zero, and (x, y) is equivalent to (x', y') if and only if $xy' = x'y$. (Details can be found in many texts, for example Griffiths [1], Ch. 22.) We must then show that the structure $\mathbb{Q}$ so obtained has the properties we require of it, as we did in the case of $\mathbb{Z}$.

Exercise

Give definitions of addition and multiplication in $\mathbb{Q}$ in terms of equivalence classes of ordered pairs of integers.

It will be seen that in this way we prove, as a statement about $\mathbb{N}$, the *existence theorem* that systems (constructed from $\mathbb{N}$) *exist* satisfying the axioms given earlier for $\mathbb{Z}$ and $\mathbb{Q}$. (We have here, of course, ignored questions of uniqueness.) There are many places in mathematics (as here) where an existence theorem is a rather specialised affair, whereas the consequences of it are of general interest; thus one takes for granted the existence theorem and comes to its proof when the interest is strong enough. This repeats the historic development of the subject. Another example of this procedure is with the logarithmic function (to base ten). Here is a function $f: \mathbb{R}^+ \to \mathbb{R}$ from the positive real numbers to the reals, which is a bijection (i.e. one–one and onto) and which satisfies the rules

$$f(x \times y) = f(x) + f(y), \qquad f(10) = 1.$$

There exists exactly one such continuous function and it is customary to denote f by $\log_{10}$, and $f^{-1}: \mathbb{R} \to \mathbb{R}^+$ by 'antilog' (the latter exists because f is a bijection). Many pupils are (rightly) introduced to the existence and properties of $\log_{10}$ long before they can prove its existence; in their early courses, the existence of $\log_{10}$ is a pedagogical axiom.

Indeed, many students at university have suffered at the hands of enthusiasts who have given, as the first topic in their analysis course, a proof that the set of real numbers $\mathbb{R}$ exists. This is one of the hardest parts of the analysis course and has been presented to students before they have possessed either the techniques or the motivation. For, unlike the construction described above to obtain $\mathbb{Q}$, the constructions yielding $\mathbb{R}$ cannot be readily explained by means of simple models. It is particularly convenient at this stage, then, to take the *existence* of the field $\mathbb{R}$ of real numbers as a pedagogical axiom and to use the axiomatic method to define (i.e. to describe) $\mathbb{R}$. Then one can pass quickly to the theory of limits, continuity and so forth that need the properties of $\mathbb{R}$ for their proof. By practising on these easier theorems, pupils can gain the strength eventually to go through a proof that $\mathbb{R}$ exists. The most convenient definition of $\mathbb{R}$ is that which says that $\mathbb{R}$ is a *complete, ordered field*, although the notions of field, order and completeness need, in any case, patient discussion. For a more detailed treatment see, for example, Armitage [1], p. 103, and then Griffiths [1], Ch. 24.

3. THE SET-THEORETIC APPROACH TO NUMBER

The axiomatic descriptions of number systems that we have so far given, are primarily designed for the 'aesthetic' side of mathematics, especially for giving clear proofs about matters involving numbers. On the 'practical' side, however, it is necessary to use numbers for modelling such aspects of real life as comparing and measuring. But, as Russell and others pointed out, such axiomatic descriptions are of no use for these applications. Indeed, it is not until we have made an intuitive acquaintance with the system $\mathbb{N}$ – by counting and working out sums with simple numbers – that we can even see that Peano's axioms really describe something significant. For these reasons, the set-theoretic approach to number was devised by Cantor, Russell, Zermelo, Fraenkel and other logicians. That approach is based on Cantor's mathematical formulation of cardinal arithmetic (see, for example, Griffiths [1], Ch. 7) and it uses the *style* of the axiomatic method in its mode of exposition.

In Cantor's theory, one defines the cardinal of *any* set X by

$$\text{card } X = \{A \mid A \text{ matches}\dagger X\}.$$

and one expresses an ordering relation between cardinals by

$$\text{card } X \leqslant \text{card } Y,$$

provided there exists an injection $X_0 \rightarrow Y_0$ for some $X_0 \in \text{card } X$ and $Y_0 \in \text{card } Y$. One adds cardinals by the rule

$$\text{card } X + \text{card } A = \text{card } (X_0 \cup A_0)$$

for any $X_0 \in \text{card } X$ and $A_0 \in \text{card } A$ with $X_0 \cap A_0 = \varnothing$. (This definition is independent of the choices of X_0 and A_0, and such sets always exist.)

These definitions yield a theory, even when the sets are infinite. In fact, the theory supplies the language that enables us to say when a set is infinite:

The set X is infinite, provided X contains a *proper* subset Y (i.e. $X \neq Y$) such that X matches Y.

For example, $\mathbb{N}$ is infinite since $\mathbb{N}$ matches its proper subset of even numbers.

Any set which is not infinite is said to be *finite*.‡ For example, the empty set $\varnothing$ is finite, since it has no proper subset, and its cardinal is denoted by zero,

$$\text{card } \varnothing = 0;$$

then the singleton $\{0\}$ is finite, with cardinal denoted by

$$\text{card } \{0\} = 1;$$

then $\{0, 1\}$ is finite, with cardinal 2, $\{0, 1, 2\}$ is finite, with cardinal 3, and so on.

Thus, for example, $\{0, 1, 2\}$ is a member of the cardinal – which is a class – denoted by 3, while $\{0, 1, 2, \ldots, 1000\}$ is a member of the class which is denoted by 1001, and so on.

Russell's development of arithmetic, from set theory, *constructs* Peano's $\mathbb{N}$ by defining it to consist of all cardinals of *finite* sets, with the successor of x corresponding to the cardinal number $x + 1$. He then verifies the Peano axioms for these 'numbers': thus in Russell's theory the Peano axioms become *theorems*, and Peano's undefined

---

† A is said to match X if there is a bijection $f : A \rightarrow X$, that is, if f is an injection (one–one) – $a \neq b$ in A implies $f(a) \neq f(b)$ in X – and a surjection (onto) – for all x in X, there exists a in X such that $f(a) = x$.

‡ We note that, strangely, it is 'infinite' which is defined positively.

terms are then *constructs*. Similarly, one can repeat the development, to have Russell's axioms (those of set formation and set theory) occurring as theorems of some more basic theory (for example, that of Lawvere (see MacLane [2]), based on categories).

For us, Russell's theory is especially important. It comes equipped with a clear, precise, and essentially simple language, which we shall show later (Chapter 18) to be a useful tool for developing approaches to number for young children. The traditional approach for them has used ordinary language, and since that was too clumsy even for geniuses like Cantor and Russell to talk clearly about number, small wonder that it failed with most schoolteachers and their pupils!

However, once the arithmetic of $\mathbb{N}$ has been taught to children, they still need to learn about fractions and negative numbers. At the appropriate stage, the existence of $\mathbb{Q}$ or $\mathbb{Z}$ can be asserted as a pedagogical axiom, in order to describe the rules for working with such numbers. Of course, the notion of an axiom cannot be used with young children, and there is a teaching problem about *describing* the notation and the rules so that children understand. But the axiomatic method suggests *what* should be described, whereas the traditional methods merely let the child pick up some idea, with much room for confusion.

Exercise

Find a traditional textbook of arithmetic and summarise its explanations of what numbers and/or fractions are. Compare these with its explanations of addition, multiplication, etc. Are the explanations clear to you? Would they be clear to a child? Do the arithmetical rules given follow from the explanations? What would a reader actually learn from the book?

4. CATEGORICITY

We close these two chapters in which we have looked at axiom systems, by considering a particular property of such systems – categoricity.

With any set of axioms, there arises the question: how many objects satisfy this set of axioms? This immediately brings up the notion of *isomorphism* between objects of that kind. For example, we could say that $\mathbb{N}$ is the unique system satisfying the axioms given earlier, because we could formulate a notion of isomorphism which allowed us to neglect any further differences between any two systems satisfying the axioms. Strictly speaking, then, we should say that there is a unique *isomorphism class* of objects satisfying the axioms for ℕ, and the question about how many objects should be phrased 'how many isomorphism classes?'. Nevertheless, while it is an abuse of language to confuse an isomorphism class with one of its objects, it is often convenient to do so. (In a similar manner we often do not distinguish between an equivalence class and one of its objects – a rational number and a fraction, a vector and a directed line-segment.)

Some sets of axioms do not define objects uniquely. For example, there are many non-isomorphic groups (although if we were to add a further group axiom that 'the underlying set should have cardinal 7', then this would no longer be the case, since we know that all groups of order 7 are isomorphic). Again, cf. p. 233, if we drop Playfair's Axiom – or its equivalent – from our set of axioms for Euclidean geometry, then the resulting geometry would not be unique. Any results which we could deduce from this weakened axiom system would hold not only in Euclidean geometry but in the non-isomorphic non-Euclidean geometry of Bolyai and Lobachewsky. The geometry defined by this system is called *absolute* geometry (see Coxeter [1], Ch. 15).

Clearly, it is important to distinguish between the two cases: where the axioms define objects uniquely and where they do not. We say in the former case that the axiom system is *categorical*. The group axioms, therefore, are non-categorical.

To know that an axiom system is categorical – as in the case of those systems we referred to above for ℕ, ℤ, ℚ and ℝ – is a great achievement. It is, in the words of Moise [1], a 'sort of arch of triumph' for it means that we have a complete understanding of the essential properties of the structure described by the system. Thus, we know that the essential features of ℝ are that it is a field, is complete, and is ordered. These three properties serve to characterise it uniquely. Of course, to characterise it is not to know *all* about it: Peano's axioms characterise ℕ, but we still do not know whether every even number greater than 4 is the sum of two primes (Goldbach's Conjecture).

This is not to say, however, that categorical axiom systems are always desirable – the lack of the categorical property can be useful in that the propositions we derive from the axioms will apply to non-isomorphic structures. Thus, for example, it follows from the group axioms that if a group G is finite with m elements, then for any x in G, $x^m = 1$, where 1 denotes the unit element of G, and x^m denotes the product $x \circ x \circ x \circ \ldots \circ x$ of x with itself m times,† using the group operation $\circ$. Apply this to the particular group $\Phi(n)$ of positive integers prime to, and less than, n with $\circ$ interpreted as multiplication mod n. Then we have Fermat's Theorem: $a^{\phi(n)} \equiv 1(n)$, where $a \in \Phi(n)$ and $\phi(n)$, Euler's totient function, denotes the number of elements in $\Phi(n)$. (In particular, when p is a prime, we obtain $a^{p-1} \equiv 1(p)$, whenever a is prime to p (cf. p. 245.) Again, it follows from the axioms that if $h: G \to H$ is a homomorphism of groups and h is also a bijection, then its inverse $h^{-1}: H \to G$ is also a homomorphism (which is why group isomorphism is a symmetric relation). Apply this to the bijection $\log_{10}$ described on p. 252, where we prescribed it to be a homomorphism from the multiplicative group $\mathbb{R}^+$ of positive reals to the additive group $\mathbb{R}$ of real numbers. Then

$$\text{antilog: } \mathbb{R} \to \mathbb{R}^+$$

is also an isomorphism, so

$$\text{antilog } (x + y) = \text{antilog } x \times \text{antilog } y.$$

Of course, $\mathbb{R}$ and $\mathbb{R}^+$ are *not* isomorphic if we take other aspects of their structure into consideration.

Exercises

1 What binary operation $*$ on $\mathbb{R}^+$ will make the function $x \to \sqrt{x}$ an isomorphism from $(\mathbb{R}^+, +)$ to $(\mathbb{R}^+, *)$? (See Baker [1].) Why do we choose to talk about $(\mathbb{R}^+, +)$ rather than $(\mathbb{R}^+, *)$ if the two are isomorphic?

2 Show that there are just two (isomorphic classes of) groups having exactly four elements.

3 Write out the multiplication and addition tables of the rings $\mathbb{Z}_n$ of residues modulo n for some special values of n. Check from your table that $-1 \times -1 = 1$ in each system considered. What difference do you observe between the cases n prime and n composite? (See Griffiths [1], Ch. 10.)

4 What is grey and has four legs and a trunk?
Answer: A mouse going on holiday.
   What does this riddle tell us about the way in which children think, and about categoricity?

5 Discuss the way in which categoricity is involved in detective stories and also how axiom systems appear in Science Fiction (e.g. Asimov's laws of robotics).

SOME FURTHER READING

Dubisch [2], Goodstein [1], Moise [2]

† More precisely, we can use the method of inductive definition (p. 248): $x^1 = x$, while $x^{n+1} = x \circ x^n$ for all $n \in \mathbb{N}$.

18

School Approaches to the Number Systems

Minus times Minus equals Plus:
The reason for this we need not discuss.†

In the previous chapter we explained some of the language developed, by Russell and others, to explain the number system to mathematicians. We shall now consider how the tools they left us can be adapted to explain the number system to pupils in school. In the first nine sections we fill in the details of an approach suggested in CCSM [2]. We then discuss other approaches.

1. STATEMENT OF A PROBLEM IN MATHEMATICAL EDUCATION

To orient the reader, our *object* here in following the CCSM proposals is to devise a way of telling children about numbers so that they can count with them, and use negative numbers; but we want them also to add, subtract and multiply with *real* rather than just with whole numbers.

Our *constraint* is that through reasons of lack of language, interest or stamina, and intellectual maturity, we cannot give the 'official' mathematical descriptions of ℝ available to adult mathematicians.

We shall be *satisfied*, therefore, if the pupils end with a 'pre-mathematical' acquaintance with the system ℝ of real numbers, enabling them to use numbers, but without their being able to say what numbers 'are'; yet they must be able to reason about what they do rather than parrot cookbook recipes of the 'two minuses make a plus' kind. In fact, those children who survive a traditional education in mathematics *can* use numbers and do problems with them, but they cannot describe the number system they use and do not know, for example, why $-1 \times -1 = 1$ – an equation traditionally imparted only by the rule of authority. Many schoolteachers have only a similar sort of pre-mathematical notion of number,

† A mnemonic learned at school by W. H. Auden (see *The Observer*, 20 May 1973).

258

which their pupils acquire (if at all) by osmosis from them. Had they a more explicit model of number, fewer pupils might fail to do mathematics.

Our strategy will be to supply the children with a model of the number system, which they can use for applications. The pupils' manipulations with the model will in principle be translatable into the 'official' language of $\mathbb{R}$, thus minimising later unlearning and errors. By basing our procedures on the axiomatic method (although the pupils will not realise it) we shall be exploiting the categoricity of the axioms for $\mathbb{R}$ (p. 256) to ensure that the system used by the children is isomorphic to $\mathbb{R}$ – but explained in their language rather than ours. We also hope to make the manipulations seem reasonable, not arbitrary, so that reasoning becomes possible.

Our exposition is an attempt (by mathematicians) to explain to schoolteachers the mathematics underlying what they may already be doing in the classroom. The language is adapted to their vocabulary. Occasionally, the teacher has to think in the vocabulary of his pupils, as he plans ways of illustrating the material. Sometimes there is a commentary placed in parenthesis, $\{\dots\}$, to show how the language used for explaining to the teacher is adapted from the language of mathematics. Thus *three* languages are being used; those of mathematician $\leftrightarrow$ mathematician, mathematician $\leftrightarrow$ teacher, and teacher $\leftrightarrow$ child. This should illustrate the complexity (and intellectual appeal) of mathematical education, and can be usefully compared with the situation faced by computer software specialists, who also need to consider three languages, but with 'child' replaced by 'computer' and 'teacher' by 'programmer'.

The mathematical educator is therefore faced with the problem of translation – translation of a piece of mathematics from one person's conceptual framework to that of another, which is often less complex and less mathematically sophisticated.

Exercises

1 In what follows try to distinguish between the three languages in use.

2 Give other examples in mathematical education of this problem of translation and of attempts to solve it.

3 Clearly it would be desirable for a primary school teacher to have read and understood the previous chapter, but to what extent is such comprehension necessary?

4 In any proposals for a school course one must take precautions against its being taught in the wrong spirit. How might one ensure that the message

communicated by the following presentation would not differ substantially from what was originally intended?

5 Discuss the difference between enriching a student's conceptual framework (e.g. by giving him mathematics of 'situations') and communicating new mathematics to him in terms of his existing framework.

6 Technical students are often familiar with the concept of a 'black box'. Devise a unit on functions for such students which is isomorphic to the usual one but which substitutes for the function formula $f: a \to b$ the scheme $\xrightarrow{a}\boxed{f}\xrightarrow{b}$ where a and b represent the input and output of the box f. (See SMP *Book 2* and Howson [6], which gives illustrations from the calculus, including the chain rule for differentiation.)

2. MATCHING

The basic notion in the underlying theory is that of 'matching' two sets A and B. We say that A *matches* B if we can pair off each element a of A with some b in B with none left over; if a and a' are different elements of A, then they are to be paired off with different elements of B. {Such a pairing-off is a bijection between A and B: see Griffiths [1], Ch. 2.}

Thus, if X is a set of married couples in a monogamous society, the set A of husbands matches the set B of wives. In a box of unbroken eggs, the set A of eggs matches the set B of places they fill. If we say that there are seven days in a week, we mean that the set A of day-names 'Monday, Tuesday, Wednesday, Thursday, Friday, Saturday, Sunday' matches the set B of number-names 'One, two, three, four, five, six, seven'. This last example illustrates how mathematicians count: first we have within $\mathbb{N}$ the subsets $\mathbb{N}_1 = \{1\}$, $\mathbb{N}_2 = \{1, 2\}, \ldots, \mathbb{N}_k = \{1, 2, 3, \ldots, k\}$, and so on, with $\mathbb{N}_k$ the set of the first k positive whole numbers inclusive, and then we say that a set A has k members if we can match A with $\mathbb{N}_k$. This matching process (see Fig. 1) is called 'counting'; we pair off elements by pointing to some element of A and pairing it with the word 'one', pointing to some other element of A, and pairing it with the word 'two', and

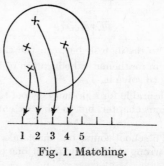

1 2 3 4 5
Fig. 1. Matching.

so on until we stop with k. We are really matching A with the set B_k of the first k number-names, because B_k is *designed* to match $\mathbb{N}_k$; and obviously if one set matches a second, while the second matches a third, then the first set also matches the third. {'Matches' is a transitive relation because two bijections compose to give a third bijection.} For brevity, k is called the *cardinality* of A (and of $\mathbb{N}_k$), and our remark about matching three sets says that

$$\text{if } A \text{ matches } B_k, \text{ then card } A = \text{card } B_k = k.$$

The number-names, and the conventions for forming them, are learned because we need to be able to refer to the sets $\mathbb{N}_k$ by name. Consequently, very young children must first learn some number-names by various means, including rhymes such as

> One, two, three, four, five,
> once I caught a fish alive.
> Six, seven, eight, nine, ten,
> then I put it back again.

Only then can they begin to 'count' objects. Later, with experience the child realises that the 'pairing-off' operation and, hence, the cardinality of a set of objects, is unaffected by a physical reshuffling of the objects. He also learns that the cardinality of a set does not depend on time or on the way in which its objects are counted. Of course, we all have counted sets and obtained different answers, but we learn by experience to recognise when there is a mistake: the invariance of cardinality {Griffiths [1], Ch. 8} actually provides the clue to our mistakes. Here is where the teacher uses her knowledge of children to devise suitable exercises in the mathematics expressed in pupil language.

The next stage in the child's development is his association of number-symbols, or numerals, with number-names. This association is reinforced by the use of 'number-lines', 'number-strips', rulers or tape-measures, on which the child finds the numerals from 1 to 10, 1 to 20 (and later, say, -20 to 20) equally spaced, always in their 'natural' order and ordered from left to right. At first, such number-lines can be used as a tool for use in matching sets; later they have other uses.

Ultimately, of course, we want pupils to have the pre-mathematical experience that the number-line can, in *principle*, be as large as we wish. Certainly, it is not long before children realise that they can lengthen any given number-line and before they ask 'Is there a biggest number?'. Here is a hint of the notion of an infinite set.

For the moment, however, we can keep to a number-line L of fixed length, say, 1–20. It then has 20 subsets of the form $\mathbb{N}_k$, which can be used for matching familiar sets.

L, however, has additional structure, for we can 'shunt' along it rigidly (see Fig. 2). That is to say, if we have a set A of objects

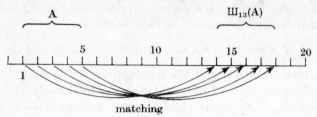

Fig. 2. The number-line and shunting.

spaced out in order on the first k numbers of L, then we can shunt A to the right through b steps (say), lifting each of the k objects bodily through b steps to the right. We assume that L is long enough (i.e. b small enough) for L to hold the shunted version of A. An important fact is clear to us: *this shunted version of A, matches A*. Thus, if we denote the operation of shunting b places to the right by $Ш_b$ (using the Russian letter $Ш$ = 'Sh'), then

$$A \text{ matches } Ш_b(A)$$

for all A and b appropriate to the size of L. The statement holds also for sets scattered in different ways on L.

Exercises

1 Explain how early sorting experiences can lead to the formation of the concepts of 'set', 'membership', 'inclusion' in children of Infant School age.
(Bristol BEd, 1971)

2 Show that $Ш_1(n) = n + 1, Ш_m(\mathbb{N}_k) = \{m + 1, \ldots, m + k\}$, and $\mathbb{N}_b \cap Ш_b(\mathbb{N}_b)$ is empty.

3. ORDERING

As well as 'counting' sets we also compare their 'sizes'. This is a basic activity with children, who complain that a friend's 'set' of things – sweets or toys – is 'bigger' than their own. The number-line L can be used to check such comparisons; for if we wish to compare (finite) sets A and B, we match A with $\mathbb{N}_a$ on L and B with $\mathbb{N}_b$. Perhaps $\mathbb{N}_a$ matches $\mathbb{N}_b$, that is $a = b$. If not, suppose $\mathbb{N}_a$ is contained in $\mathbb{N}_b$, as in Fig. 3 (written† $\mathbb{N}_a \subset \mathbb{N}_b$). We then say that the number a

† Here we use $\subset$ to denote 'a *proper* subset of' and $\subseteq$ to denote 'a subset of'.

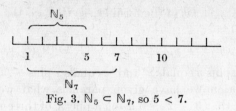

Fig. 3. $\mathbb{N}_5 \subset \mathbb{N}_7$, so $5 < 7$.

is *less* than b (and b is *greater* than a), written $a < b$ ($b > a$). {This comparison test acquaints us with the important 'Law of Trichotomy': that for any two numbers a, b, either $a = b$, $a < b$ or $a > b$, and the possibilities are mutually exclusive. The comparison process defines bijections $A \to \mathbb{N}_a$, $\mathbb{N}_b \to B$, and the composition $A \to \mathbb{N}_a \subset \mathbb{N}_b \to B$ is an injection $A \to B$.} The teacher will also need to associate comparison of numbers with such relations as 'older than'. 'larger than', etc., in which children have a strong emotional interest. By explaining adult usage of these terms, we help those children who have picked up the terms unconsciously and inaccurately.

4. ADDITION

A basic operation with sets is to combine two separated piles of objects, X and Y, to make a combined pile Z, denoted by

$$Z = X \cup Y \quad \text{(the } union \text{ of } X \text{ and } Y\text{).}$$

This operation corresponds to addition in arithmetic, as we now explain.

If we have counted X and Y, then we have found sets $\mathbb{N}_x$, $\mathbb{N}_y$ on the number line L so that

$$X \text{ matches } \mathbb{N}_x, \qquad Y \text{ matches } \mathbb{N}_y.$$

Hence, also, Y matches $\text{Ш}_x \mathbb{N}_y$, the shunt (see p. 262) of $\mathbb{N}_y$. Thus Z matches $\mathbb{N}_x \cup \text{Ш}_x (\mathbb{N}_y)$, as can be seen from Fig. 4 (which suggests a formal description of this matching).

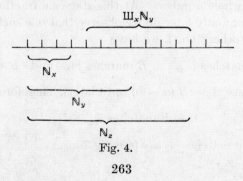

Fig. 4.

263

THE CURRICULUM IN THE SMALL

Now $\mathbb{N}_x \cup Ш_x (\mathbb{N}_y)$ is of the form $\mathbb{N}_z$, so that z is the cardinal of Z. We write

$$z = x + y, \qquad \mathbb{N}_{x+y} = \mathbb{N}_x \cup Ш_x(\mathbb{N}_y),$$

and say that 'x plus y equals z' and 'z equals x plus y'.

This description we have given parallels what we do mentally when we check that $3 + 2 = 5$ using a pile X of three peas and a pile Y of two beans: we match the peas with the set {one, two, three} of number-names, and the beans with the set {four, five}, which matches the set $(Ш_3 \, \mathbb{N}_2)$ on L.

We could have reversed the roles of X and Y in the above discussion, for clearly (by equality of sets)

$$Z = X \cup Y = Y \cup X,$$

whence we have the non-trivial fact that addition is *commutative*, i.e. that

$$x + y = y + x.$$

Similarly,

$$(X \cup Y) \cup Z = X \cup (Y \cup Z),$$

so addition is associative, i.e.

$$(x + y) + z = x + (y + z).$$

5. SUBTRACTION

Historically, men not only augmented flocks of sheep, but, by stealing, removed one flock from a larger one (thus providing a need for counting, if only to keep watch on what was happening). Thus from a set A we may remove a portion (or subset) B of A to leave a remnant denoted by $A - B$. The problem of counting $A - B$ leads us to the *subtraction* of whole numbers. {At this stage subtraction is independent of addition; only later do we observe that one such operation is inverse to the other.} On L we have

$$A \text{ matches } \mathbb{N}_a, \qquad B \text{ matches } \mathbb{N}_b \quad \text{and} \quad b < a$$

(leaving the case $A = B$ $(a = b)$ until later). Therefore,

$$A - B \text{ matches } \mathbb{N}_a - \mathbb{N}_b,$$

and this last set matches $\mathbb{N}_c$ say (see Fig. 5).

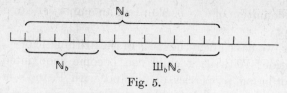

Fig. 5.

Thus, card $(A - B) = c$ and we write

$$c = a - b, \qquad \mathbb{N}_{a-b} \text{ matches } \mathbb{N}_a - \mathbb{N}_b.$$

The definition of $+$ leads us to observe (see Fig. 4) that

$$\mathbb{N}_b \cup \text{Ш}_b(\mathbb{N}_c) = \mathbb{N}_a$$

so $b + c = a$.

This last equation may be written

$$a = b + c = c + b = (a - b) + b.$$

It does *not* allow us to write the (true) equation

$$(a + b) - b = a$$

since, until the equation is established, we do not know whether it is permitted to move the brackets. However, this follows at once from the obvious fact for sets that if A and B have no members in common, then

$$(A \cup B) - B = A.$$

Exercise

Explain the validity of addition and subtraction using Cuisenaire rods.

6. ZERO

To simplify the discussion of subtraction, we assumed when removing the subset B from A, that B was not the whole of A ($B \neq A$: B is a *proper* subset of A). However, if B were all of A we could still remove it, and the result of such a removal is an example of the empty set $\varnothing$; a set with cardinality *zero* or 0. {Mathematically, 'the' empty set is unique, but this fact is a subtle one to understand. It is perhaps better not to press this point at this stage and if necessary to allow children to speak of 'an' empty set.} Clearly, for any set A

$$A \cup \varnothing = A = A - \varnothing \quad \text{and} \quad \varnothing = A - A,$$

so $\qquad a + 0 = a = a - 0 \quad \text{and} \quad 0 = a - a,$

provided we agree that $\mathbb{N}_a - \mathbb{N}_a (= \varnothing)$ is also $\text{Ш}_a(\varnothing)$ as needed in

265

the above definition of $a - b$. Shunting an empty set can never make it non-empty!

Thus far, we have given the basis of the arithmetic of $\mathbb{N}$, excluding multiplication. With it, a child can become acquainted with the essential ordering properties of $\mathbb{N}$, and with Peano's notion of a successor – of always being able to pass from a to $a + 1$. Peano's axiom 4 (p. 247) appears in the guise that we can always fill a gap on L between a and b (if $b > a + 1$) by inserting $a + 1$, $a + 2$, ..., and *that we shall actually reach b*. These facts are, of course, not *proved* for children, but nevertheless they will begin to learn what these statements mean, in a 'pre-mathematical' way.

7. MULTIPLICATION

Multiplication is rather more difficult to explain by means of the number-line than is addition. True, multiplication of natural numbers can be treated as repeated addition (indeed, this is the basis of Peano's definition), but when we do this and think of, say, 3×5 as

$$5 + 5 + 5,$$

we lose the symmetry of the operation. Furthermore, any explanation based on repeated addition will not carry over to $\mathbb{Z}$, let alone $\mathbb{R}$.

It is better, therefore, to model multiplication in two dimensions and to base our arguments on the *Cartesian product* $\mathbb{N}_a \times \mathbb{N}_b$ consisting of the 'lattice points' (i, j) in the plane, where i runs through $\mathbb{N}_a$ and j through $\mathbb{N}_b$ (see Fig. 6). Many well-known games have been

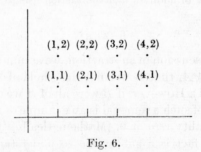

Fig. 6.

devised to make children familiar with the coordinate notation, so that they can locate a point such as (10, 6). At a more abstract level, there are also several ways (see, for example, Nuffield – CEDO [1]) of introducing children to the Cartesian product $A \times B$ of two finite sets.

It is then easy for pupils to check that by defining $a \times b$ to be the cardinality of $\mathbb{N}_a \times \mathbb{N}_b$, we obtain a 'product' with the required properties. For, clearly, $\mathbb{N}_a \times \mathbb{N}_b$ consists of a 'vertical rows' (i.e. columns), each of the form $\{(i, 1), (i, 2), \ldots, (i, b)\}$ and so $\mathbb{N}_a \times \mathbb{N}_b$ has cardinal $b + b + \cdots + b$. On the other hand $\mathbb{N}_a \times \mathbb{N}_b$ consists of b horizontal rows, each of the form $\{(1, j), (2, j,), \ldots, (a, j)\}$, and so

$$a \times b = \underbrace{a + a + \cdots + a}_{b \text{ times}} = \underbrace{b + b + \cdots + b}_{a \text{ times}}.$$

Thus, for example, 3 rows of 5 cabbages can also be seen as 5 rows of 3 cabbages. Of course, if we have *no* rows of b cabbages we have an empty set of cabbages; as also with a rows of no cabbages, so

$$0 \times b = a \times 0 = 0.$$

A great simplification in multiplication arises because of the *distributive law*

$$a \times (b + c) = a \times b + a \times c,$$

because the RHS consists of one addition of two products while the LHS consists of one product with one addition. Since multiplication on computers is usually much more expensive than addition, the distributive law helps to cut costs (amongst other advantages). To see why the law should hold, we need only observe that the RHS corresponds to a rows of b cabbages added to a rows of c cauliflowers (say), so that we can think of them as a rows of $b + c$ greens (see Fig. 7).

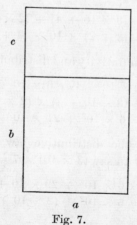

Fig. 7.

A prime consequence of the distributive law is for calculations involving the 'place' system of notation for numbers. Using the traditional base 10, we can split any finite set A, into, say, k blocks each containing 10 objects, together with a block B with $0 \leqslant$ card $B < 10$. Thus

$$a = \text{card } A = 10k + b, \text{ where } b = \text{card } B.$$

We now repeat this process using a more systematic notation to get (with $b_0 = b$)

$$\begin{aligned} a = 10k + b &= 10(10k_1 + b_1) + b_0 & 0 \leqslant b_1 \leqslant 9 \\ &= 10^2 k_1 + 10b_1 + b_0 \\ &= 10^2(10k_2 + b_2) + 10b_1 + b_0 & 0 \leqslant b_2 \leqslant 9 \end{aligned}$$

and so on, until the ks decrease below 10 and we have

$$a = 10^r b_r + 10^{r-1} b_{r-1} + \cdots + 10b_1 + b_0,$$

each b_j lying between 0 and 9 inclusive. Thus we denote a in the decimal scale of notation by

$$a = b_r b_{r-1} \ldots b_0.$$

Exercise

Carry out this process by actually splitting up a pile of counters. What happens if you change 10 to some other number like 2, or 12?

Since it is necessary to be able to add and multiply integers efficiently, many algorithms have been devised. We shall illustrate these, rather than attempt to describe them in general.

(i) $$\begin{aligned} 34 + 27 &= (3 \times 10 + 4) + (2 \times 10 + 7) \\ &= (3 + 2) \times 10 + (4 + 7) \end{aligned}$$

– by associativity, commutativity and distributivity –

$$5 \times 10 + 1 \times 10 + 1 = 6 \times 10 + 1 = 61.$$

(ii) $$\begin{aligned} 34 \times 27 &= (3 \times 10 + 4) \times (2 \times 10 + 7) \\ &= 6 \times 10^2 + 21 \times 10 + 8 \times 10 + 28 \end{aligned}$$

– multiplying out by the distributive law and using commutativity and associativity to say $(3 \times 10) \times 7 = 21 \times 10$ –

$$\begin{aligned} &= 6 \times 10^2 + 29 \times 10 + 28 \\ &= 6 \times 10^2 + (2 \times 10 + 9) \times 10 + \\ & \qquad\qquad (2 \times 10 + 8) \end{aligned}$$

$$= 8 \times 10^2 + 11 \times 10 + 8$$
$$= 9 \times 10^2 + 1 \times 10 + 8$$
$$= 918.$$

In these two examples we have, for clarity of explanation, put in much detail that one normally leaves out. When examples are presented like this we see the desirability of having short cuts, like the usual tricks of 'carrying ones' and using columns. Also it is quicker to memorise tables of 'little' sums, of the form $a \times b$ or $a + b$, when a and b lie between 0 and 9. Unfortunately the *reasons* underlying the tricks and tables are frequently forgotten – witness the continued learning of tables up to 12 – and the result is traditional, stupid arithmetic. This is why there are now moves to postpone 'tables' until there is a clear need for memorising.

Note that the methods of the above examples correspond very closely to the practical models in set theory whether or not the sets used consist of pebbles, beads on an abacus, or 1 cm Cuisenaire rods. These methods apply without change to other bases, such as 7 or 12; only the notation alters.

Exercises

1 Do the examples above in the scale of 8.
2 Write out explanations of the 'carry one' method suitable for (a) adults and (b) children.
3 It is argued that children are capable of memorising tables long before they can see any mathematical reason for doing so, and that they enjoy this activity which, although pointless at the time, is of value later. Discuss.
4 'The time to learn tables is when technique demands it.' What series of problems could eventually demonstrate to a child that it would be easier to learn tables than work out every multiplication from scratch?

8. REAL NUMBERS

So far, we have been working with a model corresponding to arithmetic in $\mathbb{N}$. Rather than passing to $\mathbb{Z}$ and then to $\mathbb{Q}$ (as in the formal development of Chapter 17) we now jump to a model of the real numbers $\mathbb{R}$; the reasons will emerge as we proceed.

The first point to stress is that the mathematics of the reals is considerably more complicated than that of the natural numbers, the integers or the rationals, and that although traditionally the reals are introduced several years after the natural numbers and the positive rationals, the intellectual maturity of pupils when they first encounter them is not sufficient to permit a detailed discussion of their properties.† Our aim then must be to present a model of the reals which will enable pupils to calculate with them with confidence and which will constantly stress those properties which characterise them – namely, that the reals form a field which is ordered and complete.

The difficulties of presenting a suitable model for arithmetic in $\mathbb{R}$ are seen when we try to extend the model we have used for $\mathbb{N}$. First, we observe that instead of the basic sets $\mathbb{N}_k$ of 'lattice' points on the number-line L, we have to use an entire segment; that is we think of the spaces between the dots as filled in. L is changed from a row of dots to a stripe of paint (with the dots retained to act as 'milestones'). In order that the number 1 be represented by a segment, we push the stripe backwards to zero. Thus we have a model of the positive real numbers in which the number x is represented by the segment S_x, consisting of all points of our number-line between 0 and x inclusive (see Fig. 8). If x happens to be a whole number, S_x is a

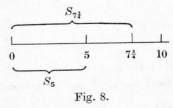

Fig. 8.

picture (in primary school terms) of a Cuisenaire rod; but by allowing x to be arbitrary we are allowing rods of arbitrary size, not just in centimetre lengths.

Having obtained a representation of each positive real number, we now try to extend our earlier 'shunt' operation. The shunt $Ш_b$ was defined in terms of movement through b steps, where b was the cardinality of the set of points on L which represented the natural number b. In our extended model, however, b is no longer represented by a set having finite cardinal, and it follows that if we wish to define $Ш_b$ when b is not a whole number, then we shall have to resort to other methods. The method we use is based on a geometrical construction. {It is, in fact, a particular example of a construction that can be employed in more general circumstances; see, for example, O'Hara and Ward [1], Ch. 7, Blumenthal [1], Chs. 4 and 5.} To

† This view is not held by all. See, for example, Papy's approach, summarised in Section 12 below.

do the drawing, children need first to be shown how to construct a line parallel to a given one, using a set square. In order to carry out the construction we have to go into the two-dimensional plane, and to employ a convenient 'copy' of our number-line L (see Fig. 9) parallel to L.

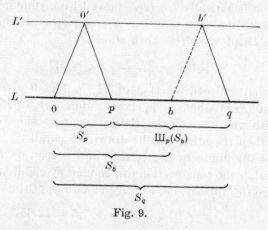

Fig. 9.

The construction we use is this: to obtain $b + p$ we choose O' arbitrarily on L', join $O'p$ and draw bb' parallel to OO'. We then construct $b'q$ parallel to $O'p$. S_q then represents the real number $p + b \ (=q)$.

The set S_b does not merely match the segment $O'b'$, it can be moved rigidly to be superimposed upon $O'b'$; and similarly $O'b'$ can be moved rigidly and superimposed upon the segment pq. The rigidity requirement follows from the properties of parallel lines, but can be demonstrated empirically to children by actually moving stiff rods.

We denote the segment pq in Fig. 9 by $\text{Ш}_p(S_b)$, and regard it as the 'shunt' of S_b. At once we see from Fig. 9 that

$$S_{p+b} = S_q = S_p \cup \text{Ш}_p(S_b),$$

analogously to our earlier definition (p. 263)

$$\mathbb{N}_{x+y} = \mathbb{N}_x \cup \text{Ш}_x(\mathbb{N}_y).$$

By rigidity, our new definition of shunt has the same effect on a set of the form $\mathbb{N}_x$ (i.e. when $x \in \mathbb{N}$) as did the original shunt of Fig. 2. Therefore, for example, $2 + 3 = 5$ whether we are thinking of 2 and 3 as naturals or reals. Thus the previous operations for adding, subtracting, and comparing size still make sense, and can be carried

271

out using segments S_x instead of the sets $\mathbb{N}_k$, and with the new definition of the shunt $ш_b$.

{This construction explains why we choose a *straight* line as a model of the number-line and why we 'space out' the numbers evenly. If this were not done, the shift functions would be hard to define. Mathematically we have moved from thinking of $\mathbb{R}$ as a set of points, to the view that $\mathbb{R}$ is a vector space complete with a (isomorphic) set of affine transformations

$$ш_b\colon \mathbb{R} \to \mathbb{R}, \quad x \mapsto b + x.\}$$

Observe that the construction of $ш_b$ in Fig. 9 gives the same result, regardless of where O' is chosen. Moreover, if OO' is known, together with any two of the three remaining sides of triangles $OO'p$, $bb'q$, then the other can be drawn by parallels. Hence, if we know *two* of the numbers b, p, q the third is read off from Fig. 9. This means that the construction can be used by children not only to add, but also to subtract, i.e. to solve equations of the form

$$3\tfrac{7}{8} + x = 10\tfrac{1}{11} \quad \text{or} \quad y + 7.8 = 12.65,$$

for x or y, provided they are given (or can make) suitably accurate scales. Such problems can be attempted long before the conventional rules about common denominators are understood.

The new operation of addition is commutative and associative, as can be checked by adults using coordinate geometry. Children, however, can at least find the statement plausible by testing special cases. Notice that these methods make mathematics accessible to children who can see and draw, but who may not possess the academic ability of abstract reasoning.

Our operations, however, have so far only been defined for positive reals, and we must now go on to see what happens when x is negative.

First we consider how to represent S_x when x is negative. Here we recall that when x and y are positive the rule for asserting $x < y$ was to check that $S_x \subset S_y$. One way in which we can extend our model in order to preserve this rule is to let S_x denote the whole *infinite* segment from $-\infty$ to x. Apart from the verbal and mathematical difficulties that might ensue from the introduction of the symbol '$-\infty$', this would also necessitate a slight recasting of our definition of addition for positive reals. Another possibility is to 'orient the line' and say that x is less than y ($x < y$) if x is to the left of y. (This means that if x is negative then S_x will still be a finite segment, and will still have endpoints x and 0; the only difference is that 0 will now be the endpoint on the right.) This approach does not in fact demand that a child should *know* which is left and which is right, although it could be a helpful step towards establishing that knowledge. In particular, then, x is 'negative' if it is to the left of 0 on the number-line L: otherwise x is 'positive'.

The addition of two numbers, one of which is negative, now presents little difficulty, since the earlier geometrical construction for Ш_p still makes sense if p or b (or both) is negative (see Fig. 10). But

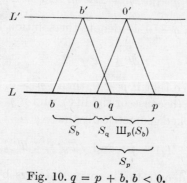

Fig. 10. $q = p + b,\, b < 0.$

we cannot always express S_{p+b} in the older form

$$S_{p+b} = S_p \cup \text{Ш}_p(S_b)$$

using set union, since $\text{Ш}_b(S_p)$ is subtracted from S_p in Fig. 10. For this reason it is perhaps better now to make changes in our notation and to write

$$q = p + b = T_p(b).$$

{We note that in doing this we are looking at addition not as a binary operation, $\mathbb{R} \times \mathbb{R} \to \mathbb{R}$, but as a set of unary operations $\{T_p \colon \mathbb{R} \to \mathbb{R}\}$. This is the usual primary approach to addition, i.e. one learns to 'add 3' and to 'add 4', rather than to 'add'. See also Hirst [1].}

The constructions make sense whether b or p are positive or negative, so addition is now always defined, and, as remarked above, is commutative and associative. Also, as the figures indicate, if $p = 0$, then

$$0 + b = T_0(b) = b$$

{so T_0 is the identity function on L}, and if $b = 0$, then

$$p + 0 = T_p(0) = p,$$

273

so we have addition with a true zero. As we remarked when considering only positive numbers, the construction also tells us how to find b if q and p are given; since we first draw $b'q$ to find b', and then construct $b'b$ parallel to $O'O$. Thus children can now find by construction the solutions of such equations as $3\frac{1}{8} + x = 1\frac{5}{16}$. The resulting operation is the function T_p^{-1}, inverse to T_p, because it satisfies the equation

$$T_p(x) = q \quad (\text{i.e. } p + x = q, \text{ so } x = T_p^{-1}(q)).$$

In particular, when $q = 0$ we denote the solution by ^-p; thus

$$p + {}^-p = 0.$$

Geometrically we see that if $p < 0$, then $^-p > 0$, that p and ^-p are equidistant from 0 (because of rigidity), and if $p > 0$ then $^-p < 0$.
{It is readily checked geometrically that

$$T_b^{-1}(q) = p \quad \text{and} \quad T_b^{-1} = T_{-b}.\}$$

If, for clarity, we denote the negative integers by $^-1$, $^-2$, $^-3$, ..., then Fig. 11 shows that $3 + {}^-2$ is $3 - 2$ in the older sense of Section 5. It is clear by Cartesian geometry (or to the children by drawing)

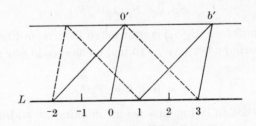

——————— construction for $3 + {}^-2$.
- - - - - - - construction for $^-2 + 3$.
Both constructions yield '1' as the result

Fig. 11.

that $^-2 + 3 = 3 + {}^-2$. Hence, the construction extends the notion of subtraction in a reasonable way to make sense of $x - y$ when either $y \leqslant x$ or $y > x$: thus $x - y$ *means* $x + {}^-y$. In particular, ^-x (with $x > 0$) is the same as $0 - x$, normally written $-x$.

The construction for $x - y$ also makes sense whether or not x or y is positive. In particular, since $0 = x - x$ for *all* x, then always $-x$ is that (unique) element y such that $x + y = 0$; so

$$\text{if } x > 0, \text{ then } -x = {}^{-}x,$$
$$\text{if } x < 0, \text{ then } x = {}^{-}y \text{ and } {}^{-}x = y,$$
$$\text{if } x = 0, \text{ then } -x = 0.$$

For these reasons

$$-(a + b) = -a + (-b)$$

(because the RHS when added to $b + a$ is zero). In particular, if we take $b = -a$, we find $0 = -0 = -a + [-(-a)]$, so $-(-a) = a$.

9. MULTIPLICATION OF REALS

We now wish to define multiplication in our new model. To define, say, $6 \times (-3)$ we could copy our procedure on p. 267 and write

$$6 \times (-3) = \underbrace{(-3) + \cdots + (-3)}_{6 \text{ times}},$$

but then we could not say that this was equal to -3×6 because we cannot add 6 to itself '-3 times'.

A possible solution – which we shall adopt here – is to use a geometrical construction, to go into two dimensions, since that seems to be where the multiplication of numbers most naturally arises. We use similar triangles as in Fig. 12 to construct the parallel lines

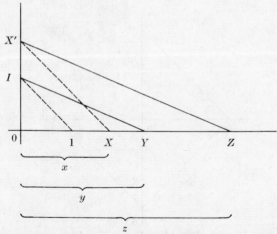

Fig. 12. Multiplication by x dilates the triangle IOY by a factor x.

YI, $X'Z$ shown, where $OI = 1$, $OX' = x$, $OY = y$ and (hence)

$$OZ = xy.$$

275

By coordinate geometry we know that, whatever the signs of x and y, z always comes out at xy. Thus we have a rule for multiplication which *always* gives an answer. The rule is easily described to children, and we shall see in a moment how they may see it to be reasonable without having the reason explained.

{*Note.* More generally one can take I to be any point on the vertical axis and construct z as shown in Fig. 13.}

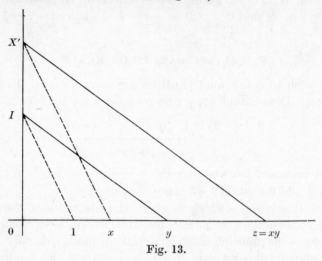

Fig. 13.

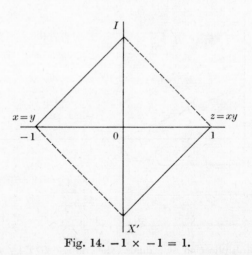

Fig. 14. $-1 \times -1 = 1$.

Fig. 14 shows at once that $-1 \times -1 = 1$, a most difficult equation to explain simply by other means. Moreover, the rule is sensible, because it agrees with the 'repeated sum' definition when x is an integer n:

$$ny = \underbrace{y + \cdots + y}_{n \text{ times}}$$

as is clear from Fig. 15 (where y is shown positive, but a reflection in the y-axis yields the case $y < 0$). Further, children can *see* the agreement in these cases, and one is introducing them informally to the notion of *extending the domain of a function*.

Since ny means the same as repeated addition, then $nm = mn$ when y is the integer m. Thus the new multiplication is commutative and associative when it applies to integers, from which children may suspect that this holds for arbitrary x and y. If they were so interested, diagrams could be devised to convince them.

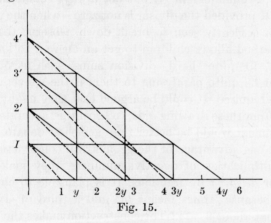

Fig. 15.

This distributivity law $x(y + z) = xy + xz$ can be seen by the children either by measuring from the diagram, or by observing that if we construct $y + z$ by the method of Fig. 10, then multiplying by x (using Fig. 12) simply dilates the addition figure, and yields the addition of xy to xz.

We now have all the arithmetic and ordering properties other than the completeness of $\mathbb{R}$ in our model. At the same time the children can see how geometrical insight may help the calculations, and they learn empirically some of the facts of Euclidean geometry – similarity, congruence, etc. Once the constructions are seen to give reasonable results, the work can be codified so as to gain technique with such problems as calculating

$$(13 - 15) \times (27 - 41),$$

277

(saying that it is $-2 \times -14 = 28$, more briefly than using the distributive law

$$13 \times 27 - 15 \times 27 - 41 \times 13 + 15 \times 14).$$

But of course the constructions work for arbitrary xs and ys, although calculations begin when the xs and ys are actual rational numbers, or – if got by measurement – integral multiples of some unit length like a millimetre. Here is where decimals can be introduced.

Division first occurs here as the inverse of multiplication: one traces Fig. 12 backwards from z and y to get $x = z/y$. Again, this process can be compared with the divisibility properties of whole numbers, to show how one cannot always divide in $\mathbb{N}$ (e.g. 3 divided by 2 leaves a remainder in $\mathbb{N}$) but one always gets a unique result (here $1\frac{1}{2}$) in $\mathbb{R}$ provided the divisor is not zero – when the geometrical construction is clearly seen to break down. Observe that the construction method allows children to get an answer, to known limits of accuracy, to quite 'hard' division sums, such as $55\frac{1}{2} \div 23\frac{1}{3}$ – which might be quite paralysing to them if the conventional algorithms were required. It could be argued that they might then never be weaned from these drawing aids, to learn the 'accurate' methods. Here, experiment would be better than armchair debate.

However, one advantage of the geometrical constructions is that, as with multiplication of negative numbers, they make plausible the rules for *dividing* by fractions. It is plausible to children that $\frac{1}{3} \times 6 = 2$ because 'times means of' and a third of six is clearly two. But what about $6 \div \frac{1}{3}$? The construction makes the 'inverting' rule clearly reasonable.

Once familiarity is gained, the teacher may judge the time to be ripe to 'idealise' the constructions by extrapolating the idea that there is an ordered field, $\mathbb{Q}$ or $\mathbb{R}$, given by the usual axioms. This statement would then merely codify, as rules of a game, what the children knew full well to be possible on 'their' number-line L, and which they are likely to feel is capable of infinite extension in each direction. This vague feeling is put into words by the axioms, since these are known to be categorical, as discussed in Chapter 17.

Exercises

1 Show by coordinate geometry that z in Fig. 13 has coordinates $(xy, 0)$. Hence prove that the construction *must* give a multiplication that is commutative, associative and distributive.

2 The proofs in the previous question are for mathematicians. Draw diagrams that will convince children visually that multiplication has the 'right' properties (cf. Fig. 15).

3 If x is kept fixed in Fig. 13, show that xy can be obtained from a straight-line graph on the same diagram once two points are known.

10. SOME OTHER APPROACHES TO THE NUMBER SYSTEMS

In the preceding sections we considered an approach to the various number systems based upon the recommendations of the Cambridge Conference (CCSM [2]). There are, of course, many other alternatives, and we shall now briefly describe some of these.

Clearly, most of the new methods of teaching number in the primary school are based on the set-theoretic approach described in Chapter 17 and, as we saw in Section 8, Cuisenaire rods provide a concrete model of this approach. (The authors know of no school approach to number based directly on the Peano axioms.)† Traditionally, children progress from the study of the natural numbers to the study of the positive rationals ('sharing numbers'). At this stage the teacher – perhaps unwittingly – is apt to base her approach on the construction of $\mathbb{Q}$ described on p. 252. For the first thing children consider are fractions which they see to consist of two whole numbers, the numerator and the (non-zero) denominator. {This corresponds to our considering an ordered pair belonging to $\mathbb{Z} \times (\mathbb{Z} - \{0\})$.} The children learn that certain fractions are equivalent, e.g. that $\frac{1}{2} = \frac{2}{4} = \frac{3}{6}$ {i.e. an equivalence relation is (implicitly) defined on $\mathbb{Z} \times (\mathbb{Z} - \{0\})$}. At this point the two paths diverge and are at different levels of abstractness. The mathematician refers to the equivalence class as a rational number and frames his definitions of addition and multiplication in terms of representatives. On the other hand the child learns to replace a fraction by an equivalent fraction in order to make these operations more meaningful to him: thus to add $\frac{1}{2}$ to $\frac{1}{3}$ he will replace $\frac{1}{2}$ by its equivalent $\frac{3}{6}$, and $\frac{1}{3}$ by its equivalent $\frac{2}{6}$, and he will now be faced with the 'simpler' problem of adding $\frac{3}{6}$ to $\frac{2}{6}$. The general rule which he deduces from this type of problem, that

$$\frac{a}{b} + \frac{c}{d} = \frac{ad + bc}{bd},$$

† They are, however, quite willing to believe that amongst the plethora of projects there will be at least one GPP (Grand Peano Project). Servais [1] proposes that the Peano axioms should be in a 'modern secondary school syllabus in mathematics for the scientific stream', but gives no indication as to what, if anything, he would do with them.

is, of course, the starting point for the mathematician's definition of addition of rational numbers.

Most teachers are content to leave the approach to rationals at this level, but some (see, for example, Mansfield and Thompson [2]), believe that 12-year-olds can profitably be introduced to equivalence classes and to the definitions of operations on quotient sets. Another approach is to use the mathematician's construction as a source for examples on equivalence relations to be attempted further up the school (see SMP *Advanced Mathematics Book 1*, Ch. 1).

The two different levels of approach can also be observed when one studies the means by which the negative integers are introduced. Again, for example, Mansfield and his associates introduce the integers as equivalence classes of ordered pairs of naturals. Others choose to model this approach without using the technical language of equivalence relations. Thus, for example, the SMP approach in its CSE series (SMP *Book C*, Ch. 2) uses what is basically the following model. We have a 'large' block of flats with its floors numbered to match the natural numbers in their natural order. It is then possible to give instructions of the type '2 floors up' or '3 floors down' (the latter kind of instruction cannot be given from all floors but can always be given from some). Here the instruction is independent of the floor on which one enters the lift. We are in fact considering ordered pairs of floor numbers {i.e. elements of $\mathbb{N} \times \mathbb{N}$} and making use of an obvious equivalence relation defined on these pairs to define the 'shift numbers' $^+2$ and $^-3$. Consideration of the model will yield an obvious definition of the operation of addition. The model will not, of course, yield an obvious definition of multiplication. One possible path to multiplication is

(i) to get as far as one can with multiplication as repeated addition,

(ii) demand that multiplication is commutative (as it is for $\mathbb{N}$),

(iii) consider what value for $^-x \times {}^-y$ would preserve the number patterns and graphical patterns already provided by consideration of $x \times y$ and $x \times {}^-y$.

{Once again the children are tackling the problems of extending the domain of a function, in this case from $\mathbb{N} \times \mathbb{N}$ to $\mathbb{Z} \times \mathbb{Z}$.}

If this 'constructive' approach to $\mathbb{Z}$ is followed, then it should be clear to the child – at least from plotting graphs of functions like $y = {}^-3 \times x, y = {}^-1 \times x$, etc. – that $^-1 \times {}^-1 = 1$ is a consequence of the definition of multiplication.

A primary school approach to the integers based on pairs that will balance on the 'equaliser' – a simple balance on which weights can be hung at multiples of a unit distance from the fulcrum – can be found in Nuffield – CEDO [2].

11. COMPLETENESS OF ℝ

As we have stressed earlier, the construction of ℝ is considerably more difficult than that of ℤ or of ℚ, especially if it is desired to discuss the completeness of ℝ. Certainly the authors cannot envisage a school approach beginning with a discussion of Dedekind cuts or Cantor Fundamental Sequences. Here one is forced back onto what is basically an axiomatic approach, although within that approach there is considerable ground for manoeuvre – indeed there are few topics in secondary school mathematics which are treated in so large a variety of ways. The approaches of the many modern projects vary from 'the least said, soonest mended' attitude of, for example, the CSE texts of the SMP, to the axiomatic treatment given by Papy.

Certainly, some would argue that it is sufficient and less confusing, merely to convey the idea that the reals can be modelled by a line on which every point represents a number – this implicitly introduces the notions of ordering and completeness – and that these numbers can be added and multiplied in some way that is consistent with the earlier rules for ℚ. It can be readily shown that all rational numbers are represented by recurring and/or terminating decimals and conversely. The next step of associating a number with every decimal is one that is accepted readily by pupils, who will willingly attach the name 'irrational' to those numbers represented by decimals that do not recur or terminate. The age, or level of maturity, at which pupils can follow the classical proof that $\sqrt{2}$ is irrational, is a matter for debate, or rather, for experiment. Certainly they are willing to accept that $\sqrt{2}$, π and e cannot be expressed as recurring or terminating decimals, long before they are capable of following the proofs that these numbers are irrational. The question of how one attaches a meaning or a decimal expression to, say, $\pi/\sqrt{2}$ is not one which seems to arise frequently in a classroom, but of course a bright child is quite capable of raising it in a class where discussion is encouraged.

At a slightly more formal level one has the treatments such as that to be found in the SMG texts (*Book 3*, Ch. 3). Here there is talk of the 'system' of 'the real numbers' and of the 'structure' of the system. The pupils are told that the structure 'is the same as that of the rational number system, but the proof of this cannot be given at this stage'. Apart from the emphasis on the field structure of the reals and the willingness to introduce technical terms, the approach is not dissimilar to that described in the preceding paragraph.

The next level of abstractness entails a more explicit mention of the notion of completeness. The approach of the SSMCIS (Draft Book I, Course II) was based on the 'lub principle'. The writers consider the length of a segment (which is clearly bounded above) and show that it may happen that there is no rational number which represents the least upper bound of the monotonically increasing sequence of rationals 1, 1.5, 1.52, 1.528, . . . which approximate to the measure of the line segment. They then postulate the existence of a new ordered field $\mathbb{R}$ which contains $\mathbb{Q}$ and which satisfies the *completeness property* that every non-empty subset of $\mathbb{R}$ which is bounded above possesses a least upper bound (see, for example, Armitage [1], p. 113). The writers then go on to *construct* a set which satisfies these properties, by means of an argument based on Dedekind cuts. It is a piece of mathematics which many (including the authors of this book) would feel ill-suited for a school course.

A more 'humane' attempt to deal with completeness is to be found in the SMP *Additional (O-level) Mathematics Book*, Part 1, Ch. 2. Here the author seeks to characterise the essential differences between $\mathbb{Q}$, $\mathbb{R}$ and $\mathbb{C}$ (the field of complex numbers). He notes that all three are fields and that $\mathbb{C}$ differs from the others in that it is not ordered. Is there a similar property which will serve to distinguish between $\mathbb{Q}$ and $\mathbb{R}$? The writer then goes on to use the 'Chinese Box' (see, for example, Armitage [1] and p. 285 below) or 'Cantor nesting' property to show how one can have nesting intervals defined on $\mathbb{Q}$ which have no rational number common to all intervals, whereas if one forms within $\mathbb{R}$ a nest of infinitely many intervals, each lying completely within its predecessor, then such a nest will always yield at least one number which is common to all the intervals. This latter approach aims then to make explicit the previous assumptions about completeness, rather than to make completeness a starting point for an axiomatic treatment of $\mathbb{R}$.

12. PAPY'S APPROACH

Perhaps the most formal of the approaches to the real numbers is that by Papy, to be found in his series *Mathématique moderne*. Here the approach is completely axiomatic – yet, it must be stressed, it is not as abstract as that to be found in, say, the draft SSMCIS texts. Moreover Papy's treatments are always enlivened by the use of imagination, humour and colour – qualities often lacking in certain texts. Papy studies the reals using a geometrical model not dissimilar to that discussed in Sections 8 and 9. The outstanding feature of Papy's treatment is the way in which he coordinates the study of geometry with that of the reals. It is his belief that 'the most fundamental and central topic of the secondary school programme is, without doubt, vector spaces' and as a natural consequence, therefore, his approach to geometry is to exhibit Euclidean space as a vector space over $\mathbb{R}$ with the natural inner product. For this reason it is necessary to study the field $\mathbb{R}$ of coefficients. The study of $\mathbb{R}$ and of Euclidean space is undertaken by means of what Papy describes as the 'progressive axiomatic approach', namely: 'State clearly that which is accepted; do not say everything at one time; state certain accepted things, little by little.'

The result of this approach is a unified course spread over the first three years of secondary education which, whatever its shortcomings, holds great mathematical interest and shows evidence of being the outcome of a considerable amount of thought and expertise.

Papy begins by studying the plane and the equivalence relation of parallelism. (To establish the transitivity of this relation one must invoke Playfair's Axiom (p. 233).) Other geometric considerations are used to introduce the notions of 'order', 'parallel projections' and 'equipollency' (i.e. given four distinct points A, B, C and D in the plane, we say that the couples (A, B) and (C, D) are equipollent if and only if $(AC \parallel BD$ and $AB \parallel CD)$. As with parallelism, the axioms already introduced are insufficient to show that the relation under consideration is an equivalence relation. In order then to ensure that equipollency is transitive we must add this as an axiom. {A weak form of Desargue's Theorem.} Once this is done, one can define a 'translation' to be an equivalence class of equipollent couples and show that under the operation of composition ('vector' addition) the set of all translations forms a commutative group. Moreover, if one has a fixed point 0 in the plane as an origin, one can establish a bijection between the group of translations and the additive group of vectors. Multiplication of vectors by integers is now straightforward, but before one can obtain a vector space structure one must define multiplication by a real number. {Multiplication by rationals would, of course, be sufficient to provide an example of a vector space structure, but it would not yield the vector space at which Papy is aiming.} It is now necessary, therefore, to establish some of the properties of the reals, and this is done in the following manner.

A line is taken in the plane and two points, labelled 0 and 1, are taken on the line. This induces a natural ordering on the line, $0 < 1$, and allows one – using the theory on equipollency – to construct certain points on the line in one–one correspondence with the integers. So as to simplify later work, Papy uses the binary notation for the integers. Thus we have the construction shown in Fig. 16 (cf. Figs. 9

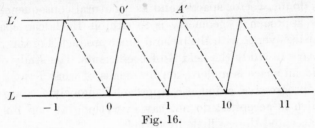

Fig. 16.

11). (As before, $0'$ is an arbitrary point on L'. A' is found by drawing $1A'$ parallel to $00'$, 10 by drawing $A'(10)$ parallel to $0'1$ and so on.)

It is now necessary to introduce Archimedes' Axiom (cf., for example, Griffiths [1] p. 393) in the form: every point on the line is now labelled or lies between two labelled points.

The labelling process is then extended to include 'finite' binary decimals. Here we make use of the theorem (known in France as the 'Small' Theorem of Thales) that parallel projections preserve midpoints. Thus to fix the point to be labelled 0.1 (binary notation for $\frac{1}{2}$) we find the point of intersection of $0A'$ and $0'1$ and draw a line through it parallel to $00'$. This will cut the segment 01 at its midpoint; the point to be labelled 0.1. In this way, cf. Fig. 17, we can add further labelled points to our line.

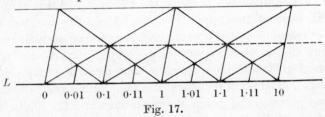

Fig. 17.

Corresponding to any 'infinite' decimal, for example, 1.010010001 . . ., it is now possible to form an infinite set of closed binary segments $\langle 1, 10 \rangle$, $\langle 1.0, 1.1 \rangle$, $\langle 1.01, 1.10 \rangle$, $\langle 1.010, 1.011 \rangle$, . . ., each one of which is properly contained in its predecessor. Papy now introduces his 'Axiom of Continuity' (cf. the 'nested interval' argument on p. 282): 'The intersection of every infinite set of such "Chinese" segments is a singleton.'

Thus every infinite 'bicimal' defines a unique point on the line and every point on the line is represented by at least one infinite bicimal. Only points which are labelled after a finite number of subdivisions have two different representations as infinite bicimals. In such cases one says the two infinite bicimals are equal. (Cf. the British Post Office Union leader's claim that 'The number of members backing the Union is 99.9 recurring per cent. Next time it will be 100 per cent' (quoted in *Math. Gazette* **56** (1972), p. 40).)

The infinite bicimals are regarded as descriptions of the objects known as real numbers. Hence, every point on the line corresponds to a real number, and conversely. The natural ordering of the line induces a natural ordering of the reals.

To complete the characterisation of the reals, one now has to define addition and multiplication and to show that the reals form a field. The two operations are defined using parallel projections in a manner similar to that described in Sections 8 and 9. Thus we have

(a) *addition*

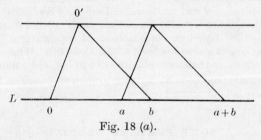

Fig. 18 (a).

(b) *multiplication* (here L' is an arbitrary line through 0 and I an arbitrary point on L').

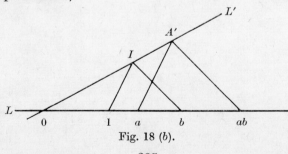

Fig. 18 (b).

285

Multiplication of a vector by a real number can now be defined and it can be shown that with these operations the plane, with the choice of origin 0, is a vector space over the reals.

By this means, therefore, Papy contrives to bind together what he sees as the three main themes of school mathematics, the number system (in particular, the reals), geometry (both affine and Euclidean) and algebraic structures (in particular, vector spaces).

Exercises

1 Devise constructions for determining ^-a and a^{-1} using Papy's model.

2 Demonstrate the various commutative, associative and distributive laws using Papy's model.

3 Provide examples of 'multiplicative' number patterns and other graphical examples that can be used to motivate the definition.
$$^-x \times {}^-y = xy.$$
(See p. 280).

4 In SMP *Additional Mathematics Book* Part 1 the naturals are referred to as 'counting' numbers, the positive rationals as 'sharing' numbers, and the positive reals as 'measuring' numbers. Describe the appropriateness or otherwise of the use of 'measuring' to describe the reals. Can you suggest a more suitable adjective?

5 Study some of the ways in which negative integers are introduced using ordered pairs of naturals (see, for example, Nuffield – CEDO [2]). What do you see as their advantages (if any) over more traditional methods? (Consider both pedagogical and mathematical aspects.) How would you attempt to test whether or not a particular method using ordered pairs was superior to, say, a method based on consideration of temperatures below freezing point, or the height above sea level of the Dead Sea? What criteria would you use? Remember that some children live in hot, flat countries.

6 What objections could be made to using arguments based on Dedekind cuts in a secondary school course?

SOME FURTHER READING

Dienes [1, 3], Durell [1], NCTM [4], Nuffield – CEDO [1, 2], NSSE [1]

19
Applied Mathematics

1. WHY TEACH APPLICATIONS OF MATHEMATICS?

In Chapter 2 we stressed the need to teach mathematics because of its practical aspects and we indicated, in Chapter 10, the vast range of applications of the subject in today's society. However, even when one has elected to teach topics in mathematics because of their practical value, one still has the option of either presenting mathematics without specific reference to its applications, or of actually including the applications as part of one's teaching. The former method has traditionally been followed in some countries which have left applications to specialists (e.g. physicists); it is also the way in which many mathematics courses are given to scientists at university level. In Britain, however, there is a tradition of teaching from applications – at least at the school level – and this method of teaching has much to commend it, for reasons which we outline below.

Perhaps the principal motive for teaching applications is that, by so doing, we better prepare pupils to take their place in society; for even those who will not use mathematics in their later jobs will still need as citizens to reckon, to estimate, to make decisions, to weigh probabilities, and so on.

There are, however, other important reasons for teaching applications and these might be listed as follows:†

(a) *For motivation*: many pupils are not moved by curiosity about mathematics but they can be persuaded to learn mathematics as a tool for solving 'practical' problems which appeal to them more.

(b) *For cultural reasons*: applications of mathematics, such as Newtonian mechanics, are part of our cultural heritage and of the

† The reader may well notice similarities between the views and examples to be found in this chapter and those expressed in Chapter 7 of the UNESCO publication *New Trends in Mathematics Teaching* Vol. 3 (1972 publ. 1973). This is because the principal author of the latter chapter is one of the writers of this book. In this respect, the present chapter represents 'later thoughts' on the topic.

human activity of mathematics. To learn calculus without under-standing what led to its development and how it was used by Newton and others, is like learning to play scales on the piano without being shown any compositions.

(c) *For fear of 'something worse'*: The mathematics used in physics lessons (for example) is often crude, unsuitable, or wrong. Formulae are obtained by mysterious processes, strange rules are laid down and the whole mess has to be learnt by heart, since it could hardly be reproduced by understanding.† If the mathematics is taught by the mathematics teacher it is more likely to be understood, but even so this solution is unlikely to be ideal unless there is cooperation between the two teachers concerned.

(d) *To teach recognition of structure in the presence of 'noise'*: it is one thing to learn the group axioms, but quite another to recog-nise a group structure *within* a piece of mathematics, let alone chemistry or physics. Consequently it is desirable to give pupils opportunities to search for and identify mathematical structures in a variety of situations. This is related to the art of model-making, as we shall see below, and is an extension of such activities as estimating the volume of a tree by regarding it as the frustum of a cone. The difficulty of the problem is exemplified by traditional classical mechanics in which students will follow a theoretical treatment of force, energy, etc., but will fail to recognise these entities in a specific problem.

These, then, are four mathematical and pedagogical reasons for teaching applications. But, as we mentioned earlier, these are pro-bably less important than the social reasons. At every stage of his life a man is called upon to take decisions, many of which can be taken more rationally if approached quantitatively. At the personal level, a driver entering a motorway must have decided on whether he has enough petrol to get him to the next service-station, whether he has enough cash to buy supplies, etc.; faulty decisions will involve him in penalties. He must be able to decide whether to buy a car one way rather than another – bank loan or hire-purchase. Is it better to rent or buy TV? What kind of central-heating will suit him and his pocket best? Less personally, communities and firms must decide whether to act, or spend money in one way rather than another, and so on through the highest parts of Government. Thomas Tate in his

---

† A non-mathematical former colleague of one of the authors was overheard telling a class of electronics students always to remember that 'in this type of example i^2 equals plus 1, but don't ask me bloody well why'!

Principles of Geometry (1848) wrote of the need to teach 'estimation' in an attempt to prevent the pupil from later rendering 'himself penniless, with all the moral evils resulting from hopeless and irretrievable pecuniary embarrassment'. Such Victorian highmindedness seems faintly comical today, but nevertheless mathematics teachers still have similar social obligations to fulfil.

2. THE FIELD OF APPLICATIONS: CRITERIA FOR SELECTION

In Chapter 10 we described in general terms some of the many ways in which mathematics is applied. For more specific accounts the reader is directed to such writers as M. S. Bell [1] (for operational research), George [1] (geology), Haggett and Chorley [1] (geography), and Maynard Smith [1] (biology). The vast field of probability and statistics has already yielded many problems suitable for examples in school discussion, which can be found in most modern texts.

It is impossible to provide here a more detailed account of all such applications, but we shall endeavour in our discussions below, even when we cannot describe the material in detail, to show how the applications we mention meet the criteria listed in Section 1. We shall, in particular, adopt the viewpoint that applications should be selected and treated *as if decisions were to be taken upon them.* 'What would you do if . . . ?' is a question which strongly influences one's approach to a problem in applied mathematics, as we shall see later. (At its most extreme, someone's life may depend on the approach; for example, one must calculate whether certain scientific experiments will kill the observer, see Wallis [1].)

Of the many examples we give, the reader should distinguish between those which are solved by observations of the form 'what goes in must come out', and those which depend on more subtle concepts, such as Force. In traditional British treatments the latter type has been emphasised and the former neglected, either through academic snobbery or ignorance (pedagogical or mathematical). We have therefore laid some stress on the first type. We have omitted material on topics such as linear programming and probability which are now well covered in many texts. Some classical material has been included both for those readers who have not met British 'Applied Mathematics' and also for those who may have coped successfully with the topic but failed to see its relevance. Several of the more advanced exercises are based on the SMSG lectures of Pólya [3] and Schiffer [1], and the treatments of these that we give are necessarily condensed and omit the discursive questioning of the originals. Both these authors have devoted imaginative effort to showing how physical laws came to be formulated. Thus Pólya avoids what he calls 'a bland statement of the Universal Law of Gravitation' and in so doing makes the reader realise the boldness of Newton's hypothesis that the law really is universal (see the Exercise on p. 309). This is in marked contrast to the conventional 'According to Newton, the law of attraction between two gravitating bodies is . . .'. The reader is recommended to study both these books in detail. We also omit stressing the need to check and even to guess formulae using dimensional considerations. Such considerations, we assert, go without saying.

We have also explained earlier the great influence on the applications of mathematics that the computer has already had. Many problems which were previously unassailable can now be attacked. Because of the importance of the computer, many school programmes now include the elements of computing – the construction of flow charts and an introduction to such languages as FORTRAN and ALGOL (see p. 117). Such topics, however, are not *applications* of mathematics so much as examples of a mathematical way of thinking. They are often valuable examples, but a mathematics curriculum should not be regarded as treating applications of mathematics solely because it includes computation.

3. MODELS: SOME EXAMPLES

To allow more precise discussion, we need now to use the notion of a mathematical model. We recall that we gave a 'literary' description of the term 'model' in Chapter 1, pointing out that a thought could be expressed using different sorts of language – metaphorical, mystical or mathematical. If the language used is mathematical, then we have a mathematical model of the thought. This model will not, in general, be unique, for, depending on how much detail we choose to consider, several models will exist. Thus, we are not using the term 'model' in the sense only of an *isomorphic* copy, nor in the sense of mathematical logic. Reasons for making models are so as to *analyse* a situation or to *communicate* a thought. In both cases, the deliberate suppression of detail is frequently essential for comprehension; the kind of detail suppressed will be related to the choice of language for the model.

We have met examples of models for communication in Chapter 18, where we saw how, in the primary school, knowledge of the number system is imparted through the (physical) model of Cuisenaire rods. Here, a physical model is an aid to the understanding of a related mathematical model (cf. the way in which Riemann gained a deeper understanding of the Theory of Functions through studying the physical analogues of potential theory). We shall now consider the use of models for purposes of analysis.

To avoid a difficult technical discussion of the word 'model' in the abstract, it will suffice for our purposes to say that 'A is modelled by B', provided that A is a system (i.e. a set of elements between which certain relations hold) as is B, and there is a correspondence $a \mapsto a'$, $R \mapsto R'$ between the entities (elements and relations) of A and those of B, such that if a and x are in a relation R in the system A, then a' and x' are in the relation R' in B. The correspondence is not required to be one–one (in order that we may suppress detail).

In the simplest situations, the correspondence *is* one–one (an *isomorphism* of the structures of A and B). For example, a class of children can be so structured as to act as a digital computer. Here the computer is A, consisting of memory, adder, etc., with mutual links, while B is the set of children, each performing his allotted role. Similarly, a performance of a play by one company models any other performance of a play by any other company, provided that they follow the same script. A well-known geometrical example is that underlying Exercise 4 on p. 243, where the concrete object is a model of a certain finite projective geometry; and in this case, vice versa also. The Cuisenaire rod example, mentioned above, models addition sums in $\mathbb{N}$, if we make each number n in the sum correspond to a train of rods n cm. in length, and if each $+$ sign between numbers corresponds to joining together the corresponding trains. An example where the correspondence is not one–one is seen in the way in which an aeroplane is modelled by a plastic toy; such features as engines, wings, etc., in the real aeroplane are reproduced in the toy by corresponding lumps of plastic placed in appropriate positions, but other parts having a vital function are ignored.

Two especial types of modelling occur in mathematics. First, *when we have a 'concrete' mathematical realisation (B) of something 'abstract' (A)*. For example, we have described in Chapter 16 how the axioms of Euclidean geometry are modelled by Cartesian geometry, and in Chapter 17 how those of the system $\mathbb{Z}$ can be modelled by $\mathbb{N}$ and some set theory; we indicated that those of $\mathbb{R}$ can be modelled by $\mathbb{Q}$ and some set theory. In these cases, B is regarded as being more familiar or accessible than A, though whether one is more 'abstract' than the other is a matter of taste.

Second, *when we have a mathematical description* (B) *of a physical situation* (A). We shall give examples of this second type, which is the more appropriate to applied mathematics, below.

Now, the basic activity of applied mathematics (in the general sense of Chapter 10) has three stages:

(1) One is interested in some problem in the world – a problem of physics, or biology, or scheduling, or decision-making, etc. – and to think about it more easily, one selects a model of the problem-situation built from mathematical components.

(2) One studies the mathematical features of the model, using mathematical techniques (which may suggest working with a mathematically simpler, more tractable, model). For this stage, pure mathematics is important, to supply models and techniques. However, one must still keep the physical problem in mind, so as not to be led astray by the pure mathematics.

(3) Finally, one relates the newly discovered features of the model to the situation where one started, by going back to the real world. Depending on one's confidence in the 'goodness-of-fit' of the model, one then makes appropriate conclusions about the real world; or one rejects the model as being too crude, and one tries again!

Example 1. On leaving a store, we have to pay the bill. To compute it, the cashier uses a model essentially (and usually unconsciously!) selected from the arithmetic of integers, in which certain integers correspond to the prices of goods ordered. (These prices are in multiples of some basic coin – cents in the USA, half-pence in the UK.) She then arithmetises, and arrives at an integer called the total. This integer is in the model (the system $\mathbb{N}$) and corresponds to an amount of cash which has to be handed over, before the real-life problem is settled. In cases of dispute with the buyer about the total, the usual cause of the trouble is found to be that the prices were not modelled agreeably in the cash register when the cashier set up the model. We all have such confidence in arithmetic as a model of our mental processes of counting, that pricing and monetary policy are usually designed in terms of the model itself. Certainly the decision-making criterion of Section 1 is very evident here.

Example 2. In the eighteenth century, Euler solved the famous problem of deciding whether or not there was a 'unicursal' route through Koenigsberg. A rough map of the town, with its bridges and islands is shown in Fig. 1, and a unicursal route would be one which

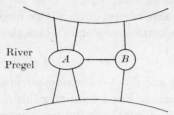

River
Pregel

Fig. 1. Map of Koenigsberg.

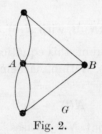

Fig. 2.

crossed each bridge once and only once. The rough map is already a model of the geography of the town, with many irrelevant details suppressed. Euler threw away even more detail by realising that it sufficed to look at the linear graph G shown in Fig. 2. And now he modelled G by a function $f: V \to \mathbb{N}$, where V is the set of nodes of G, and for each node v, $f(v)$ is the number of arcs of G ending at v. Euler then showed, within mathematics (see, for example, Ore [1]), that a (connected) graph H has a unicursal route *if and only if* the corresponding function f takes either 2, or no, odd values. Since this is not the case for G, no unicursal route exists in G, and hence none exists in Koenigsberg town. Notice that Euler models G in the system $\mathbb{N}$ considered as a set of odd and even integers, ignoring any other relations between them. Such suppression of detail is typical of modern topology; a more subtle, but still elementary, example, is given in the excellent article by Crowell [1], who shows how to distinguish between certain knots in loops of wire, by using a function that maps each knot onto the spokes of a wheel in a certain way. It is important to realise that with such examples, if one did not suppress information the model would be exactly as complicated as the original, so the art is to make progress by losing unmanageable complexity.

293

Example 3. In questions concerning population growth or decline (say with a population of bacteria or atoms) it often seems to be an appropriate hypothesis that the number N present at time t is proportional to the rate of change of N. One model of this situation is the 'discrete' mathematical model, in which there holds the equation

$$kN(t_{n+1}) = [N(t_{n+1}) - N(T_n)]/(t_{n+1} - t_n)$$

where k is constant and $N(t_n)$ is the number in the population at the time t_n of the nth census. For a computer this may be the best model to take, but in classical mathematics one passes to the 'continuous' model, in which N is a differentiable function of t satisfying the differential equation

$$kN(t) = \mathrm{d}N/\mathrm{d}t, \text{ with } N(0) = M \text{ given.}$$

By the Mean Value Theorem, this equation approximates the earlier one to within $o(t_{n+1} - t_n)$. Within mathematics, the equation has the unique solution

$$N(t) = M\mathrm{e}^{kt}.$$

If $k > 0$, this tells us that N increases, and will have doubled its initial size when $t = \ln(2/k)$; also, for any times t_1, t_2,

$$k = \ln[N(t_1)/N(t_2)]/(t_1 - t_2).$$

These predictions can be checked experimentally, and if they agree with what is observed, confidence in the mathematical model is built up. If there are discrepancies, the initial hypothesis might have to be reviewed as being too simple and ignoring some other factors. No completely satisfactory model has yet been found that will enable a government to take firm decisions about its future population trends (but see Exercise 1 on p. 306). On the other hand, the above model (considered as modelling a population of atoms) is sufficiently good for it to be used as a basis for deciding whether employees have been subjected to dangerous doses of radioactivity. For further information on this kind of problem, see H. T. Davis [1, Ch. 5], Rosen [1], Newman [1], etc. For other material on modelling in biology, see the fundamental book *On Growth and Form* by D'Arcy Thompson; a recent upsurge of interest in this work among topologists has arisen because of the additions due to Thom [3] and Zeeman [2], who use very powerful tools of algebraic and differential topology.

Example 4. To investigate the vertical fall of a stone in a vacuum, under gravity, Newton's Laws of Motion lead to a model of the situation in which the stone's height at time t corresponds to a point x on the real number-line $\mathbb{R}$, where x and t are related by a smooth function $x = f(t)$ satisfying the differential equation $d^2f/dt^2 = $ constant (C). By pure calculus, with no reference to stones, height, earth or Newton, we know that there is a unique solution to this differential equation, of the form $f(t) = A + Bt + Ct^2$ where A and B are two further constants. These can be evaluated once we can model two states of the stone (say, its position at two times t_1 and t_2). The actual numerical model of C, if needed, is inserted after an experiment in the real world (and some hard thinking).

The question of just which parameters can be measured conveniently, is a factor which will affect the mathematical treatment of the model. With the falling stone, we could determine A and B if we know the stone's position and velocity at $t = 0$ – an important economy if we had only Galileo's timepieces, and the initial velocity were zero. (In conventional 'Applied Mathematics' examination questions, one is often asked to prove a relationship between parameters, where the ones to be inferred from the 'data' are just those which in practice would be easier to measure than the data.) Given position and velocity at $t = 0$, we might be led to write the equation $d^2f/dt^2 = C$ as $v\,dv/dx = C$, where v is the velocity and $x = f(t)$; the solution is now the familiar formula $v^2 = 2gx$, which gives us a convenient way of determining v in terms of the more easily measured x. The new formula is an instance also of the Law of Conservation of Energy, because it says that the gain in Kinetic Energy ($\frac{1}{2}mv^2$) equals the loss in Potential Energy (mgh); it was the accumulation of such instances, using abstract concepts suggested by terms like $\frac{1}{2}mv^2$ in the mathematical models, that led to the later more far-reaching models of general mechanics and physics of Lagrange, Hamilton, Maxwell, Einstein and others.

Returning to our falling stone, suppose we wished to model, not a vacuum, but wind resistance to the stone. Then we should need to choose another differential equation: observation suggests something like $d^2f/dt^2 + k(df/dt)^2 = C$, where k is another (negative) constant, and the extra term is the 'model' of the wind resistance. The mathematics of the new equation is more difficult than that of the old, at the price of a more accurate model. Similarly the well-known model for a pendulum is the equation $\ddot{\theta} = -A^2 \sin\theta$, to which we are led from Newton's Laws; but to handle this equation mathematically is difficult, so mathematical technique suggests that we change to the more tractable equation $\ddot{\theta} = -A^2\theta$. Checking with the original physical situation, we expect the new model to be adequate, at any rate if θ is 'small', because of the mathematical fact that $|\theta - \sin \theta| \leqslant |\theta|^3/6$.

All the material of this example is familiar to anyone who has taken a traditional applied mathematics course. However, Stage 3 of the activity we described earlier is neglected in those courses, but not by scientists. Rarely is it checked whether the predictions of the models are confirmed in the physics. Naturally, then, many courses in applied mathematics are regarded as inferior forms of pure mathematics! Sometimes the modelling aspect of applied mathematics is so forgotten that experiments are sometimes designed using apparatus that is calibrated from the model; thus Ohm's Law is sometimes 'verified' using ammeters calibrated on the assumption that the law is true! A more subtle instance is to use the formula $T = 2\pi\sqrt{(l/g)}$ for the period of a pendulum, in order to find g. This formula is derived from the approximating equation $\ddot{\theta} = -A^2\theta$ (here $A = g/l$), and it is not evident mathematically that it approximates the period of the solution of $\ddot{\theta} = -A^2\sin\theta$ (or even that the latter equation *has* a period).

Here is a situation where the physics suggests an interesting mathematical investigation *within mathematics:* to show that the required period exists and to estimate the discrepancy due to the approximation. For a discussion of this last point, see Pollak [1] and Brauer [1].

Defenders of existing treatments are likely to say that Newton's Laws are *known* to give a good model of mechanics, apart from the situations of relativity theory (at large distances) or quantum theory (at small distances); certainly the flights of astronauts have shown this. Also, verifying predictions cannot be done within the mathematician's world of theory, pencil and paper, so it is a waste of time (they say) to bother about verifications. The extreme view against the performing of experiments is the argument of Todhunter's that 'if he (the pupil) does not believe the statement of his tutor – probably a clergyman of mature knowledge, recognised ability and blameless character – his suspicion is irrational, and manifests a want of the power of appreciating evidence, a want fatal to his success in that branch of science which he is supposed to be cultivating'.†

† Todhunter was neither against genuine experiment nor resolutely opposed to laboratory work, but he insisted that the latter should come out of the time allocated to organised games, since they were an equal waste of time!

In spite of these arguments, it is found that applied mathematics is not popular, that it is in 'decline' (Heading [1, 2]) and that students of the subject are frequently bad at formulating models in new situations. Some remedial action appears to be needed, and more needs to be done to help teachers acquire 'The Art of Teaching the Art of Applying Mathematics' (see Lighthill [1]).

Exercises

1 Find out how Galileo surmounted the difficulties caused by the crudity of the available apparatus.
2 Read the article by Crowell, mentioned above, and develop it into teaching units.
3 Look at some examination questions in mechanics, to see whether they are realistic about the measurement of data.

4. A CHOICE OF MODELS

One's choice of model may well depend, as in Example 4 above, on the need to render the mathematics tractable. In other instances, however, the choice of model may hinge on other considerations. This problem is concealed in classical treatments of mechanics, for Newton's model is normally taken as a starting point and other models are not considered. However, by considering the theory of light, for example, one can more easily track the evolution of different models. There was much dispute as to whether Euclid's 'optics' or Newton's particle theory was correct, until this was resolved in favour of Euclid and Fermat (see Schiffer [1]). Later there was dispute as to whether particles or waves gave the 'right' model, a disputation only resolved by the quantum theorists, who showed that each was valuable in its place. Again, Gauss and Lobachewsky are reputed to have attempted to carry out experiments to see whether Euclidean or their newly discovered non-Euclidean geometry provided the 'right' model of space. The question of choice of model is therefore an important one and could be discussed with benefit at school. Examples that could be used there are Steiner's theory of voting systems [1] and the choice of different probability functions to describe the same situation (see, for example, Hirst [1]).

An early and delightful example on the choice of models is that pointed out by Schiffer [1], and concerns light. Euclid's axioms of optics state:

(i) *rays of light are straight lines,*

(ii) *when reflected from a plane surface, the incident ray and the reflected ray each make the same angle on opposite sides of the normal to the surface.*

(Note that (i) really says that straight lines are *models* of light rays: if we say *are*, we find it harder to ask for a mechanism of transmission – a question that was vital for the later development of physics. Indeed, much of the controversy about non-Euclidean geometry was due to the initial belief that the Euclidean model of space *was* real space, that a system was identical with its model.)

Heron simplified Euclid's model of light rays by postulating the single axiom:

(iii) *rays of light always take the shortest path from one point to another,*

which implied both (i) and (ii) above (see Question 3 in the Exercise on p. 299).

Heron's axiom, however, did not seem to yield a law to account for refraction of light at an interface (e.g. between air and water). The law (usually known as Snell's Law) could hardly be inferred from experimental results, see Schiffer [1]; but if we take Fermat's version of (iii) – suggested by it, but more far-reaching –

(iv) *rays of light have finite velocity and always take the least time to pass from one point to another;*

then (iv) implies (iii) and easily yields Snell's Law by the use of calculus (see the same Exercise). Fermat's 'law' (iv) was the forerunner of the later theories of minimum energy that have influenced modern physics so profoundly, and required the insight that the speed of light is not infinite, but a constant for each medium (air, glass, water) through which it passes.

It may happen that the mathematical model of a physical process may be too difficult fully to analyse mathematically, yet a preliminary analysis may show that it also models some quite different physical process which is easier to study by 'analogue' methods; the information about the 'easy' process is then used to make conclusions about the more difficult one. For example, the 'travelling salesman' problem, to find the most economical itinerary for a salesman who travels from city to city, is most easily solved for each salesman by setting up a system of strings and weights, and reading off certain parameters from the resulting equilibrium; the simplest mathematical model is too difficult to solve mathematically. Similarly, the shortest network of roads between towns can be found by constructing a certain soap-bubble (see Courant [1, Ch. 7]) because the road problem and the bubble problem – of finding equilibrium with minimum energy – have the same mathematical model, but it is far harder to solve than to make the bubble and take measurements from it.

Exercises

1 Read the article on voting in Kemeny [2] and then incorporate it into an account of Steiner's theory [1].

2 A man with hobnailed boots stamps on a flea. If the area of his boot is A, that of the nails is B, and that of the flea is negligible, what arguments would you adopt for believing that the flea has a probability of $(A - B)/A$ of survival?

3 Use calculus to show that Fermat's axiom (iv) above implies Heron's axiom (iii), the axioms ((i) and (ii)) of Euclid, and Snell's law of refraction, as follows.

   (a) Take the x-axis as the line joining two points O, A. The distance from O to A along a smooth curve C of the form $y = f(x)$ is $\int_0^a \sqrt{(1 + y'^2)}\, dx$ (where $A = (0, a)$, $a > 0$). Prove that this is always greater than a if C is not the straight path OA.

   (b) Suppose a ray of light from A is reflected at O on a mirror, to a point B, where AB is parallel to the mirror. Let OM be the normal at O to the mirror (see Fig. 3). To establish Euclid's axiom (ii), using Heron's axiom (iii),

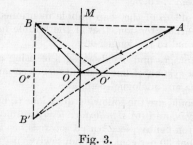

Fig. 3.

it is necessary to prove that $\angle BOM = \angle AOM$, knowing from (i) that AO and OB are straight lines. Let B' denote the reflection of B in the mirror, so that triangles OBO'' and $OB'O''$ in the figure are congruent. Let AB' cut the mirror in O'. If $O' \neq O$, then $AO + OB' > AB'$ so

$$AO + OB > AO' + O'B$$

since $O'B = O'B'$. Thus the path AOB is not the shortest that the ray of light could take, so $O = O'$ by contradiction. Now complete the proof that $\angle BOM = \angle AOM$. (This sort of argument can be used in other problems: see Courant [1, Ch. 7].)

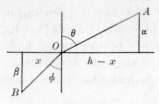

Fig. 4.

(c) For Snell's Law, the total time taken from A to B (see Fig. 4) is $T = AO/v + OB/w$, where v and w are the velocities of light in the media above and below the interface. (NB. We have to assume light does not have infinite velocity, and this can be experimentally determined.) We then find that since $OB = \sqrt{(x^2 + \beta^2)}$, etc., then

$$\mathrm{d}T/\mathrm{d}x = 0 \quad \text{when} \quad \frac{h - x}{v \cdot AO} = \frac{x}{w \cdot OB},$$

i.e. when $\sin \theta / \sin \phi$ equals v/w and is constant, for the two media. Moreover, $\mathrm{d}T/\mathrm{d}x$ is increasing, so the actual path taken requires minimum time.

4 Write an essay, giving a historical account of the way in which Snell's Law was discovered and the related discussions to show that light travels with finite velocity.

5 It is required to build a road from A to B on opposite sides of a motorway. Assuming the cost of the bridge over the motorway is independent of its position, and the cost per mile of road is x units on A's side and y units on B's side of the road, use the refraction problem above to find the cheapest route for a road from A to B.

6 Investigate the problem of the soap bubble mentioned above. Show from the definition of surface tension, that if the bubble has an edge, then its energy is proportional to the length of edge.

7 Consider the advantages and disadvantages of using a system of variable resistors to solve a system of linear simultaneous equations. (The electrical system acts here as an 'analogue' computer.)

8 Discuss how you could use the following questions to introduce the concept of (a) *null hypothesis*, and (b) *significance level*, to middle-school pupils.
(i) 'Is the new ten-penny piece biased towards heads or tails?'
(ii) 'Is it true that 30 per cent of all housewives use Fairy Snow washing powder?' (Bristol BEd, 1971)

9 A 10-year-old boy is collecting 'Historic Car' coins, which are given free with petrol. The boy needs 5 coins to complete his set of 20, and his father gets 3 coins when buying petrol one day.
(i) What is the probability that all the coins are new ones for the boy?
(ii) What is the probability that none of the coins is a new one for the boy?
(iii) Devise a series of a few lessons suitable for explaining such calculations to the boy.
(Use any model that seems reasonable.) (Bristol BEd, 1971)
[Cf. Fletcher [2], p. 213.]

300

5. TEACHING APPLICATIONS OF MATHEMATICS: PRIMARY SCHOOLS

In this section and the next, we shall discuss ways in which applications of mathematics can fit into a mathematics curriculum. These therefore leave out of account those highly 'abstract' curricula such as the 'K–6' curriculum of p. 187, and others based on an appeal only to a child's intellectual curiosity; the latter are sometimes a reaction against a too materialistic attitude in society, but we would not agree that it is materialistic or incurious to ask questions about the real world.

We described in Chapter 11 how, building on those methods of teaching associated with the names of Froebel and Montessori, the Nuffield Project and others have attempted to involve young children in practical mathematics. Thus children weigh, measure, pour liquids, estimate areas by counting squares, etc. The numerical data thus obtained are then manipulated by the techniques of arithmetic, for various purposes. Some measurements cannot be made directly and instead are worked out from scale drawings. Thus practical geometry is developed, leaving the child with a knowledge of the physical facts of geometry – similitude, area, shape, tiling, symmetry – that can be organised theoretically at the secondary level. Conclusions are drawn from frequency charts about traffic passing school, frequency of birthday, structure of families, etc.

Mathematical models are constantly made at this level, although few are suitable for explicit discussion (cf. Example 1, above). However, there is much scope here, because they do provide excellent material for discussion and so reinforce prevailing trends in primary schools towards an 'open' approach (see p. 213).

Such old problems as that of the bath with taps and outlets can serve as rough models of water conservation in reservoirs. Children can find out about their local system of water supply (they often make such surveys as part of the widespread use of project work); then they can discuss to what extent the bath with source and sink is a good model of the system, and how water engineers might make plans for changes in the weather. And then the old problem, of m men taking d days to dig s units of earth, is good for discussion. When is it unrealistic? Why is it structurally the same problem as that of the laundry that needs so much time and so many workpeople to deal with a given work-load? How are overtime payments calculated? How can a manager decide in advance to take on extra workers (for example on a farm during the fruit-picking season)?

A further technique of modelling comes through approximate arithmetic. That is to say, one may have an arithmetical model of a situation in which awkward numbers occur, leading to (say) awkward division sums, like $531 \div 47$. What features of the problem allow one to be satisfied with $530 \div 40$, or $530 \div 50$, or a decimal answer to several places? Such features affect the amount of necessary arithmetical labour, and also can act as checks to calculations. For example, if $531 \div 47$ represents the average speed of a runner in kilometers per hour, one will reject 83 as an answer (because of an alternative crude model of the problem, containing the generalisation 'no human can run faster than 40 km/hr'). 'Clock arithmetic' (i.e. arithmetic modulo n for a specific n) is now common in primary schools, and can also be used to check calculations: for example '$531 = 47 \times 13$' is wrong because in arithmetic modulo 3 it becomes '$0 = 2 \times 1$', although in modulo 10 it says '$1 = 7 \times 3$', which is correct. Here, one is using the natural homomorphisms $h_3 : \mathbb{Z} \to \mathbb{Z}_3$ and $h_{10} : \mathbb{Z} \to \mathbb{Z}_{10}$ to map the calculation in $\mathbb{Z}$ into calculations within $\mathbb{Z}_3$ and $\mathbb{Z}_{10}$ respectively. The model in $\mathbb{Z}_{10}$ is not good enough to pick up the error that is here picked up within $\mathbb{Z}_3$.

At all stages, one cannot make a good model of a situation without possessing a good intuitive experience of the situation. Thus a particular model cannot be discussed until the pupils are judged by the teacher to be ready. Intuitive preparation may be necessary beforehand. (This applies to all stages of the course, and at the beginning one needs to make sure that the children have appreciated such notions as that of conservation of volume – e.g. when liquid is poured from one vessel into another of different shape. Also terms need to be defined; children do not automatically know what 'greater than' means. All this has been pointed out by Piaget [1, 2].) Especially is this true of probability, where children can understand the problem long before the solutions. For example, when cards come with packets of tea, what is the chance of completing a set? The frequency definition of probability is quite accessible in primary school, and various teaching units have been written to develop this (e.g. by Minnemast, ESI in Boston, etc.).

We have indicated, then, many fields of application of mathematics in the primary levels. The teaching in a good primary school often tends to be highly integrated, with such subjects as reading, writing, geography, history, transport, marketing, etc., often used one within the other. Applications of mathematics thus seem to be natural to this stage, and may greatly facilitate the teaching of 'pure' mathematics. Such teaching must still be provided for, of course, to cultivate the faculty of abstraction. Frequently the impression is given that the practical work of pouring and measuring *is* mathematics, which is not true.

Exercises

(The following exercises are intended to give ideas to teachers; the material may be too difficult for some age-groups of children.)

1 Investigate the relationship between the water-tank with taps, and the flow of electricity through circuits, using the following observation. Let the water-tank have volume V units and suppose that a tap empties or fills the tank in t minutes. Then we may call the tap a 't-tap', with t positive or negative according as the tap fills or empties. The rate at which the t-tap empties or fills the tank is V/t units per minute. Hence show that (if $V \neq 0$) an s-tap and an r-tap working together are equivalent to a t-tap, where $1/t = 1/r + 1/s$, provided $t \neq -s$. This corresponds to connecting resistances in parallel, but allowing negative resistances; it also corresponds to the combination of thin optical lenses, by the usual formula for calculating focal lengths. What arrangement of water-tanks corresponds to connecting resistances in series? Devise explanations suitable for primary schools. Relate this to flows of cash (income/expenditure).

2 A class of children notes the number-plates of cars passing their school in one day (relays of working parties must be organised to do this). What inferences would you expect might be made about traffic in the school district, concerning peak use, age of car, county of origin, etc.? How would the conclusions vary as between a port and an inland town?

3 If the linear dimensions of an object increase by a factor k, its surface and volume increase by factors k^2 and k^3 respectively. Devise ways of preparing primary children to appreciate this.

4 Devise ways of introducing children to the growth of such functions as 2^n, n^2, n^3, and $n!$ For example, if a joke is passed on by each hearer passing it on to two different listeners each day, how many people will have heard it by the end of a week? What if 'two' is replaced by 'one', or by 'three'? Might this explain the rapidity with which children's rhymes can spread over a country (see Opie [1]). Has this problem any relevance to the rabbit-breeding problem that generates the Fibonacci numbers? (NB. Since population problems are of obvious social importance, these questions are related to the criteria we gave in Section 1 of this chapter.)

5 Anticipate the kind of answers an examiner might receive to the problem, set in an examination in 1908: 'Given three shillings and sixpence, plan a meal for four' – a most unusual type for those days, and soon to disappear. Could the question be structured to be made less open-ended and therefore easier to mark, without 'closing' it completely?

6 Suppose a room is papered with patterned wall-paper. How will the size of the pattern affect the amount of paper wasted from the rolls? Is elaborate planning (of how to cut the pieces) going to make a worth-while difference to the final cost?

7 Devise an arrangement of jobs for children in a class, so that the whole class operates as a computer/data-processor. (For assistance see Bolt [1].)

8 It is required to paint a notice with lettering that can be read at 100 feet. Determine a practicable minimum size for the lettering.

9 Estimate how many woman hours are necessary to produce a school dinner. How many child hours and teacher hours are necessary (i) to produce a Christmas play, (ii) to produce a class project, (iii) to teach decimals?

10 There are now some attempts to teach mathematics using various games. Investigate some of these, to find out what knowledge of the world they might impart to a child (e.g. 'Monopoly' and finance, 'Bridge-it' and plane topology, chess and logic). How might such games help a child's commitment to learning, regardless of its verbal skills? In what sense are such games 'applied mathematics'?

6. TEACHING APPLICATIONS OF MATHEMATICS: SECONDARY SCHOOLS

For ease of reference, it seems better to discuss applications at secondary level by subject within mathematics. This unfortunately might seem to perpetuate the traditional subdivision into 'compartments', and hence we stress that *we have always in mind a unified approach to mathematical ways of thinking*. Also, we shall write at first as if secondary education did not have to take into account the various public examinations; discussion of assessment is postponed until Chapter 20.

6.1. *Arithmetic*

As technique increases, harder applications are possible, such as problems on heating a house, see SMP Book 5. When pupils are old enough to appreciate the terminology, examples involving money – credit, investment, depreciation (how long should one keep a new car?) – become possible. Such items were always the staple material of the traditional texts (e.g. Pendlebury [1]) but they were closed-ended, to train the future clerks whose mental processes would be used as mechanical aids are used today. Nowadays, speculation must be built-in. Thus, hypotheses can be constructed to account for fluctuations in data: for example, given the nightly receipts of the only cinema in a town, make inferences about the social patterns of the town. First-order approximations can also be worked out to such problems as : how much food does a city need in a week? How much rubber flies into the atmosphere from car tyres, per year? Questions of pollution and the environment, or running down of natural resources, can well be raised here, with their impact on decision-making.

Adding-machines can be useful at this stage, partly to give insight into the nature of routine calculation, partly to perform iterative calculation, and partly as a 'second chance' for those less able children who failed to understand arithmetic earlier. Problems involving costing need to be structured gradually; see the articles 'How much does it cost to keep a dog?' (Tammadge [1]) and on building flats (Wallis [1]).

Exercises

1 If the population of a region is N, and the average life-expectancy is n years, one might expect that about N/n babies will be born each year, in the region. Is this true of your region? If so, and if there are M people in your town, there should be about MT/n schoolchildren in the town, where T is the average stay in school. Check this against the actual figures. Given the allowed expenditure per child, estimate the total educational expenditure by the town. How does it compare with other towns, or other years? Does it seem to keep pace with inflation?

2 Many manufactured items are never repaired; if they are faulty they are thrown away and renewed. Take an example of such an item, and compare the cost of doing this (i) to the manufacturer, (ii) to the community at large, which has to pay for treating the rubbish created.

3 Deaths in aircraft crashes are normally stated in terms of passenger-miles. What is the effect of expressing them in terms of the total of take-offs and landings?

4 Compare the cost of transporting a passenger by different vehicles, using the energy of one gallon of petrol.

5 A coal-mine employs 1000 men and loses £N per year. If it were closed down, suppose 750 of the men would not expect to find another job, and would have to live off Social Security payments. What value of N makes it cheaper to the State to keep the mine open? Is this a good way of thinking about the problem?

6 Look at some O-level arithmetic questions set by an examining board. Discuss which of them are (a) realistic, (b) sensible, (c) artificial, (d) silly.

7 Look at some O-level papers on Principles of Accounts. Would you say that the questions were better than the Arithmetic questions of the Mathematics papers? What about the Commercial Arithmetic papers?

8 An O-level question begins 'Find without tables . . .'. Is there a justification for doing elaborate arithmetic without using tables or mathematical aids?

9 A question gives the dimensions of a large petrol tank and asks for the depth of petrol if the number of gallons contained is known, and 1 cu. ft = 6.23 gall. Is this question realistic?

10 Estimate how many man hours are necessary to build a house (i) on site, (ii) taking materials and transport into account. Compare your figures with those for (i) a school, (ii) a public building. (Many interesting problems on estimating costs are to be found in the 1848 *Geometry* by Tate.)

6.2. *Algebra*

Here the notation is commonly used to express functional relationships, such as $A = l \times b$ for the area of a rectangle, $V = IR$ (Ohm's Law), $v^2 - u^2 = 2gh$, etc. Immediately, questions arise as to those variables that might be best to measure, and those easiest to calculate (with the necessary transposition of variables). Some problems in operational research are solvable by this kind of simple algebra: see Singh [1] for problems on railways and the siting of bridges, and Gale [1] for the 'Jeep' problem (of depositing supplies across a desert, using a vehicle whose range is less than the distance across the desert). The notation of the algebra of sets is often learned before this stage, and perhaps that of the algebra of propositions. In any case, Boolean algebra can be contrasted with ordinary algebra, and used to solve simple logical problems and to design switching circuits (see, e.g., Kemeny [2], Jeger [2]). The calculus of propositions and predicates can itself be shown as a model of language and thought-processes; see Griffiths [1].

Matrices begin at this stage, as shopping lists, descriptions of geometrical transformations, and as aids in solving linear equations. At a more advanced but still elementary level, they can be used in a treatment of Special Relativity, as in Schiffer [1]. Linear inequalities and simple problems of optimisation are now possible, while theory of simple games can be introduced. The notation of algebra having been developed, the formulae of probability and statistics (with natural applications) are ready to be discussed. When matrix groups are available, an interesting application outside mathematics is that to certain primitive marriage-systems (due to A. Weil: see Kemeny [1] and White [1]). Groups also can illuminate the calculation of permutations and combinations, via the theory of operators and the class formula (see Herstein [1]). Fletcher [2] is a rich source of examples on linear algebra in general.

Exercises

1 In the population problem considered in Example 3 (Section 3), the crudest 'discrete' model can be expressed in the form $N_{n+1} - N_n = pN_n$, using unit intervals of time and allowing N_n to denote the population after the nth interval of time, while p is constant. Thus $N_n = q^{n-1} N_1$, so the population would increase geometrically if $q > 1$, with the dire consequences predicted by Malthus. In the 1850s the Belgian Verhulst was dissatisfied with this law (as were many others, filled with humanitarian horror at what Malthus

predicted and the callousness his predictions were inducing in some employers of labour). Verhulst proposed instead a law of the form

$$N_{n+1} = qN_n - rN_n^2 \tag{i}$$

where the term $-rN_n^2$ represents the 'braking effect' due to competition between pairs of individuals. It is mathematically easier to change the model slightly to

$$N_{n+1} = qN_n - rN_nN_{n+1} \tag{ii}$$

so that

$$y_{n+1} = q^{-1}y_n + r/q, \quad \text{where } y_n = N_n^{-1}.$$

Find α so that, if $y_n = z_n + \alpha$, then $z_{n+1} = q^{-1}z_n$. This reduces the mathematics to that of Malthus' law, but the social consequences are quite different when we turn from the model to the situation modelled by (ii). Show that, now,

$$N_{n+1} = q^nN_1/[1 + rN_1(q^n - 1)/(q - 1)], \tag{iii}$$

approximating to q^nN_1 if r is small, and to $(q - 1)q^n/r(q^n - 1)$ if n is large and $q > 1$. What are the degrees of approximation?

Thus N_{n+1} tends to $(q - 1)/r$ as n tends to ∞, and this limit is an upper limit, independent of N_1 (which is surprising). Plot a graph of N_n against n.

Given the population statistics of a country, how would you find q and r in (iii)? Verhulst in 1850 predicted population trends for several countries, and was amazingly accurate for some countries; his estimate for the USA was only 1,000,000 out (i.e. about $\frac{2}{3}$ per cent) in 1940. Curiously he was most inaccurate about Belgium's population: that country had changed from an agricultural economy to an industrial one during the nineteenth century, so r and q *were not constant*; they fluctuated much more than those of other countries. To do full justice to these considerations, one must read the account in Schiffer [1], which gives a plausible argument for introducing the square term in (i).

Estimate the error in using equation (ii) rather than (i). (If N_n' refers to (i), write $N_n' = N_n + U_n$, and use the earlier approximation that $h - \epsilon < N_n < h = (q - 1)/r$ for any $\epsilon > 0$ if n is sufficiently large.)

2 Look at some GCE algebra questions based on applications set by an examining board. Discuss which of them are (*a*) sensible, (*b*) artificial.

3 Find or construct a good question on inequalities based on an application of mathematics.

4 The routine calculations occurring in practical (and other) problems can often be summarised in 'commutative' diagrams of the following kind.

(*a*) Fig. 5 tells us that if a and b are positive numbers, and if we form the product ab, followed by the logarithm, then the result is the same as taking

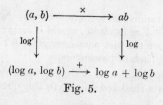

Fig. 5.

the separate logarithms and adding. That is to say, if the various operations of multiplying, adding, and taking logarithms are as indicated by the arrows in the diagram, then the way we go from the top left to bottom right is independent of the route taken; in functional notation, $\log \circ \times$ (corresponding to one route) is $+ \circ \log'$ (corresponding to the other). Thus the diagram 'commutes'. We have written it using variables rather than their domains. From a practical viewpoint, the diagram tells us that the way to find $\log ab$ is to go round the diagram 'down and across'.

(b) Similarly, the formula $1/a = 1/b + 1/c$ for combining resistances in parallel can also be expressed by means of a commutative diagram. In Fig. 6 the 'across and down' route says 'take reciprocals and then sum';

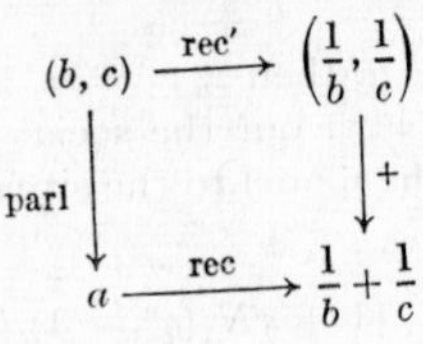

$$(b, c) \xrightarrow{\text{rec}'} \left(\frac{1}{b}, \frac{1}{c}\right)$$

$$\text{parl} \downarrow \qquad\qquad \downarrow +$$

$$a \xrightarrow{\text{rec}} \frac{1}{b} + \frac{1}{c}$$

Fig. 6.

the other route says 'form the equivalent resistance to b and c in parallel [summarised by the operation 'parl'] and then take the reciprocal value'. Since the two routes yield the same result by the theory of electrical circuits, we have a practical prescription for finding a ($=$ parl(b, c)): it is rec$(+ (\text{rec}'(b, c)))$ since in this case we can reverse the bottom arrow because $1/(1/a) = a$.

Construct such diagrams for (i) computing simple interest, (ii) computing compound interest, (iii) finding the roots of a quadratic equation, (iv) illustrating the matrix equation $(M^T)^{-1} = (M^{-1})^T$ (T denotes the transpose).

For further examples, see Fletcher [3].

5 Look at those questions on mechanics or applied mathematics, on a GCE paper, which involve an algebraic formula. Which elements in the formula would be easiest to measure? Express the others in terms of these. Did the examiner do the same?

6 Each of n universities has a single professor of (say) Greek, and requires the professor from another of the n universities as an external examiner. In how many ways can the examining be done, if no professor examines at more than one university? Show that this problem has the same model as that of asking in how many ways n letters can be inserted into n addressed envelopes without a single one being correctly placed (the Bernoulli–Euler or Montmort problem). What happens if the n universities have a rule allowing an examiner only three years of office? In how many ways can the second round be operated?

7 How would you programme an automatically operated lift so as to minimise the use of power? For example, would it return to ground level when empty, or remain where the last passenger got out? Consider whether this minimises the waiting time of passengers.

8 Would you introduce the idea of a singular matrix to children at lower secondary level? Give a reasoned answer to this question, considering it in

relation to two of the following approaches: (a) through coding; (b) through transformations of the plane; (c) through the solution of linear equations in two unknowns.

(Bristol BEd, 1971)

6.3. *Mathematical physics*

Using simple algebra, ideas of astronomy can serve to introduce pupils to the marvellous work of Kepler and Newton. We write it as a sustained exercise for the reader, for brevity.

Exercise

Suppose a particle describes a circular path of radius r with velocity v cm/sec. Let us assume the result that the particle's acceleration is v^2/r cm/sec^2 towards the centre (i.e. its acceleration is centripetal) – there seems to be no way of making this result plausible without a limiting process. Then we can introduce pupils to the classical results of gravitation by using only algebra, and the following outline is suggested by the discussion in Pólya [3].

First, let the *period* of the particle be T, so it takes T seconds to make one revolution round the circle.

Step 1. Prove that the particle's centripetal acceleration is

$$a = 4\pi^2 r/T^2. \tag{i}$$

If the particle happens to be a planet moving round a central 'sun' (as earth round sun, moon round earth, or the moons of Jupiter round Jupiter), then Kepler's observations of these cases led him to enunciate that

(A) *the area swept out by the radius was proportional to the time of sweep*, while

(B) T^2 *is proportional to* r^3.

Step 2. Show that (A) implies that the angular velocity, and hence v, are constant. Show also that from (i), (B) implies

$$a = \frac{4\pi^2}{c^2} \cdot \frac{1}{r^2}, \tag{ii}$$

where $T^2 = c^2 R^3$ and c is a constant depending on the planet and its 'sun'.

Now consider ourselves in Newton's position. Equation (ii) tells us (with Newton's Law that force = mass × acceleration if mass is constant) that the force on the planet is of the form

$$F = km/r^2 \tag{iii}$$

(a form guessed by others before Newton did something about it).

Step 3. Conversely, then, since $km/r^2 = mv^2/r$ establish Kepler's Law (B): $T^2 = c^2 r^3$. The constant c thus obtained does not quite agree with observation because our model does not take into account the fact that the earth's orbit is an ellipse with the sun not stationary.

Step 4. Show that

$$k = 4\pi^2 r^3/T^2. \tag{iv}$$

In particular, for the moon and earth, let R denote the radius of the moon's orbit, and k_1 the corresponding constant. Thus k_1 can be computed from (iv)

309

since R and T can be measured. Next, if we consider the mass of the earth to be concentrated at its centre, then the same theory applies to a particle on the earth's surface if we think of it as a planet. Thus, ρ being the radius of the earth and k_2 the corresponding constant, we have from (ii) that the centripetal acceleration γ on the particle due to the earth is

$$\gamma = k_2/\rho^2.$$

But γ is conventionally denoted by 'g' and can be measured experimentally using the formula $2\pi\sqrt{l/g}$ for a pendulum (in which Newton was rightly quite confident, in spite of the criticisms expressed on p. 296 above). Once ρ was measured with sufficient accuracy, Newton could then calculate k_2, and it turned out that this agreed with his computation of k_1. This presumably then led him to infer that k in (iii) is *independent of the planet*.

However, for a planet P and sun S, the force on the planet due to the sun is, by (iii), $F_P = k_P m_P/r^2$, while that on the sun due to the planet is $F_S = k_S m_S/r^2$. Another of Newton's laws then tells us that $F_P = F_S$: action and reaction are equal and opposite. Hence we may write the common magnitude of the force as

$$F = G m_S m_P/r^2, \tag{v}$$

where G is independent of r. Mathematically we would expect G to be a function of m_S and m_P, say $G = G(m_P, m_S)$. Physically we would expect that if the planet or sun could suddenly have another mass stuck to them, then the resulting force would be the sum of the original force and one due to the new mass. Thus $G(m_P, m_S)$ is *bilinear*, i.e.

$$G(um_P + vm', m_S) = uG(m_P, m_S) + vG(m', m_S)$$

for all numbers u, v: and similarly for the m_S variable. It was truly astounding that Newton, on the basis that $k_1 = k_2$ as described above, staked years of work on the assumption that G was actually constant, that it was a truly *universal* 'constant of gravitation'. Had he allowed for the possible variations in g, and other crudities of his model, he would surely have been too inhibited by choice, to hit on the idea.

With G in (v) assumed constant, Newton could 'weigh the planets'. For, by (iv) we now have

$$Gm_S = 4\pi^2 r^3/T^2,$$

so if r', T' refer to a planet orbiting round a sun S', then

$$m_S/m_{S'} = (r/r')^3/(T/T')^2.$$

Step 5. Show which planets we should observe, to express the masses of earth and Jupiter in terms of that of our sun.

The assumption that G is constant, implies that it can be measured by experimental work on earth. For details of these experiments the reader is referred to books on physics.

Step 6. How would you estimate the density of Jupiter?

Step 7. Two satellites orbit the earth of radius ρ at heights $h < h'$, with constant speeds v, v' and periods T, T'. Show that

$$v' = v\sqrt{\frac{\rho + h}{\rho + h'}}, \quad T' = T\left(\frac{\rho + h'}{\rho + h}\right)^{3/2};$$

and if $h = 0$, then $v = \sqrt{(g\rho)}$.

For other simple and interesting uses of mathematics in physics, see one of the newer texts in physics for schools, such as Jardine [1]. This latter emphasises basic principles of conservation from which other laws are deduced.

6.4. *Geometry*

Since the overthrow of Euclid's *Elements* and derivative texts, geometry has been taught in British schools mainly as a branch of physics (see p. 237) with great emphasis placed on its applications. As we mentioned earlier, not all countries are satisfied with such a 'physical' approach, but there are few, if any, in which the first approaches to geometry are not visual and tactile.

Thus examples on scale-drawing, perspective, surveying and navigation can be found in many texts, along with applications of trigonometry to the calculation of heights, bearings, etc.

More recent applications to find a place in the texts are examples taken from biology, on changes of scale – for example, how a 10-fold increase in the linear dimensions of an animal may produce a 1000-fold increase in its internal heating system and only a 100-fold increase in cooling through its skin (see, for example, Haldane [1], Maynard Smith [1], Thompson [1]) – and problems on networks solved by means of simple linear graph theory (see SMP *Book 2*, Ore [1]).

The introduction of different 'metrics' can generate interesting discussions about suitable models: thus we have the 'flat rate' bus metric (suitable for the 'Red-Arrow' services in London)

$$\mathrm{d}(P, P) = 0,$$
$$\mathrm{d}(P, Q) = 5, \quad \text{if } P \neq Q,$$

and the 'Manhattan taxi-driver's metric'

$$\mathrm{d}(P, Q) = |\, x(P) - x(Q)\, | + |\, y(P) - y(Q)\, |$$

(see, for example, Noble [1]), which is appropriate for cities laid out in a grid pattern. Questions arise concerning the optimal arrangement of shops etc., which reflect trends in using topological techniques by geographers and town planners (see, for example, Haggett and Chorley [1]). Consideration of cities with one-way traffic systems will lead to distance measures that are no longer symmetrical, i.e. for which one has $\mathrm{d}(A, B) \neq \mathrm{d}(B, A)$.

An example of a 'metric' used in a biological application (due to Marczewski and Steinhaus) can be found in the Exercises below.

The introduction of geometric transformations to school courses has meant that problems on reflection of light etc. can be tackled more readily by geometric means (see, for example, Yaglom [1], Jeger [1], Courant [1] and Schiffer [1]).

Once analytic geometry is available, problems can be solved using the intersection of pairs of graphs. The tracing of curves, the areas under graphs, and the gradients of curves, all have well-known applications. Certain aspects of conic sections are applicable before calculus is available: for example, a source of light or heat placed at the focus of a parabolic reflector produces a parallel beam, with applications to car lights, radar antennae and solar furnaces. Parabolas are also needed for the discussion of the flight of projectiles with constant acceleration. (See Pólya [3] and Schiffer [1].)

Exercises

1 If one ship is twice as big (linear dimensions) as another, what can you say about the energy requirements of the larger if its speed is to be twice that of the smaller? (See Thompson [1].)

2 If animal A is k times the size of animal B, would you expect the gestation time of A to be k^3 times that of B?

3 The 'metric topology' mentioned above can be explained as follows. The usual requirements of measuring, whether on the flat or on a curved surface, lead us to say that we can measure distances between points A, B of a set X provided we can at least specify (by some process) a number $\mu(A, B)$ such that $\mu(A, B) \geqslant 0$ and

  (i) $\mu(A, B) = \mu(B, A)$;

  (ii) $\mu(A, B) = 0$ if and only if $A = B$;

  (iii) (the triangle law) $\mu(A, B) \leqslant \mu(A, C) + \mu(C, B)$ for all points C in the set X.

Thus, although for some ways of measuring distance we might be able to say more, we *always* require that (i) the distance between A and B shall be the distance between B and A, (ii) distinct points are at a strictly positive distance apart, (iii) in any trio of points, the sum of a pair of side-lengths is not less than the remaining one. Now μ above is a special kind of function from the set $X \times X$ of ordered pairs (A, B) of points of X, to the real numbers; and any such function satisfying (i) to (iii) is called a *metric* on X. The prime example is the Euclidean metric in three dimensions

$$\rho[(x, y, z), (a, b, c)] = \sqrt{[(x - a)^2 + (y - b)^2 + (z - c)^2]}.$$

  (*a*) Prove that ρ is a metric.

  (*b*) Prove that the 'Red-Arrow' and 'taxi-driver's' functions, given above, are metrics.

  (*c*) Under what conditions will a railway timetable give a metric, if we define $\mu(A, B) =$ time taken to travel from A to B?

  (*d*) Where should a shop be placed, so as to be equidistant from three points, in a town with a grid pattern of equally spaced streets?

  (*e*) The following problem was suggested by biologists:

312

Suppose two forests A, B, have a and b different species of trees, respectively, of which w species are common to both forests. Is there a good numerical measure of 'distance' between A and B? Interestingly, one can define a metric on the set of all forests by the rule

$$d(A, B) = (a + b - 2w)/(a + b - w).$$

Prove that this function is a metric, assuming that two forests are regarded as equal if $a = b = w$. The formula allows plenty of practice in computing $d(A, B)$ in specific cases. Many questions are raised: what if $d(A, B) = 1$? For what C do we have $d(A, B) = d(A, C) + d(C, B)$?, etc.

(f) If μ satisfies rules (ii) and (iii) for a metric on X but μ is not symmetric, show that if we define

$$d(A, B) = \max[\mu(A, B), \mu(B, A)]$$

then d is a metric on X.

4 Prove the statement in the text, about the parallel beam from a parabolic reflector.

5 Let two planets of the sun have angular positions α, β measured from a fixed radius in the plane of themselves and the sun, assuming this plane and the sun to be fixed. Suppose the planets have constant angular velocities ω_1, ω_2 about the sun. Find when they can become collinear with the sun. (Such a 'synod' is used by astronomers for observing the period of one of the planets with accuracy.)

6 Particles are injected into the air from a source, with constant velocity v at the rate of N per second. Show that the number per cc, at distance r from the source, will be approximately proportional to $1/r^2$. Use this crude model to show why the intensity of light, of smells, and of paint sprays, falls off as $1/r^2$. Could this be a basis for the various inverse square laws of physics?

7 Assuming the earth to be a perfect sphere, repeat the calculations of Eratosthenes (280–195 BC) as follows (see Fig. 7). He knew that Alexandria (A)

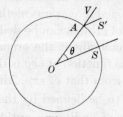

Fig. 7.

and Syene (S), both on the Nile, were on the same meridian, since the Nile flows there almost exactly north–south. By watching the sun's rays hitting the surface of a deep well at noon on a midsummer's day in Alexandria, he could measure angle VAS' as $7° 12'$ between the vertical AV and the sun's rays AS' (assumed in a parallel beam containing OS). Knowing the measured length of the arc AS, find the radius of ρ of the earth.

(Here is an example where we have two metrics for the earth's surface, since we can use either the arc-length AS or the straight length AS as a measure of distance. Clearly the arc-length has the greater practical significance here.)

8 (i) Give a vector proof of the theorem that the angle in a semi-circle is a right angle.

   (ii) Discuss in detail the background knowledge of vectors that a pupil would need in order to justify his use of this proof.   (Bristol BEd, 1971)

9 Discuss the view that one should approach geometry through topology, since such notions are more 'primitive' (hence easier to understand psychologically) than the ideas of metrical geometry.

6.5. *Calculus*

Starting with elementary kinematics and the problem of calculating gradients, it is fairly easy to teach calculus as a manipulative subject, outside rigorous analysis, after the manner of Mercer's curriculum (p. 171). This method has been widely adopted in Great Britain for over half a century. We must admit that those who learn calculus in this way find 'analysis' hard, later: on the other hand, such treatments are accessible to those who would not aspire to a more formal treatment.

Simple problems of maxima and minima can soon be attempted. Next comes the problem (not considered by Mercer but common in traditional treatments) of introducing the logarithmic and exponential functions, and an excellent opportunity arises for applications to motivate the theory. (One can of course simply ask within 'pure' mathematics for a function having itself as derivative.) Problems of radioactive decay, or of population growth, lead, as in Example 2 of Section 3, to the differential equation $\mathrm{d}N/\mathrm{d}t = kN$. The solution can be obtained by various methods within pure mathematics, or its properties of growth can be deduced from the equation and inferences can be made about the original population that led to the model. The inverse function of *exp* is the logarithm, whose behaviour can be deduced from that of *exp*. Alternatively, the concept of 'function' allows *log* to be defined by the integral $\int_1^x t^{-1}\,\mathrm{d}t$, as a response to the question of finding an antiderivative of t^{-1}; its inverse function is *exp* (see, for example, SMP *Advanced Mathematics Book 3* and, for a criticism of this approach, Matthews [1]). The logarithmic function can also be approached via its functional equation, obtained from suitable models, as in Engel [2].

The notion of integration is, in British treatments, introduced in a way suited to physical applications. Thus, a variant of the discussion in Mercer's curriculum (p. 174) is to *assume* that the integral $A(x)$ exists as an area under the graph of $y = f(x)$ on an interval $a \leqslant x \leqslant b$. One has the approximation $\delta A \simeq f(x)\delta x$, and then passes to the differential equation $\mathrm{d}A/\mathrm{d}x = f(x)$ from which A arises as anti-derivative of f. Of course, one assumes (implicitly at first) that $A(x)$ exists and depends smoothly on x. This process occurs in the formulation of many differential equations in physics, such as those involving changing mass (rocket motors, chains sliding off tables, vehicles refuelling while in motion), where Newton's Law must take the form that $\mathrm{d}(mv)/\mathrm{d}t = $ Force. Once the process has been seen to be useful and interesting, a more rigorous discussion within pure mathematics is appropriate; and even then, various levels of rigour are possible, as shown in Griffiths [1], Armitage [1] and Quadling [2].

Once integration and the exponential function are known, pupils can be introduced to differential equations. Cooling bodies lead to the equation (Newton's Law of Cooling) $\mathrm{d}\theta/\mathrm{d}t = \theta - \theta_0$. Populations of predators and prey lead to systems of equations linking the number $N(t)$ of predators with the number $M(t)$ of prey:

$$\mathrm{d}N/\mathrm{d}t = aN + bM, \quad \mathrm{d}M/\mathrm{d}t = pN + qM,$$

where a, b, p, q are taken to be constant. The type of problem being modelled gives rise to discussions of the roles of the constants θ_0, a, b, p, q within the respective models. The simultaneous equations here offer a good opportunity for solution by linear (matrix) methods, while their orthogonal trajectories offer many applications of other (non-linear) techniques. See, e.g. Kemeny [1], and the articles of Richardson [1] and Lanchester [1] on a theory of conflict. Second-order equations occur in the theory of electrical circuits and in mechanical systems with damping, and there is an interesting analogy between them; each is a model of the other (see Exercise 2 below).

By writing acceleration as $v\,\mathrm{d}v/\mathrm{d}x$ ($v = \dot{x} = $ velocity), many problems in mechanics involve non-linear equations of the form

$$v\,\mathrm{d}v/\mathrm{d}x = f(x, v)$$

which can be integrated elementarily for special types of function f. With each integration should come a discussion of the graphical form of the solution and the consequences for the situation being modelled. See Pólya [2], pp. 175–205, and Exercise 2 below, involving the terminal velocity of a falling body.

The motivating examples so far mentioned have been applications within classical physics. For biological examples see Rosen [1], who considers such problems as the optimal angle for a vein to branch into two, and Maynard Smith [1], who discusses, for example, problems of muscular movement. Several mathematics programmes now exist, in which probability theory plays a strong motivating role (e.g. the SSMCIS and SMP texts); applications of calculus – especially of integration – are developed within them.

The traditional treatment of calculus tends to gloss over the various infinite processes involved. However, many applications of mathematics require discussion of iterative processes and the related error terms. For this reason, there have been approaches based on computing (see, for example, G. S. Young [1]).

Exercises

1 Given that k is a real number and f a continuous function, show that the solution of the first-order differential equation

$$(D - k)\,y = f(t), \qquad y(0) = y_0$$

(where D stands for the operator d/dt) is *unique*, and is given by

$$y(t) = e^{kt}\left(y_0 + \int_0^t e^{-ks} f(s)\,ds\right).$$

Hence show that the second-order equation

$$(D - l)\,(D - k)y = f(t), \qquad y(0) = y_0, \qquad (Dy)\,(0) = y_1$$

has a unique solution, and find it. (Solve $(D - l)\,u = f(t)$, then $(D - k)\,y = u$.)

2 The equation of motion of a particle of mass m, subject to a constant driving force F and retardation proportional to its velocity, is of the form

$$m\,dv/dt = F - rv, \qquad r \geqslant 0, \tag{i}$$

with v and dv/dt prescribed at $t = 0$. Also $v = dx/dt$ where x is the displacement of m at time t. Now, the current i in an electric circuit driven by an e.m.f. E satisfies Lenz's Law

$$L\,di/dt = E - Ri, \qquad R \geqslant 0, \tag{ii}$$

where L is the inductance and R the resistance.

Show that

$$\begin{aligned} v(t) &= e^{kt}(v(0) - F/r(e^{-kt} - 1)) \qquad (k = -r/m) \\ &= e^{-rt/m}(v(0) + F/r) + F/r \end{aligned}$$

so that v settles down to the terminal velocity F/r as t gets large. Noting the correspondence $m \leftrightarrow L,\ r \leftrightarrow R,\ F \leftrightarrow E,\ v \leftrightarrow i$, show that

$$i(t) = e^{-Rt/L}\,(i(0) + E/R) + E/R$$

which settles down to E/R, as if only Ohm's Law were in force.

Find v in terms of x in (i), by writing $dv/dt = v\,dv/dx$. What is the analogue for (ii)? Investigate periodicity of the solutions of (i) and (ii) when F in (i) is $-w^2x$ with w constant and $v = dx/dt$. What is the electrical analogue?

3 An SMP A-level question reads as follows:

'For a body whose temperature is $\theta(t)$ at time t, Newton's Law of Cooling may be expressed in the form

$$\theta'(t) = -A[\theta(t) - \sigma(t)],$$

where $\sigma(t)$ is the temperature of the body's immediate surroundings and A is a positive constant. Verify that

$$\theta(t) = e^{-At}\left\{\theta(0) + A\int_0^t \sigma(u)\,e^{Au}\,du\right\}$$

satisfies the differential equation.

The body is in a laboratory which itself cools according to the above law with a constant B in the place of A, and with $\sigma(t)$ denoting the external air temperature. On a winter's day when the external temperature is zero, the body and the laboratory are initially at the same temperature $\theta(0)$. Find a formula for the temperature of the body at any subsequent time.

Prove what you would expect, namely that this temperature is greater than that of the body if it had been cooling outside the laboratory.'

Write out a solution to the question, and then draw a flow diagram to indicate how a calculus course might be planned, to enable pupils to be able to do the question and understand the manipulations involved.

[Southampton BSc, 1969]

4 The 'continuous' analogue of Verhulst's equation (i) in Question 1 on p. 306 is

$$dN/dt = qN - rN^2, \qquad N(0) = N_1.$$

Find a formula for $N(t)$, using partial fractions to evaluate $\int \{dN/N(q - rN)\}$. (N.B. you must assume $N \neq q/r$ and distinguish between the cases $N < q/r$, $N > q/r$, since $\log x$ is defined only when $x > 0$.) How does $N(t)$ compare with the solution of the discrete equation (ii) on p. 307? Show that $N(t)$ converges to the same limit as the discrete solution as $t \to \infty$; and sketch the graph of $N(t)$. For a fuller account see Schiffer [1].

5 Write the predator-prey equations on p. 315 in the form

$$(D - a)N - bM = 0, \qquad pN - (D - q)M = 0 \qquad \left(D = \frac{d}{dt}\right)$$

and, as with simultaneous algebraic equations, obtain two second-order differential equations $LM = 0 = LN$ where L is the operator $(D - q)(D - a) - bp$. Find conditions on a, b, p, q, so that if $dM/dt = dN/dt = 0$ at $t = 0$, then (M, N) always lies on an ellipse, centre $(0, 0)$. Show that this corresponds to an equilibrium state of the two populations, in that the prey increases, leading to an increase in prosperity of the predators, leading to a scarcity of prey, then a decline in the predators, and so on. What happens if a, b, c, d are such that these fluctuations cannot occur? Discuss the special case when $aq - bp = 0$.

6 It is often found that a person does not acquire knowledge of a subject (especially mathematics) in steadily increasing fashion. After a long, slow start, he 'takes off', learns rapidly and eventually levels off again. Would Verhulst's model of population (Question 1 above) be a suitable one for modelling this type of learning? What are the implications for teachers of the subject?

7 After a heater is placed near a wall to heat a room, a grubby mark is eventually seen on the wall, and it rapidly seems to get worse. Explain this exponential effect by using the assumption that the number of par-

317

ticles of carbon already deposited on the wall is proportional to the rate of increase in their number. Is this a physically reasonable assumption?

8 The forces on a particle P of mass m in the surface of a rotating fluid may be assumed to be mg vertically downwards, and R perpendicular to the tangent at P to the meridian through the surface (see Fig. 8). If P is a distance x from the axis of rotation and its angular velocity is w, show that

$$\mathrm{d}y/\mathrm{d}x = w^2x/g$$

where $y = y(x)$ is the equation of the meridian. Hence show that on the moon, a cup of tea when stirred will have a much steeper hollow in it than on earth. Show that the surface of the fluid is a paraboloid of revolution, its steepness increasing as we stir faster.

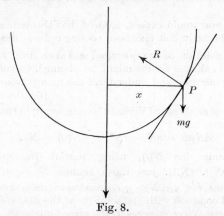

Fig. 8.

9 The equation of motion of a particle subject to a central inverse square law of force is $\ddot{r} - r\dot{\theta}^2 = -\gamma/r^2$ where $r^2\dot{\theta} = h = $ constant, because of the conservation of angular momentum. In this mathematical model, put $r = 1/u$, so $\dot{r} = -h(\mathrm{d}u/\mathrm{d}\theta)$, $\ddot{r} = -h^2(\mathrm{d}^2u/\mathrm{d}\theta^2)$ and $\mathrm{d}^2u/\mathrm{d}\theta^2 + u = -\gamma/h^2$. Show that this last equation is satisfied by the point (r, θ) on an ellipse with polar equation $lu = 1 + \mathrm{e}\cos\theta$, where $u = 1/r$ and $1/l = -\gamma/h^2$. Hence show that the particle has an elliptical orbit. Calculate the eccentricity. When is the orbit approximately a circle?

10 In order to escape completely from the earth's gravitational field, a rocket must have sufficient kinetic energy to overcome the work done on it by the earth's gravitational force. If v is its 'escape' velocity and m its mass, show that

$$\tfrac{1}{2}mv^2 = -\int_\rho^\infty (GMm/r^2)\mathrm{d}r$$

where M is the earth's mass, and ρ its radius. Hence show that $v = \sqrt{2}$ times the 'orbital' velocity discussed in Step 7 (p. 310).

7. APPROXIMATION, IDEALISATION AND STRUCTURAL STABILITY

Applied mathematicians and engineers frequently make their work difficult to understand, because they neglect various terms in mathematical arguments, often without giving explicit reasons. Such reasons can be interesting and wholly appropriate to a mathematics lesson, as can be seen in the discussions in some SMP texts. As a simple example: in some problems, the earth is taken to be flat, in others it is a spherical ball, in others pear-shaped, and in yet others it is regarded as a point-mass. The reasons usually depend on the ratio of the earth's diameter to the other parameters involved in the problems. If these ratios are estimated, it then becomes clear and reasonable why various terms may be neglected, and why others may not. Sometimes, of course, a non-linear term may be omitted because the mathematics is made simpler, but the consequent change of model should be remembered. Similarly with the moon: a recent lunar landing met difficulties because the model used for calculation did not take account of fluctuations in lunar gravity. In general, simplifications may lead to interesting mathematics; but the beauty of the mathematics should not obscure the fact that the model may then be without physical interest. If this rule is neglected, then extreme violations lead pupils to adopt the scientist's view – which is a grave comment on past teaching of mathematics – that 'stupid computations are mathematics, but serious reflection is science'. The 'serious reflection' almost always consists in finding a model which is both tractable and a good fit.

We ought to mention here the notion of 'structural stability', first introduced by Russian control engineers, and of great current interest among research workers. It is, however, not an esoteric idea, but of great practical importance. Essentially it starts from the fact that all physical observables can only be measured to within certain degrees of accuracy. Thus, for example, the equation of a pendulum with damping should be of the form (see Brauer [1]):

$$\ddot{\theta} = a \sin \theta + f(\theta, \dot{\theta}, t),$$

where f denotes a 'small' function that accounts for all effects that have been neglected. The general features of the graph of θ should be independent of small fluctuations in f, just as we can still tune a radio, even though dust may accumulate on the controls inside. The graph should therefore be *structurally stable*. To make the necessary analysis, greater mathematical technique is required, and especially a knowledge of the theory of functions of several variables. Nevertheless, in elementary treatments, the role of inaccuracies of observation and model need not be passed over in silence. This attitude concerning structural stability is rather different from the 'uncertainty principle' of quantum physics, which is nowadays mentioned in school physics courses.

We conclude this section with a set of examination questions from English textbooks and examinations. They are chosen here because they are typical of traditional applications; in view of the previous discussion their virtues and vices should be apparent, and we invite the reader to try them and then comment on them.

Exercises

1 The following table gives data concerning flight:

	Length of wing (metre)	Beats per sec	Speed of wing tips (metre per sec.)
Stork	0.91	2	5.7
Pigeon	0.30	6	5.7
Bee	0.01	200	6.3

Assuming that a beat is half a complete oscillation, that the speed given is the maximum speed, and that the motion of the wings is simple harmonic, find the amplitude of the vibration of the wing tips.

If this interpretation of the figures is correct, the angle swept by the wings in each beat is approximately the same in each case. What is this angle, approximately, in circular measure?

2 A corridor of width h opens into a passage of width b, which runs at right-angles to the corridor. A girder of length l is to be moved on rollers along the corridor and into the passage. Neglecting the thickness of the girder, show that its maximum possible length, in order that it may just clear the corner, is

$$lb^{2/3} + h^{2/3}l^{3/2}.$$

3 (i) A pack of 52 cards contains 4 Aces and 4 Kings. Three cards are taken at random from the pack. Find the chance that they should be

(a) 3 Aces,

(b) 2 Aces and 1 King.

(ii) The chance of any one engine of a four-engined aeroplane failing on a long journey is 5 per cent. If only one engine fails the chance of the aeroplane completing the journey is 80 per cent; if two engines fail, but are on opposite wings, its chance of completing the journey is 50 per cent. It cannot fly with two engines out of action on the same wing. Find the chance that the aeroplane will complete the journey.

(The probabilities given in this question bear no relation to those that may exist on regular air services.)

4 A chemical dissolved in a fixed volume of solvent exists in two forms, type A and type B, each of which is of the same density when dry. The rate at which type A changes into type B is proportional to the concentration x of type A in the solution and similarly the rate at which type B changes into type A

is proportional to the concentration y of type B in the solution. Prove that $dx/dt = -kx + k'y$, where k, k' are constants, and find a similar differential equation for dy/dt.

Hence prove that x satisfies the differential equation

$$\frac{d^2x}{dt^2} + (k + k')\frac{dx}{dt} = 0.$$

Solve this equation for dx/dt and hence for x, given that the initial concentration of type A is 1 part in 20 and that its final concentration is half this.

8. NEEDS AND CRITICISMS

Some of the questions reprinted in previous sections are of 'applications' of the utmost artificiality, although some, for example Exercise 3 above, are distinguished by a whimsy which can serve as a substitute for practicability so far as motivation is concerned. Others have fewer saving graces. Pollak [1] has drawn attention to such shortcomings, and has pointed out that a problem like:

'The specific weight s of water at a temperature $t°$ is given by the cubic $1 + at + bt^2 + ct^3$; find at which temperature the water has maximum specific weight'

gives rise to such questions as: Why a cubic? What is the nature of the coefficients? Certainly, it would be more appropriate in a 'real' application to find the critical point by inspecting the graph that suggested the cubic rather than by finding ds/dt. Such criticisms show, however, that even a 'bad' example can be a source for useful discussion – and even for use as a first approximation to better models.

Finding worthwhile applications is not a simple matter for the teacher, and he has been helped in recent years by the publication of such collections as those by Noble [1] and the Mathematical Association [5]. The various projects, too, have played a part in searching out new applications of mathematics and translating them into terms suitable for use in school. Examples are found in the texts of the Engineering Concepts Curriculum Project, and of the SMP and the Schools Council Sixth Form Mathematics Curriculum Project. Nevertheless, there still exists a need for many more practical examples, especially those of models simple enough for elementary use.

Again, if more emphasis is to be placed on the applications of mathematics, then there will have to be changes in examination methods. If the pupil is to show evidence of what is entailed in constructing a mathematical model, then he will have to do so outside the constraints of a conventional examination. Such explanations and the actual construction of models take time both to do and to describe, and for that reason are better suited to coursework than to timed examinations. This is perhaps why the questions traditionally associated with prescribed models have degenerated to merely manipulative ones, for example, those concerning rods of length $2l \cos \theta$ lying in smooth hemispherical bowls. These may or may not have had value within 'pure mathematics', but they have had little to do with applied mathematics or with the training of applied mathematicians. We shall, however, deal more fully with problems of examining in Chapter 20.

Finally let us conclude this chapter with a further quotation from Pollak [1]:

We should perhaps bring to the surface one more fundamental assumption of this brief study of applications of mathematics and their relations to mathematics teaching. We have assumed throughout that it is possible and proper to attack the problem rationally, that is, that a logical and hopefully well-reasoned discussion will contribute to the solution of the problem. It is currently fashionable at times to question the relevance of reason to matters of education. We also are forced to question the relevance of reason, but on grounds totally different from those associated with current educational disturbances. No one seriously questions the importance of applications of mathematics, and yet we have had (and continue to have) enormous difficulties in obtaining a proper place for them in the curriculum. Why? There exists at least a possibility that a fair number of people go into mathematics precisely because they wish to hide from the real world. They conceive of mathematics as a beautiful, orderly structure that has nothing to do with life; amazingly enough, it is possible to make a living at it. If this is the case with sufficient numbers and the real source of difficulty in introducing applications into mathematics teaching, then the writer is afraid there is no hope.

Exercise

Design a 'core' syllabus in applied mathematics for a particular type of pupil.

SOME FURTHER READING

Brown, M. [1], NCTM [6]

PART 7
A CLOSER LOOK AT EXAMINATIONS

20

Examinations: Instruments for Evaluation

1. CHARACTERISTICS OF EXAMINATIONS

We wrote in Chapter 3 of the evolution of the external examination system and of the way in which the oral examination of medieval times gradually gave way to written examinations of a kind which until ten years or so ago appeared to be completely immutable. In the same chapter, we briefly described the work of Bloom and his colleagues, directed at providing a classification of educational objectives. We now look in a little more detail at the techniques of examining, and at some of the problems one encounters when designing examinations.

Examinations have certain technical characteristics of great importance. We first look briefly at two of these – validity and reliability – whilst remarking that not all authors would choose to describe these terms exactly as we do (cf., for example, Bloom [1], who uses the term 'scoring objectivity' for what we refer to as reliability).

It is important when setting an examination to ensure that it tests what we wish to test – that the examination has *validity*. Far too frequently, as we shall see later by the use of specific examples, we examine only a small subset of our educational objectives, frequently no more than the ability to memorise. If our examination is to be valid, then *all* the objectives must be tested. (Here, of course, one still encounters the 'Pollock' dilemma (p. 35) – 'Can examinations be designed to test all types of objective?') The examination even though valid, would still be largely worthless, however, if the candidate's marks were to depend upon the examiner's digestive system. Would different examiners award the same paper different marks, or would the same examiner mark the paper similarly on different occasions? These questions refer to the *reliability* of the examination.

Clearly, a valid examination which is not reliable has great defects. That an examination can be reliable without being valid is easily shown by a 'pathological' example. If the candidate receives full marks for spelling his name correctly and zero otherwise, then the examination is reliable. The examination is valid, of course, only if it is testing the ability to spell one's own name.

Great attempts have been made in recent years to improve the reliability and validity of examinations.

Another characteristic of examinations is that of *beneficence*. Any examination, and particularly an external one, will bring with it a backwash of possible good and bad effects of an extraneous nature. How are we to ensure that the good effects predominate – that is, that the examination is beneficent? Certainly, those examinations that were not valid, in the sense that they tested only rote learning, have turned out to be far from beneficent.

It will help to fix ideas if we consider briefly the method of examining used by the traditional examining boards in the British GCE system. As explained in Chapter 3 these boards have been in existence for a century or so, and their methods have not changed significantly within that period. We saw how their primary concern originally was with impartiality, so they employ independent examiners to set the question papers which are very closely related to a prescribed syllabus. These papers are vetted by a 'moderator', who tries to keep the questions of the same level of difficulty from year to year, looks for possible sources of misunderstanding in the wording of the questions, and rules out any questions that are not on the syllabus. In each subject there is a chief examiner, who makes a break-down (with his colleagues) of the steps required to answer each question, and assigns a possible numerical mark for each step. This marking scheme is modified after a proportion of the candidates' answers have been marked, to take account of steps which might have caused greater difficulty than expected, and after inconsistencies between individuals on the marking team have been revealed. Marks are then assigned to all the scripts; these are 'raw' and are next subjected to a 'scaling-process' to make the resulting statistical distribution agree with that of previous years; for example, if the mean mark obtained was lower than last year's, it would be assumed that this year's paper was harder than before, and the candidates compensated by having their marks all increased proportionately. After the scaling, each candidate's performance is compared with that predicted for him by his school before the examination, to help in borderline decisions and as a check on the whole procedure. Finally, the results – perhaps coded into grades A, B, C, . . . – are issued in the form of legal documents, since they entitle the holder to various privileges within the different professions. (See Schools Council [14], Appendix D, for further details.)

The impartiality of this procedure is clear, especially if we observe the secrecy of the printing and other arrangements. As to its beneficence, the boards normally claim that any teacher may submit his own syllabus for the board to examine upon it; but it is the style of the written, timed, examination which seems to have a most important effect (for a wide variety of different syllabuses is in current use) and no teacher can expect a board readily to vary that style. If he should insist, then he is generally driven to use one of the more flexible 'Modes' offered by the less prestigious CSE boards which have only recently come into existence (see Chapter 3, Section 6).

As to its validity, cases are known where candidates have been entered for the examinations of two different boards, with markedly variant results. Discussion of possible reasons for this are made difficult by the boards' secretiveness; their officials are conservative men who, like Victorian bank-managers, wish to present a front of total reliability, jealous of any hint of loss of trust in their currency. They do not question publicly what their examinations are testing, except perhaps to assert that they test 'ability'. Nevertheless, the bigger boards are now undertaking research into other methods of examining, stimulated by responsible criticism and the working groups of the Schools Council (see Chapter 12). See also the article by the Secretary of one of the largest boards in *Where?* (1971); that board's syllabuses now contain statements of 'the objectives of the examination' for certain subjects (e.g. physics), but not mathematics.

Exercises

1 Discuss the reasonableness of the assumption, by examining boards, that the distribution of marks each year should be the same. How could this be verified experimentally?

2 If marks are numerical, then the order-relation between numbers can be used to rank the candidates in an order. Is it reasonable to compress the information into a linear (one-dimensional) scale? What would happen if marks were vectors (i.e. we use a k-dimensional 'scale', $k > 1$)?

3 What could be meant by 'maintaining constant standards'? What are 'standards'? Is there an abstract notion suitable for all subjects?

2. OBJECTIVES

If our examination is to be valid, then we must know what our educational objectives are. The traditional way of dealing with this problem is to prescribe a syllabus which describes content, and then to define the depth of treatment expected by example, that is, by setting over the years questions at what the examiners believe to be the correct level. In this way the objectives of the *examination* become quite well defined. As we have seen however, they then frequently grow away from the objectives of the *curriculum*. As many authors have pointed out, a more satisfactory solution is to describe the objectives of the curriculum in more detail and then to set an examination based on these.

One starting-point for such an attempt is the two taxonomies of educational objectives drawn up by Bloom and his fellow workers. An attempt to translate part of *Handbook I: The Cognitive Domain* into the terms of mathematical evaluation has been made by Avital and Shettleworth [1].

These two authors take the first five taxonomic levels of Bloom's classification and suggest that they correspond to three levels of thinking:

Taxonomic level	Thinking process
1. Knowledge	Recognition, recall
2. Comprehension	Algorithmic thinking, generalisation
3. Application	
4. Analysis	Open search
5. Synthesis	

Here 'knowledge', for example, is the ability to repeat something which has been explicitly taught. It is the lowest of the cognitive categories and the category given most emphasis in examinations.

At the other end of the scale, 'synthesis' demands that a pupil should be able to recognise the need for principles which, at a first glance, might seem unrelated to the problem, but which can be applied to the given data in order to reach a solution of the problem.

It must be stressed here that what might be 'synthesis' to one pupil, will be 'knowledge' to another who is already acquainted with the problem and its solution (cf. the argument about categories on p. 211). The cognitive level of a question will, therefore, depend upon the experiences of the pupil. Consequently, a lack of knowledge about the pupil's mathematical experience will often render an examination invalid.

Alternative 'cognitive levels' or 'levels of behaviour' have been set out by other research workers (see, for example, Wood [1], Husén [1], Bloom [1].) Thus, Wood used the following levels:

A. *Knowledge and information*: recall of definitions, notions, concepts.
B. *Techniques and skill*: computation, manipulation of symbols.
C. *Comprehension*: capacity to understand problems, to translate symbolic forms, to follow and extend reasoning.
D. *Application*: of appropriate concepts in unfamiliar mathematical situations.
E. *Inventiveness*: reasoning creatively in mathematics.

Whereas Manheim (see Bloom [1]) suggests:

1. Ability to remember or recall definitions, notations.
2. Operations and concepts.
3. Ability to interpret symbolic data.
4. Ability to put data into symbols.
5. Ability to follow proofs.
6. Ability to construct proofs.
7. Ability to apply concepts to mathematical problems.
8. Ability to apply concepts to non-mathematical problems.
9. Ability to analyse problems and determine the operations which may be applied.
10. Ability to invent mathematical generalisations.

The existence of such lists does not, of course, solve any of our problems. We are still faced, for example, with that of designing examination questions which will test levels E and 10 above. Where these lists help is in providing us with a way of classifying existing examination papers and checking whether or not they provide adequate coverage of our educational objectives – whether or not they are valid.

Testing in the *affective* domain is by no means so straightforward. Bloom's *Handbook 2: The Affective Domain* provides us with a classification of objectives in this domain, and Wilson, writing in Bloom [1], tries to apply these to mathematical education. He distinguishes the wish to assess attitude, interest, motivation, anxiety, and the way in which a pupil thinks of himself as a mathematician. There is also a wish to test the pupil's appreciation of mathematics. For example, Hardy [1] describes how 'The mathematician's patterns, like the painter's or the poet's, must be *beautiful*; the ideas, like the colours or the words, must fit together in a harmonious way. Beauty is the first test: there is no permanent place in the world for ugly mathematics.' He goes on to give examples of proofs which he values for their beauty. This is but one way in which we can appreciate mathematics – yet how are we to test a pupil's appreciation? The object of such testing, of course, is not to torture a pupil but to find out whether our teaching is doing what we should like it to do.

The need for more research into testing in the affective domain becomes clearer when, for example, we recognise the part which attitudes and appreciation play in a university mathematics course. Yet these are never tested by conventional entrance examinations, to their detriment as predictive instruments. Regrettably, the need is reinforced by those questions that have been devised to date. We shall quote some below and others can be found in Wilson's paper. Those that are 'reliable' are, in the main, faintly ridiculous, whereas those which seem to reach the heart of the matter appear to be far from reliable. Yet this latter might well be a price worth paying.

Exercises

1 It is often said that examinations test character and the ability to withstand the stresses of life. Is this true of conventional mathematical examinations? If so, is it a legitimate aim?

2 The following is an attempt by F. W. Land (University of Hull, 1967) to classify the abilities associated with the construction and application of mathematical models.

  (A) *Model Construction*

    (1) To search for, and find patterns common to a variety of situations.

    (2) To isolate and define the critical variables.

    (3) To describe symbolically these variables.

    (4) To determine and define the relationships that exist between the variables.

    (5) To establish the necessary validity of statements of these relationships.

    (6) To describe and perform operations on the variables.

  (B) *Model Application*

    (1) To establish that a given situation is one in which a certain model is applicable.

    (2) To establish that the model is the most appropriate (if a choice exists).

    (3) To formulate the problem in terms of variables defined in the model.

    (4) To manipulate the model in order to solve the problem within the model.

    (5) To formulate the solution in terms of the original problem.

  Describe a situation leading to a mathematical model suitable for discussion in the sixth form and attempt to describe the construction and application of the model using Land's categories.

  Comment on the aptness of Land's taxonomy.

                          (Southampton BSc, 1972)

3 Rephrase Manheim's objectives 7–10 in order to make them more exact, and give suitable examples.

4 Give examples of examinations which can be rendered invalid by an ignorance of the pupils' mathematical experience and preparation.

3. TESTING OBJECTIVES

There are still pockets (by no means small) of the British external examination system which seem completely insulated against change. Thus, for example, the reader might care to speculate which of the questions reproduced below was set in 1910 and which is taken from a 1970 GCE paper:

(1) Prove that, if α, β, γ are the three roots of the equation
$$x^3 - 21x + 35 = 0,$$
then $\alpha^2 + 2\alpha - 14$ will be equal to β or to γ.

(2) Express each of the coefficients a, b, c of the equation
$$x^3 + ax^2 + bx + c = 0,$$
in terms of α, β, γ, the roots of the equation.

Hence, show that
$$\alpha^2 + \beta^2 + \gamma^2 = a^2 - 2b$$
and find an expression for $\alpha^3 + \beta^3 + \gamma^3$ in terms of a, b, c.

By constructing a cubic equation, or otherwise, solve the equations
$$x + y + z = 2, \qquad x^2 + y^2 + z^2 = 14, \qquad x^3 + y^3 + z^3 = 20.$$

Exercises

1 What do you think the examiners are testing in the above questions?

2 Suppose a candidate gets the wrong expression for a, b, c in Question 2, verifies the equation by a lucky error, and finds the expression for $\alpha^3 + \beta^3 + \gamma^3$ by a correct method but using his wrong formula for a, b, c. Should he get no marks at all? Most mathematical marking concerns such 'botched' work; can it then be said to be 'objective'?

3 Try to think of some mathematical mistakes which you have made and from which you have profited.

A strong argument in favour of such examinations is their *economy* in examiners' energy: they are difficult neither to set, nor to mark within the system described on p. 325. Their validity and reliability are, however, open to question. Both of the questions above, for example, provide many opportunities for the pupil to make errors and be side-tracked. If this happens, then reliability of marking suffers.

The problem of reliability can be partly resolved by replacing such 'essay-type' problems† by 'objective-type' tests consisting of questions each of which can be answered without a great deal, if any, of intermediate working and which, therefore, carries either full marks or none at all.

† This nomenclature, although somewhat misleading, is commonly used to describe questions that present the candidate with alternative routes of attack.

Objective tests, nowadays, tend to be presented in 'multiple-choice' form, that is, the pupil is given five or so alternatives from which he has to choose the one(s) which is (are) correct. This method of presentation greatly increases the reliability of the examination and reduces the problem of marking. It is possible to mark such papers by computer – a great advantage when the number of candidates is large.

Doubts remain, though, as to whether such tests are as valid as the traditional ones. Can they really be designed so as to test other than the most trivial educational objectives? Many researchers believe that the answer to this last question is 'yes'. Certainly, the examples contained in Avital and Shettleworth [1] and in the two Schools Council Examination Bulletins (Schools Council [4 and 7]) show that much more is possible than might at first be supposed. Thus, for example, the following question can be used to test ideas concerning formal proof:

(I) Assuming that $\sqrt{2}$ is an irrational number, we can prove that $a + \sqrt{2}$, for any rational a, must also be an irrational number. Which one of the following statements can be used to obtain such a proof?
   (a) The difference between two rational numbers is always a rational number.
   (b) $\sqrt{3}$ is also an irrational number.
   (c) The sum of two irrational numbers is sometimes rational and sometimes irrational.
   (d) The product of a rational number, different from zero, by an irrational number is always an irrational number.
   (e) None of the above can be used to prove the Theorem.

(Avital and Shettleworth [1])

and the following question can test the ability to recognise patterns:

(II) The last digit in 4^{10} is
   (a) 0,
   (b) 2,
   (c) 4,
   (d) 6,
   (e) 8.

(Wilson, J. W. et al. [1])

In this particular example we have an instrument incapable of distinguishing between the pupil who can recognise a pattern and the one who cannot but who compensates for this lack by a delight in multiplication. This is, of course, a drawback, and only serves to emphasise the fact that to set good examples of this type is not as easy as it is to set conventional questions, and with the latter we have over a century of experience!

It is, therefore, necessary to check by experiment, using pilot groups of students, that test items do perform the job for which they are intended; and that the incorrect answers are, at least, reasonably plausible.

Setting multiple-choice questions to test objectives in the affective domain is even more difficult. Consider such items as:

(III) I would like to study about the life of Gauss
    (a) not at all,
    (b) a little,
    (c) a lot.

and

(IV) I like the problem '$359 - 574 + 6480 - 999 - 46937 + 9748 + 97483$ $= ?$'———— than the problem
'Jane is half as tall as Dick. Joe is half as tall as Jane. Mark is half as tall as Joe. Dick is 60 inches tall. How tall is Joe?'
    (a) a lot more,   (c) a little less,
    (b) a little more,   (d) a lot less.

(Wilson, J. W. *et al.* [2])

One would have reservations about their ability to tell us anything useful about, respectively, the pupil's interest in mathematical activities and his attitudes to the subject. Certainly, it is doubtful whether knowing that, say, a child prefers a pointless question to an idiotic one, will advance the cause of mathematical education.

Essay-type questions such as

(V) What do you think the poet meant who said,
    'Euclid alone has looked on beauty bare'?

(Wilson, J. W. *et al.* [2])

do, of course, raise the old bogey of reliability of marking.

In composing some of the exercises in this book, the authors have met the same difficulties. However, these exercises are self-testing; they are intended to arouse curiosity in the reader, and so present a different problem to that of formal assessment.

The normal multiple-choice *situation* in which the pupil is presented with one *situation* and asked to make a single, independent decision, can be extended to what are known as 'multi-facet' examples. In these, a mathematical *situation* is looked at from a variety of viewpoints and the pupil is accordingly faced with the task of making a number of decisions.

Thus

(VI) C and S are concentric circles with radii 3 cm and 5 cm respectively.

	True	False
(a) The area of C is $\frac{3}{5}$ of the area of S.	......	......
(b) The circumference of C is $\frac{3}{5}$ the circumference of S.	......	......
(c) The circumference of C is less than twice the diameter of S.	......	......
(d) All tangents to C are chords of S.	......	......
(e) A chord of S which is tangent to C has length 8 cm.	......	......

(Wrigley [1])

Exercise

Find the probability of obtaining full marks for Question VI, by guesswork.

Many teachers might still be doubtful about the validity of any examination procedure which used such test items as the only means of assessment. All the 'objective' questions are of necessity 'closed' in the sense that the answers are predetermined and the pupil is constrained to move along fixed lines – there is no scope for either assessing or encouraging a pupil's creativity (see Wood's level E, p. 328).

Various attempts have been made recently to remedy this deficiency. The advent of Mode 3 CSE assessment made it possible to experiment with problems set in external examinations. Thus, one London school – Abbey Wood – set open-ended, 'discovery'-type questions such as:

1. Write about the mathematics of a chessboard.
2. A polygon is to be made from a piece of string 24 inches long. Investigate the possibilities.   (1967, CSE, Metropolitan Regional Exam. Board)

Similar questions were also set at GCE O-level, for example,

3. Investigate the set of triangles which have perimeters of 12 units.
(1967, GCE O-level, AEB)

In these examinations, unlike those consisting of multiple-choice test items which are, in general, compulsory, the pupils were able to select which question they wished to answer and were then allowed a considerable time for their attempt.

Exercises

1 Try the three open-ended questions reprinted above. How would you mark the work?

2 A 'closed' version of Question 3 (but hardly suitable for O-level) would be:
'Show that the set of all triangles of perimeter 12 units is bijective with a certain convex subset of the plane π; $x + y + z = 12$ in $\mathbb{R}^3$'.
Is there anything more to be said?

Such investigations are, however, not very suited to the conditions of a timed examination and it was natural, therefore, that attempts should be made to have them carried out during term-time.

These ideas were developed in the ATM Sixth-Form Project referred to earlier (p. 131), which suggested topics for investigation by sixth-formers such as:

4. Construct Pascal triangles using numbers from modular arithmetic and investigate the properties of such triangles.

5. What shapes can be obtained as the shadow cast by a square, with a point source of light? (ATM [4])

In this case, the teacher would be expected to act as adviser – to indicate possible books and judge on blind alleys – and observer. These are great responsibilities involving a change in his relations with his pupils.

The assessment and external moderation of such individual projects pose many problems and it is obvious that their marking is likely to be less reliable than the marking of conventional examination papers. Many, however, would feel that this is outweighed by the valuable feedback which will result from such examinations, compared with the cramming and syllabus-oriented work which are associated with more traditional examination forms. It is hoped, therefore, that gains in validity and beneficence will more than compensate for any loss in reliability.

4. CAN EXAMINATIONS TEST INSIGHT?

In this section, which is largely a sustained exercise for the reader, we try to see how insight (see Chapter 15) might be cultivated even within the 'traditional' syllabus – which consists of a more detailed version of that propounded by Mercer in Case Study 1, Chapter 13. This problem arose because the mathematics panel of one examining board thought this might be a way of improving mathematics teaching which would not require a great deal of in-service retraining. Unfortunately, it could not be put to the test, since several teachers thought that insight was too difficult to cultivate; they preferred the existing style of question because their weaker pupils could do them. They admitted that the resulting 'skill' was useless to such pupils, but defended it by saying that it was a way by which pupils could get a paper qualification for a later (and usually non-mathematical) job.

Mercer, of course, had insight and wished to cultivate it in his pupils, partly for strictly vocational reasons. Thus we see how the plans of one generation can be misunderstood and garbled by the next.

We now reprint a set of English A-level Mathematics questions, with our own criticisms of a few of them, and with suggested alternatives based on the original idea in the examiner's mind – who, in the nature of the system, needs considerable insight and cunning to produce questions which client teachers will regard as 'fair' according to their own conventions. The reader is then challenged to say whether or not he thinks they are good, to say what they seek to test, and to produce better ones if he can. He should also try them out on his friends, and he will be surprised at how much practice in the mathematics he himself gets in the process. Many teachers of mathematics have consolidated their own knowledge and skills by inventing and modifying questions: this is the nearest that conventional training gets to allowing a student to do creative work – but he does it after full-time training is over, rather than during it. Most students are found by the authors to be curiously reluctant to set their own questions, but perhaps this is because one's first attempts always lead to questions that one would fail to do in a real examination. Patience and humility help, with the knowledge that one is not unique in one's difficulties! The setting of some questions can lead to interesting technical problems, to choose the numbers, for example, so that the result works out nicely. For information on such matters, see Sawyer [1, p. 34] and Macdonald [1, 2].

5. SOME A-LEVEL QUESTIONS

(1) (a) If $y = e^{xy}$, show $\dfrac{dy}{dx} = \dfrac{y^2}{1 - xy}$.

    (b) Evaluate (i) $\displaystyle\int_0^{\pi/3} \cos^2 x \, dx$,

           (ii) $\displaystyle\int_0^{\pi/4} \left\{ \frac{\sin x}{\cos x} + \frac{\cos x - \sin x}{(\cos x + \sin x)} \right\} dx$

    (c) Obtain the coordinates of the points of inflexion on the curve

$$y = \frac{x^2}{1 + x^2}.$$

335

Comment

(*a*) As it stands, the question asks for purely formal implicit differentiation, requiring no real understanding of the issues involved. It suggests, however, the following:

(i) If $y = e^{xy}$, why is it correct to say that $x = (\log y)/y$? What would you do if $y = -e^{xy}$?

(ii) Sketch the graph of the curve $v = (\log u)/u$ showing its maxima, minima and asymptotes (if any). Infer from the graph that u is not a (single-valued) function of v.

(iii) Hence show that the equation '$y = e^{xy}$' is not a statement about *functions* without further amplification.

(iv) Assign a meaning to the command 'If $y = e^{xy}$, show $dy/dx = y^2/(1 - xy)$', and carry out your version.

N.B. (iii) and (iv) would be found difficult by traditionally trained candidates, but those having the elements of sets and functions should find the questions involved fairly natural.

(*b*) These are purely formal questions testing memory. A suggested alternative is:

'Many professional mathematicians forget how to integrate simple functions because they practise other techniques. Explain to one of them what basic rules you remember to work out an answer to (ii) above. Include also an estimate of the numerical value to within 1 per cent.'

(2) (*a*) If the roots of the equation $x^2 + px + q = 0$ are α and β, obtain in terms of α and β, the equation with roots p and q.

If the roots of these two equations are identical (and are not zero), determine the values of p and q.

(*b*) In the expansion of $\left(2x - \dfrac{1}{x^2}\right)^{12}$,

(i) determine which term is independent of x,

(ii) express the 10th term in its simplest form.

Comment

(*a*) This is just juggling with the facts that $-p = \alpha + \beta$, $q = \alpha\beta$. It may have some charm, but the question is unmotivated. We suggest:

'Consider the mapping $\theta: \mathbb{R}^2 \to \mathbb{R}^2$ which associates to each (α, β), the pair (p, q), where $(x - \alpha)(x - \beta) = x^2 + px + q$. Is the mapping one–one? Indicate on a diagram the image of the mapping. Show that there is a fixed point X of θ (i.e. for which $\theta(X) = X$) which lies off the axis.'

(3) Give the first 3 terms and the general term in the expansion of $\log_e (1 - x^2)$. Calculate the values of $\log_e(1 - x^2)$ for $x = 0.2, 0.4$ and 0.6, and show that the sum of the first 3 terms of the expansion differs by less than 1% from the value of $\log_e(1 - x^2)$ in the range $0 \leqslant x \leqslant 0.4$.

Hence evaluate $\int_0^{0\cdot4} \log_e(1 - x^2)\mathrm{d}x$ to within 1%.

Comment

This is quite good as it stands (but why?), except perhaps for the arithmetic. A variant (because students are weak on 'nth terms' when n is not a specified integer) is:

'Write down an expansion of $\log(1 + x^2)$ in powers of x, to n terms, with an error term E_n. Given that $|x| < 1$, how large must n be to ensure that the error is less than 10^{-6}?'

(Although 'error terms' are not normally included in the school syllabus, it is an easy exercise to derive the term here using integration by parts (see Griffiths [1], p. 526).)

(4) (a) Solve the following equations, giving all solutions in the interval $0 \leqslant x \leqslant 360°$.
  (i) $\sin x = \cos 2x$,
  (ii) $\tan 2x \cdot \tan x = 1$,
  (iii) $15 \cos x + 8 \sin x = 10$.
(b) By writing the equation in terms of $\sin x$, show that, in the range $0 \leqslant x \leqslant 360°$, there is only one solution to $\sin x \cdot \cos 2x = 1$. Obtain this value of x.

Comment

(a) Here the language of sets can clarify the problem. We suggest the modifications:

'(i) Sketch the curves $y = \sin x$, $y = \cos 2x$ on the same graph, and hence describe the set of all real numbers x such that $\sin x = \cos 2x$. For which of these x does x also satisfy $0 \leqslant x \leqslant 2\pi$?

(ii) Use the same technique to find all real solutions of the equations

$$\tan 2x \cdot \tan x = 1$$

and $$15 \cos x + 8 \sin x = 10,$$

Have these equations a *common* solution?'

(b) This is good as it stands, but because of its 'one-off' nature it might be preferable to have:

'Describe a geometrical procedure for estimating roots of $\sin x \cdot \cos nx = 1$, where n is a positive integer.'

337

Notice that a candidate could get high marks on both the original and on the revised questions, and still be demoralised in his first-year lectures in a university, by lecturers who assume that he knows what a definition is (*and why we make them*) or what a proof is, other than an application of an algorithm, or that mathematics is frequently concerned with statements of the kind 'For all $x \ldots$, $P(x) \ldots$'. In particular, no inductive proofs are present.

Exercises

1 Criticise the following questions taken from the same examination paper, saying why you think they are good or bad. You should write out the answers expected by the examiner first!

(5) Express $1/(x-1)(x-1)^2$ in partial fractions.

Expand the result as a power series in x up to terms in x^4 and state for what values of x the expansion is valid.

Sketch the graph of the function $1/(x-1)(x+1)^2$ using the first three terms of the expansion as a guide for small values of x.

(6) Determine the equation of the locus of the point which moves so that its distance from the line $x + a = 0$ is equal to its distance from the point $(a, 0)$.

Prove that the equation of the normal to the parabola $y^2 = 4x$ at the point $(t^2, 2t)$ is $y = -tx + 2t + t^3$. Show that the normal at the point $(1, 2)$ passes through the point $(15, -12)$ and obtain the other points from which normals can be drawn to pass through $(15, -12)$.

Write down the coordinates of the mean centre of the triangle formed by the three points on the parabola, the normals at which pass through $(15, -12)$.

(7) The minimum safe distance in feet, between the front bumpers of consecutive vehicles on one lane of main roads is $18 + v + v^{2/32}$, where v ft/s is the average speed of the traffic. Express, in terms of v, the greatest number of vehicles that can safely pass per hour at v ft/s, if this speed is observed. Determine the maximum value of this greatest number as v varies, stating the corresponding average speed and the distance between vehicles.

Sketch, for values of v from 0 to 80, a graph of the number of vehicles passing per hour against v and comment briefly on the significance of the results.

(8) Sketch roughly the curve $y = \sin x \cos^2 x$ from $x = 0$ to $\frac{1}{2}\pi$.

Determine the area enclosed between the curve $y = \sin x \cdot \cos^2 x$, and the part of the x-axis between $x = 0$ and $\frac{1}{2}\pi$.

If this area is revolved about the x-axis, determine the volume of revolution so formed.

Determine the x-coordinate of the mean centre of the same area.

(9) A is the point $(\sqrt{3}, 0)$ and B the point $(-\sqrt{3}, 0)$. Sketch the complete locus of the point P which moves so that angle $APB = 60°$. Obtain equations for this locus.

Obtain the equation of the common tangent to this locus nearest to the point A. If this common tangent touches the locus at C and D, determine (i) the area, (ii) the perimeter, of the region bounded by the tangent CD and the part CBD of the locus.

(10) The gnomon of a sundial (the vertical part which casts a shadow) consists of a triangular lamina ABC; AC is horizontal with A due South of C, BC is vertical, AB is l units long and angle $BAC = \lambda°$. When the sun is at an altitude $\theta°$ and shines from a direction South $\phi°$ West, the shadow of B on the horizontal plane through AC is B'. Prove

$$\tan CAB' = \frac{\tan \lambda \cdot \sin \phi}{\tan \theta + \tan \lambda \cos \phi}.$$

Consider now, instead, a gnomon in the shape of an isosceles triangle ABC, again in a vertical plane. $AB = BC = l$, the angle $BAC = \lambda°$, AC is horizontal with A due South of C. Prove that, with the sun shining from the same direction as before, the area of the shadow cast by triangle ABC is

$$\frac{\frac{1}{2}l^2 \sin 2\lambda \sin \phi}{\tan \theta}.$$

2 Suppose it was thought good to test the candidates in 'drill' questions. Separate out those parts of the questions on the paper reprinted above which might be thought to be good tests of drill.

3 Write a flow-diagram for teaching enough *geometry* to do the above paper. Is it a satisfactory test of three-dimensional geometry?

4 Write a flow-diagram for teaching enough *calculus* to do the paper. Expand your 'boxes' as in Case Studies 3 and 4 of Chapter 14 to show the detail necessary for a College student as compared with a sixth former.

5 Criticise the reply of the teachers (p. 334) concerning weaker pupils. First, find examples of jobs or College courses where it is more acceptable to have weak mathematics than nothing (they do exist).

6 Criticise the questions on applied mathematics given in Chapter 19 from the point of view of *this* chapter, and suggest alternative questions and assessment procedures.

7 Criticise the examination papers contained in the Appendix (p. 343) from the points of view of this chapter and Chapter 19.

SOME FURTHER READING

ATM [3, 4], Hooper [1], NCTM [5], NSSE [1], Schools Council [4–8, 10], Wrigley [1]

Epilogue

This book is an attempt to establish an intellectual framework for mathematical education, which we view as a subject sufficiently specialised in its own right to be distinct from other disciplines. To this end, we have summarised in the text what seem to us the important social factors that have accompanied and interacted with the growth and teaching of mathematics.

We prefer to look at the basic problem not in the succinct but vague form: 'How should we communicate this piece of mathematics?', but as 'These pupils will have such and such a function in society, within the limits of prediction.† What mathematical equipment will best help them to perform that function? How can we best supply that equipment?' This means that we must be able to look within the store of 'official' mathematics, to see what it has to offer. If we find a piece of material that might be suitable for the pupils, we must then ask whether it can be communicated in some form to them.

With any attempt at 'coding' mathematics for pupils, there arises the problem of testing them for understanding. Such testing should be robust enough to withstand the usages that society may make of it, although it should never be forgotten that its main object is to see whether the pupils have 'understood'. (The word 'understood' must, of course, be related to both the cognitive and affective domains.) This problem of communication must be related to the question 'Is the teaching process we are adopting compatible with other aspects of the society's systems?' If it is not, it may not 'take' or it will be absorbed in a garbled form.

This 'coding' problem is not simply one of replacing a refined mathematical language by a cruder (but still mathematical) one, like the conventional teaching of calculus to engineers. With all types of pupil, the final teaching language may have to take account of their *social* language: it is no good using the language of mandarins to the children of factory workers, as studies by teachers of English have

† This is quite different from fixing their role in a static order.

shown. For example, the early SMP texts T and $T4$ were written in the language of mathematical specialists, intent on getting the mathematics right. These were rewritten in the language of grammar school boys, and the resulting books 1–5 were again rewritten (with modifications) in the language of 'CSE' children, as books A–H. In France the corresponding work has not been done, and the official texts of 'mathématiques modernes' are translations from Bourbaki into the language of the Lycée Professeur with the associated mandarin conventions about choice of topic, about definitions and consistency. As such it is not surprising that they are not 'taking' among the majority of French schoolchildren, and a democratic gesture of 'mathematics for all' has misfired. Some African countries have met the same problem when borrowing European or American project materials. It has been found necessary to translate them into the language and conceptual framework of the 'borrowing' culture (see Phythian [1] and Howson [5]).

Of course, we have used several terms in the paragraphs above without proper analysis. Some are analysed in the text, but some are not and need further research: what precisely is a 'piece' of 'official' mathematics? Nevertheless, our questions make sense at an intuitive level, sufficiently to act as a guide for future work. A list of 'Problems in Mathematical Education' is to be found in Long *et al.* [1], to which we refer the interested reader, but we would add the cautionary note that they are concerned more with the 'coding' problem itself and not with the systems into which the code needs to be fitted.

Finally, what we have done is all very theoretical, yet our preference is to supply a mode of thought for the teacher who aspires to be in Stage 4 of the Beeby model (p. 62): thus, he will know the blueprints, be able to choose an appropriate one and – above all – modify it or create an improved one to suit the particular circumstances of his classroom. After all, it is his professional responsibility, acting as an individual, to make it possible for his pupils to renew both mathematics and the society that nurtured them.

Appendix

Some Specimen Examination Papers

So as to provide further examples for practice, comment and criticism, and in order to give those readers unfamiliar with all the various levels of the British educational system a better indication of the standards involved, we reprint below a selection of papers set in recent years in public examinations. Some of the rubrics (instructions to candidates) are worth noting.

The papers are:

A. Examples of the papers taken by boys of about 13 years attempting the scholarship examinations at Winchester College – one of England's leading public schools. All boys take Paper I. Paper II is one of the three or more 'alternative' papers which boys must take. The purpose of these papers is talent spotting rather than certification.

B. A Mode 1 CSE paper (see p. 41). Candidates take two papers, and specimens of both types are included. (For reasons of space, we have not copied the exact lay-out of these papers and, in particular, the 'spaces' left for the answers have been omitted.) The syllabus used is a 'modern' one based on the course to be found in the CSE books of the SMP.

C. A Mode 3 CSE paper (see p. 41). This paper was supplemented by one that was set in a traditional manner on the lines of that reproduced on p. 350.

D. A 'traditional' GCE O-level examination set on Geometry. The complete examination included similar papers on Algebra and Arithmetic and candidates would need a certain minimum mark in each paper to pass the examination in mathematics. This sort of 'tri-partite' examination is going out of favour and 'mixed' examinations (see E) are now more popular.

E. A GCE O-level examination on a 'traditional' syllabus. Candidates take two papers, both of which are designed on similar lines.

F. A 'commercial' mathematics examination at GCE O-level. Again, this is one of two roughly identical papers.

G. A 'special' examination devised for a particular school. This paper was complemented by one set on traditional lines although on a modern syllabus.

H. An A-level GCE paper set on a modern syllabus in an unconventional manner. The syllabus includes both pure and applied mathematics. Candidates also take a more traditional-looking paper consisting of 'essay-type' problems (see p. 330). This paper is taken by candidates taking 'single-subject' mathematics (i.e. those devoting about seven periods a week to mathematics).

I. An A-level GCE 'Applied Mathematics' Paper. This paper is intended for candidates specialising in mathematics in the sixth form (i.e. devoting about twelve periods a week to mathematics). The syllabus followed is one of the 'modern' ones. (Readers are reminded that such specialists would also take a paper in 'Pure Mathematics'; and one such paper – set on a traditional syllabus – is to be found on pp. 335–9.)

J. A specimen, alternative A-level paper on numerical methods and computing. Such a specimen paper, together with an examination syllabus, is always produced some years prior to the first examination.

K. An examination paper taken by those who wish to enter Cambridge University to study mathematics. The reader will notice that this examination is concerned not with certification but with classification. There are therefore few questions containing bookwork and many which demand initiative and insight.

Note. Those readers who are unfamiliar with the British system are reminded that many of the questions to be found in these papers – particularly those which examine 'traditional' material – recur in disguised form, year after year. In short, they are stereotyped and, accordingly, candidates will be well-practised in their solution, to the unfortunate exclusion of other mathematics and skills.

A

WINCHESTER COLLEGE ELECTION

Monday 22 May 1972. 2.15 *p.m. to* 3.45 *p.m.*

MATHEMATICS I

[*Time allowed:* $1\frac{1}{2}$ *hours.*]

1. X is a set containing $\{a, b\}$ as a subset, but X is not $\{a, b\}$. X is, however, a subset of $\{a, b, c, d, e\}$, though X is not $\{a, b, c, d, e\}$. How many different sets X can there be? Make a list of them.

2. [*In this question no slide rules or tables are allowed. Your method must be clearly shown.*]

   Considering only positive values, which is the greatest and which is least of: the square root of 2; the cube root of 3; the sixth root of 6?

3. If $y = \dfrac{t - x}{1 + tx}$, express x in terms of t, y.

4. (i) Find by experiment a rule which you think gives all the values of m, such that $2^m + 1$ may be exactly divisible by 3. [Try about a dozen values of m.]

   (ii) What do you think is the rule giving all the values of n such that $2^n + 1$ may be exactly divisible by 5?

   (iii) Assuming you have found correct rules (you need not prove them), explain why 15 cannot be a factor of $2^q + 1$, whatever the value of q.

5. In Figure 1, AM and BN are perpendicular to the line PQ. $AM = 1.5$ cm, $MN = 2$ cm, $BN = 2$ cm. The shaded circle has centre B and radius 1.5 cm. Take a fresh sheet of paper and turn it sideways so that a long side is the bottom. Using ruler and protractor, make an accurate version of Figure 1, placing it near the centre of the so-called bottom of your rotated sheet of paper.

345

Sketch tidily, *but free-hand if you wish,* as many circles as you can find which touch the line PQ and the shaded circle, but also pass through A.

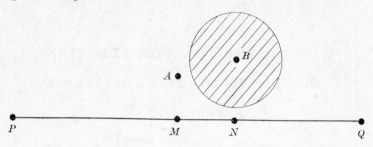

6. I have some home-made lemonade and some jars to put it in. If I put 4 pints into each jar that I use, I have 5 jars unused. If I put 7 pints into each jar that I use, I have 11 jars unused. How many jars do I have?

7. [*In this question no marks will be given for a solution which only uses accurate drawing and measurement. Calculations are required.*] In Figure 2, HE is a girder 12 metres long, hinged to the ground at H. The end E is attached to a taut cable which winds off a reel of cable R. The reel R can move on an overhead horizontal rail which is 15 metres above the ground. Originally the reel R is directly above H, and the cable is perpendicular to the girder. E is then lowered to the ground at F (which is directly under the rail), and R is moved 8 metres along the rail to S in the general direction away from F. The cable remains taut. How much extra cable unwinds from the reel?

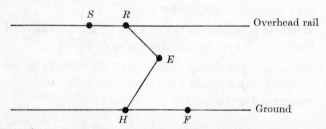

8. Figure 3 is inaccurately drawn, except that A, B, C, D, E, F, G are distinct (i.e. separate) points, and the lines drawn are straight. The marked angles ($\angle BAG$, $\angle GAF$, and the four others) are all meant to be equal. What size should they be? [Make your method clear.]

346

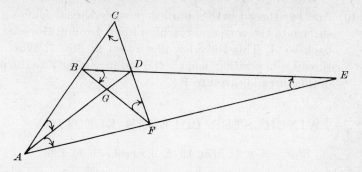

9. The ownership of the Juggernaut Construction Co. is split into 120,000 shares. Mr Moneybags owns 900 of these shares and they are valued by people who would like to buy them at £3 each.

On 3 May 1972 the Juggernaut Company made an offer to people who would own shares in it on 3 July 1972. The offer was to sell to them, on 20 July 1972, at a cost of £2 per new share, 1 new share for each 6 old shares they would own on 3 July 1972.

Mr Moneybags will have no spare cash to buy these offered shares. He has decided to sell today (22 May 1972) some of his present shares (at £3 each) to get exactly enough cash to buy (at £2 per share) all the new shares to which he will then be entitled in July.

How many shares should Mr Moneybags sell today?

10. A circle of radius 8 cm is cut into 3 equal sectors, one of which, $VABC$, is shown shaded in Figure 4. $VABC$ is then taken up, and VA is laid along VC to form a cone with circular base as illustrated in Fig. 5. B is the point on the perimeter of the base diametrically opposite to A (and C).

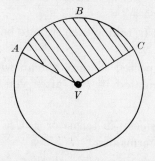
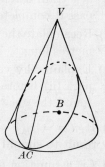

(i) Find the length of the shortest thread lying on the curved surface of the cone that will reach from A to B.

(ii) Another thread is the shortest possible thread lying on the surface of the cone and reaching from A round the cone and back to A. This thread is illustrated in Fig. 5. When this thread is in position, what is the distance from V to the point of this thread nearest to V?

WINCHESTER COLLEGE ELECTION

Wednesday 24 May 1972. 9 a.m. to 10.30 a.m.

MATHEMATICS II

[Time allowed: $1\frac{1}{2}$ hours.]

1. In a certain essay competition the essays are to be typed and each placed in its own envelope together with its writer's identity badge. In this way the judges will not know who wrote which essay. The identity badges are to be square plates of metal with sides of 3 cm. The plates will be ruled *on both sides* into 9 squares each side of 1 cm. In each of these smaller squares a hole of diameter 0.5 cm can be punched out only *exactly centrally* in a smaller square. By punching 0, 1 or 2 such holes in each plate how many distinguishably different identity badges can be produced? Show sketches of as many as you can find.

2. (i) Two trains, each 90 metres long, are travelling in the same direction on parallel tracks with steady (but different) speeds. They are observed to take 15 seconds and 10 seconds respectively to pass a certain point on the line. When the fast train catches up the slow one, how long will it take to pass it completely?

   (ii) Recalculate the answer to part (i) taking each train to be 180 metres long.

   (iii) Explain why you could be expected to know the answer to part (ii) before you had calculated it, but after you had found the answer to part (i).

3. (a) If $g(x) = 1/(1 - x)$, simplify $g(g(x))$ in terms of x. Use this result to simplify $g(g(g(x)))$ in terms of x.

   (b) f is a function operating on numbers. f is such that, for all values of x, we have $f(f(x)) = x$; for example, we might have $f(x) = -x$. Without using any numerals (except, if you

348

wish, 1) write down some other possible expressions for $f(x)$, not equal to the one given and not equal to each other. [Do not look for more than 4 more.]

4. Rectangular blocks of margarine have square ends of side a, and are of length x. They are completely wrapped in paper, but with no overlaps of paper. Each block has its paper peeled off and laid flat. The margarine in each block is then spread on its own paper to form a layer of uniform thickness, y. In the following problems a is a fixed quantity, but x may vary (thus the blocks do not all have the same volume).

   (i) What happens to the values of y if blocks are chosen with smaller and smaller values of x?

   (ii) What happens to the values of y if blocks are chosen such that x gets larger and larger without limit?

   (iii) What is y for a block that was originally a cube?

5. (i) Study the following argument closely:
$$7 \times 143 = 1001,$$
$$\therefore \begin{cases} 7 \times 0.143 = 1.001 \\ \text{and } 7 \times 0.000143 = 0.001001. \end{cases}$$
From the last two lines it follows that
$$7(0.143 - 0.000143) = 1.001 - 0.001001,$$
and hence that
$$7 \times 0.142857 = 0.999999.$$
Thus
$$\tfrac{1}{7} = 0.\dot{1}42857.$$

   (ii) Replace the first line of the above argument by
$$13 \times 77 = 1001,$$
and then, showing working of exactly the same kind as in part (i), express $\tfrac{1}{13}$ as a recurring decimal. [No marks will be given for merely doing a division sum.]

   (iii) Use $73 \times 137 = 10001$, in the same way, to find the recurring decimal for $\tfrac{1}{73}$.

6. (i) You are given that $ab > a + b - 1$ for all values of a, b such that $a > 1$ and $b > 1$. Assuming this, show that $abcd > a + b + c + d - 3$ for all values of a, b, c, d each greater than 1.

   (ii) If you can prove that $ab > a + b - 1$ for $a > 1$, $b > 1$, do so.

349

7. A circle is inscribed in a triangle ABC; that is, the sides of the triangle (none of which have been produced) all touch the circle. BC touches the circle at D, and CA touches the circle at E. Prove that

   (i) $BD = BC - CE$.

   (ii) $BD = \frac{1}{2}(AB + BC - CA)$.

8. A piece of chain is to be split into 2 pieces by cutting and destroying a single link. One of the resulting pieces (it matters not which) is to be at least 3 times as long as the other.

   (i) If the original chain has 12 links, then in how many ways can the link-to-be-cut be chosen?

   (ii) Repeat part (i) with the original chain having 93 links.

   (iii) How many more links than 93 would there need to be in the original chain before there was a larger possible number of links-to-be-cut than in part (ii)?

   (iv) Give a list of arithmetical instructions which would enable someone to calculate without any trial-and-error in how many ways the link-to-be-cut could be chosen once the number of links in the original chain was known.

B

SOUTHERN REGIONAL EXAMINATIONS BOARD

CERTIFICATE OF SECONDARY EDUCATION EXAMINATIONS

MATHEMATICS
(Syllabus R)

PAPER 1

Monday 4 May 1970: morning

TIME ALLOWED: 2 HOURS (MINIMUM), 3 HOURS (MAXIMUM).

The invigilator should read these instructions to the candidates immediately before the examination begins.

(i) All questions to be attempted. Write your answers in the spaces provided.

(ii) Any working must be done on the page opposite the question and should not be crossed out.

(iii) Do not spend too long on any questions you find difficult.

(iv) Mathematical tables, slide rules, or calculating machines may be used.

1. What will be the total value, in new pence, of the six new coins of the realm?

2. Write 5.386 correct to (*a*) 2 decimal places, (*b*) 2 significant figures.

3. Sketch the triangle with vertices at (2, 1), (5, 1) and (4, 2) and find its area.

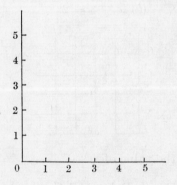

4. About how many German Marks (D.M.) can I buy for £5 if 1 D.M. is worth $10\frac{1}{2}$ p?

5. Find the value of $\frac{1}{3} + \frac{2}{5} + \frac{1}{4}$.

6. (*a*) Sketch the image of the point (3, 5) after reflection in the line $x + y = 0$.

   (*b*) Write down the coordinates of this image.

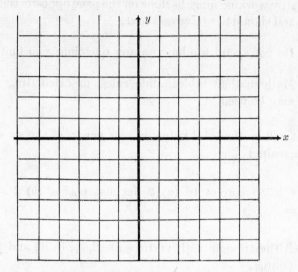

7. Compile the addition table for modulus 4 and solve the equations:

   (*a*) $x + 3 = 2$.   (*b*) $x - 3 = 2$.

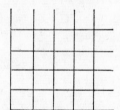

8. $2^x = 64$. Find the value of x.

9. (*a*) Add 1 to 355_6. (*b*) Subtract 1 from 100_2.

10. $A = \begin{pmatrix} 2 & 3 \\ -4 & 1 \end{pmatrix}$ $B = \begin{pmatrix} 3 & 1 \\ 2 & -2 \end{pmatrix}$.

    Complete these statements

    (*a*) $A + B = \begin{pmatrix} & \end{pmatrix}$ (*b*) $AB = \begin{pmatrix} & \end{pmatrix}$

    (*c*) $A^2 = \begin{pmatrix} & \end{pmatrix}$.

11. The numbers thrown on two dice can be expressed as sets of ordered pairs (p, q). These can be arranged in tabular form as follows:

		q					
		1	2	3	4	5	6
p	1						
	2						
	3		(3, 2)				
	4					(4, 5)	
	5						
	6						

(a) Complete this table.

(b) In how many ways can a pair of numbers be chosen whose sum is odd?

(c) What is the probability that the number 7 will be the sum of the pair of numbers thrown?

(d) What is the probability that $p - q$ will be negative?

(e) What is the probability that p/q will be equal to or less than one?

12. The square $ABCD$ is mapped onto the quadrilateral $ABPC$ by the shear S.

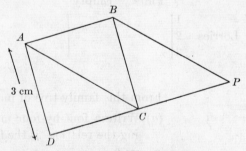

(a) What kind of quadrilateral is $ABPC$?

(b) Write down the area of quadrilateral $ABPC$.

(c) Calculate angle BPC.

(d) Calculate the length of BP, giving your answer correct to 1 decimal place.

353

(e) On the diagram, sketch the image of the quadrilateral $ABPC$ under the shear S.

13. (a) Each interior angle of a regular polygon is 140°. How many sides has the polygon?

(b) If $P = \{$regular polygons$\}$ and $Q = \{$quadrilaterals$\}$. What is $P \cap Q$?

14.

x	1	2	3	4	5	...
y	8	11	14	17	20	...

Find a relation between x and y.

15.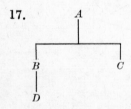

(a) Find the value of (i) tan B. (ii) sin B.

(b) Calculate the shortest distance from A to BC.

16. Use matrices A and B to write the correct figures in C.

$$A = \begin{array}{c} \\ \text{Lorries} \end{array} \begin{array}{c} 1 \\ 2 \\ 3 \end{array} \begin{bmatrix} 60 & 0 & 0 \\ 40 & 10 & 10 \\ 30 & 15 & 10 \end{bmatrix}$$

Load (crates): Beer, Cider, Lemonade

$$B = \begin{array}{c} \text{Beer} \\ \text{Cider} \\ \text{Lemonade} \end{array} \begin{bmatrix} 12 & 5 \\ 10 & 4 \\ 9 & 4 \end{bmatrix}$$

Weight (kg): Full, Empty

$$C = \begin{array}{c} \\ \text{Lorries} \end{array} \begin{array}{c} 1 \\ 2 \\ 3 \end{array} \begin{bmatrix} \\ \\ \end{bmatrix}$$

Weight (kg): Full, Empty

17.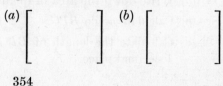

From this family tree of males,

(a) Write a four-by-four matrix describing the relation 'is the father of' and,

(b) Write a four-by-four matrix describing the relation 'is the son of'.

(a) $\begin{bmatrix} \\ \\ \\ \end{bmatrix}$ (b) $\begin{bmatrix} \\ \\ \\ \end{bmatrix}$

18. Solve the inequality $2 > x - 2 \geqslant 5$ and indicate the result on this line graph.

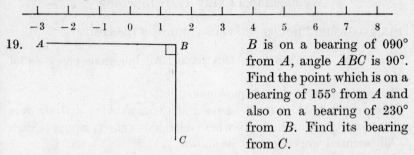

19. B is on a bearing of $090°$ from A, angle ABC is $90°$. Find the point which is on a bearing of $155°$ from A and also on a bearing of $230°$ from B. Find its bearing from C.

20. A drawer contains four blue pullovers and two white ones. John takes one pullover from the drawer. Then Bob takes one, too. Both boys make their selection without looking at the colour.

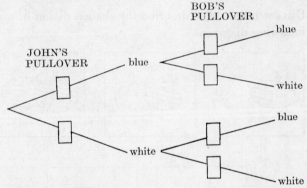

(a) Complete this tree diagram by writing the appropriate probability on each branch in the box provided.

(b) Use the diagram to find the probability that John and Bob:
   (i) each select a blue pullover,
   (ii) each select a white pullover,
   (iii) select pullovers of the same colour.

21.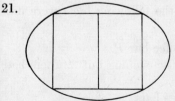

(a) State the number of regions, arcs and vertices (nodes) in this network.

(b) The diagram is to be coloured according to the rule that no two regions with a common arc are the same colour. What is the minimum number of colours required?

Show your colour scheme on the diagram.

PAPER 2

Wednesday 12 May 1971: morning

TIME ALLOWED: 2 HOURS (MINIMUM), 3 HOURS (MAXIMUM).

There are 10 questions in this paper. All questions carry equal marks.

Answer not more than 6 questions.

The answer booklet must be used for your working and answers except for questions 4 and 10 where detachable sheets are provided.

All essential working must be shown.

You may use mathematical tables, slide rules or calculating machines, but if slide rules are used this should be shown by writing (S.R.) after your answer.

1. (*a*) Use set notation to describe the shaded region in each of the following diagrams.

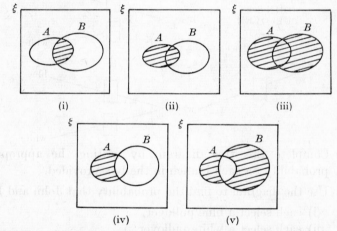

(*b*) The police were investigating the passing of counterfeit money at coffee bars P, Q and R, on the evening of 4 May. 11 people were suspected and the following facts were discovered.

5 of the suspects had visited coffee bar P.

4 remained in coffee bar R all the evening but 3 others had looked in.

These 3 had visited all the coffee bars.

6 had visited Q.

1 had been out of town all evening.

356

   (i) Draw a diagram to show the above information.

  (ii) Find (1) $n(P \cup Q)$,
               (2) $n(R')$,
               (3) $n(Q \cap R \cap P')$.

 (iii) Find the number of suspects who visited both P and Q but not R.

  (iv) Express this number in set notation.

2. A breakfast cereal called 'Krispweet' is marketed in two sizes:

   'Standard'   25 cm by 18 cm by 7 cm,
   'Family'   30 cm by 21 cm by 7 cm.

The contents of both fill the packets.
The standard size costs the manufacturer 10p made up as follows: contents 5p, packaging 1p, overheads 4p.

  (i) What percentage profit on his cost price does a manufacturer make on a 'Standard' packet which he sells to a retailer for 12p?

 (ii) The retailer makes a $37\frac{1}{2}\%$ profit on the 12p he pays for this packet. What does the consumer pay?

 (iii) How much does the cereal in a 'Family' packet cost the manufacturer?

 (iv) The 'Family' packet retails at 28p. Does a housewife get better value for money if she buys this packet? Show your working.

3. 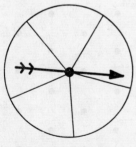 The diagram shows a fixed disc and a rotating arrow, used in a spinning game.

  (a) 5 girls draw different numbers to decide the order of play. How many possible orders are there?

  (b) The circular disc has only red and blue sectors such that the probabilities of landing on the red and blue sectors are in the ratio of $3:2$.

Draw a circle of radius 3 cm to represent the disc and show, accurately, a possible arrangement of the red and blue sectors. [Mark each sector RED or BLUE.]

(c) If the arrow is spun freely, what is the chance of
  (i) landing on a blue sector,
  (ii) landing on a red sector with two consecutive spins,
  (iii) landing on different colours with two consecutive spins?

(d) The rules of the game state, for each player's turn
  stopping on RED–score 3,
  stopping on BLUE–score 2 and spin again unless you have had 4 spins already.

List the scores which it is possible to obtain with just one turn.

4. ANSWER ALL PARTS OF THIS QUESTION ON THE DETACHABLE PAGE PROVIDED.

(a) What is the size of an interior angle of a regular hexagon?

(b) Regular hexagons can be used to form a tessellation.
Use the lattice paper provided to construct part of the pattern accurately.

Imagine that the pattern is extended to fill the whole plane and

(c) on your diagram, mark
  (i) a centre of rotational symmetry of order 6 (label it S),
  (ii) a centre of rotational symmetry of order 3 (label it T),

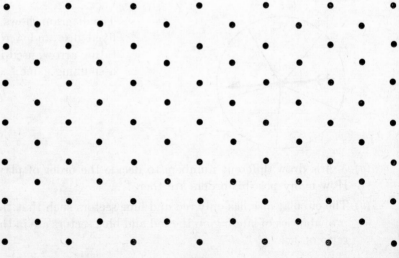

(iii) a centre of rotational symmetry of order different from that of S and T (label it U).

(d) State the order of rotational symmetry about the centre U.

(e) State the number of lines of symmetry which pass through the centre of each hexagon.

5.

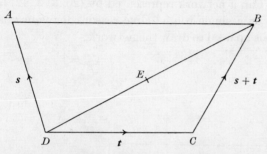

(a) If DE is $\frac{1}{3}DB$ express in terms of s and t,

    (i) $\overrightarrow{DB}$ (ii) $\overrightarrow{DE}$ (iii) $\overrightarrow{AD}$ (iv) $\overrightarrow{AB}$ (v) $\overrightarrow{CE}$ (vi) $\overrightarrow{AE}$.

(b) What does the result of (iv) tell you about the lines AB and DC?

(c) What does the result of (v) and (vi) tell you about the points A, C and E?

(d) State the ratio of $AE:EC$.

6.

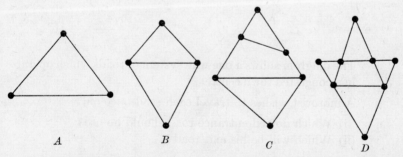

     A   B   C   D

(a) Give the letter of each of the above networks which can be traversed by a single route without travelling more than once along any arc.

(b) In which of the above networks could such a route begin and end at the same node?

359

(c) If → means 'may be represented by', we may write
Network $A \rightarrow (2, 2, 2)$
Network $B \rightarrow (2, 3, 3, 2)$.

  (i) Copy and complete
    (1) Network $C \rightarrow$
    (2) Network $D \rightarrow$

  (ii) Can a network represented by $(20, 3, 3, 42)$ be traversed by a single route? Give reasons for your answer. There is no need to draw the network.

(d)

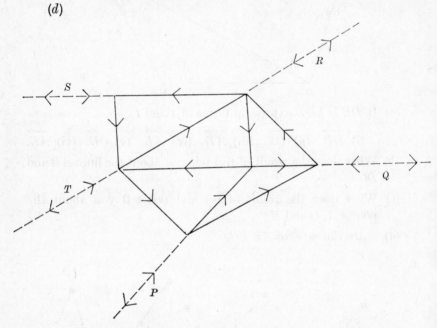

The diagram shows a one way system of roads which operates in a congested town centre.

A motorist wishes to travel each road once only.

  (i) Which dotted entrance road should he use?
  (ii) Which will be his exit road?

7. (a) Find the values of a, b, c, d, e, f in the following equations.

  (i) $\begin{pmatrix} 3 & 5 \\ 1 & 2 \end{pmatrix}\begin{pmatrix} 2 \\ 3 \end{pmatrix} = \begin{pmatrix} a \\ b \end{pmatrix}.$

  (ii) $\begin{pmatrix} 3 & 5 \\ 1 & 2 \end{pmatrix} + \begin{pmatrix} 2 & e \\ f & -3 \end{pmatrix} = \begin{pmatrix} c & 7 \\ -5 & d \end{pmatrix}.$

360

(*b*) The petrol sales in litres at 2 garages were as follows

$$\begin{array}{c} \text{4 star} \qquad \text{2 star} \\ \begin{matrix} \text{Garage A} \\ \text{Garage B} \end{matrix} \begin{pmatrix} 300 & 100 \\ 200 & 200 \end{pmatrix}. \end{array}$$

  (i) 4 star petrol was sold at 10p per litre and 2 star petrol at 9p per litre. Write this information as a 2 by 1 matrix.

  (ii) Multiply the two matrices to form a 2 by 1 matrix $\begin{pmatrix} x \\ y \end{pmatrix}$.

  (iii) What is the value of $x + y$? What does it represent?

  (iv) 4 star petrol gave a profit of 1p per litre, but 2 star petrol gave a profit of 2p per litre. Write this information as a 2 by 1 matrix.

  (v) Multiply the matrices which will give the profit made by each of the garages and state these profits.

8.

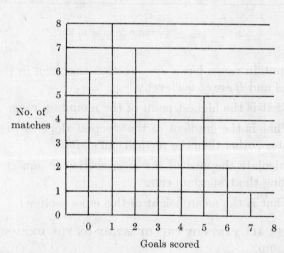

Goals scored

(*a*) The chart shows the frequency distribution of goals scored by Exton during part of the season. From the chart find

  (i) the total number of matches played.

  (ii) the mean number of goals per match.

  (iii) the probability that a spectator who watched only one match during this part of the season, saw Exton

    (1) score exactly 4 goals,

    (2) score at least 4 goals.

(b) How many goals must Exton score in the remaining 4 games if the team is to average 2.6 goals per match for the whole season?

9.

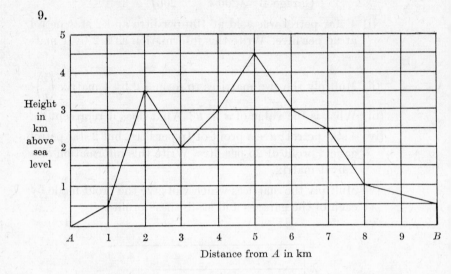

Distance from A in km

A mountain range has a cross section as shown in the diagram. Both A and B are at sea level.

(a) What is the highest point of the mountain range in metres?

(b) What is the gradient of the steepest slope in the mountain range, other than the vertical cliff at B?

(c) Calculate the area of the cross section in square kilometres using the trapezium rule.

(d) What is the mean height of this cross section?

10. ANSWER ALL PARTS OF THIS QUESTION ON THE DETACHABLE PAGE PROVIDED.

A shopkeeper buys x metres of material at £1 per metre and y metres of material at £2 per metre.

(a) If he wishes to buy not more than 250 metres of material, write an inequality connecting x and y.

(b) If he has £400 available, write another inequality connecting x and y.

(c) Represent these inequalities on the graph paper provided, shading out the regions you do not require.

(*d*) The shopkeeper makes a profit of 50p per metre on the cheaper material and 75p per metre on the more expensive material. Use your graph to find the number of metres of each which must be bought to give the maximum profit. State this profit.

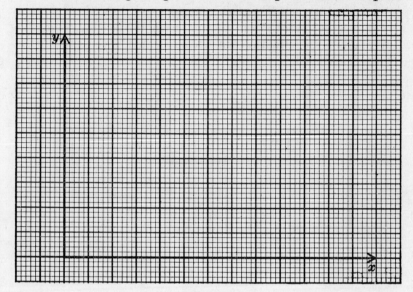

C

METROPOLITAN REGIONAL EXAMINATIONS BOARD

for the Certificate of Secondary Education

1967

PAPER II(b)

MODE 3 EXAMINATION TO BE TAKEN BY ABBEY WOOD SCHOOL

4 *May* 1967 – 9.30 *a.m.*

(No time limit)

Answer ONE question only.

1. Write about the mathematics of a chessboard (or draught board).
2. Investigate the set of numbers
   1, 1, 2, 3, 5, 8, 13, 21, 34, . . .

3. A polygon is to be made from a piece of string 24 inches long. Investigate the possibilities.

4. When there were 5 houses in the school, 3 houses were in the hall for assembly each Wednesday, in rotation. Now there are only 4 houses, 2 houses are in the hall each Wednesday. Compiling a rota for either situation can be complicated. Discuss the mathematics involved.

5.

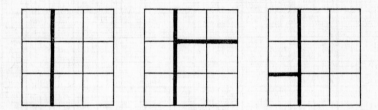

The diagrams above show dissections of a 3-unit square into rectangles. (N.B., {squares} ⊂ {rectangles}) The rules are

 (i) in each dissection the rectangles are all different;
(ii) all the edges are a whole number of units long.

The diagram below shows a similar dissection of a 4-unit square.

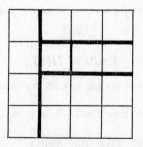

Investigate other dissections of a 4-unit square, using the same rules. (Credit will be given for a systematic approach.)

364

D

Universities of Manchester, Liverpool, Leeds, Sheffield and Birmingham

Joint Matriculation Board

GENERAL CERTIFICATE OF EDUCATION

Mathematics Ordinary Syllabus A
Paper III   Geometry

Monday 19 June 1972   9.30–12
Careless work and untidy work will be penalised.

Answer all questions in Section A and four questions from Section B.

In calculations sufficient steps in the working must be shown to make clear how the calculations have been performed.

The use of slide rules is permitted. Write the symbol SR at each point where such use has been made.

Section A   *Answer all questions in this section*

A1   (*a*) The angles A, B, C of a pentagon $ABCDE$ are respectively 90°, 100°, 110°, and the angles D and E are equal. Calculate these equal angles.

(*b*) The diagonals of a rhombus are 12 cm and 16 cm long. Calculate the length of a side of the rhombus.

(*c*) PQ is a diameter of a circle, centre O. Points R and S on the circle are such that $\angle QPR = 25°$ and SR is parallel to PQ. Calculate $\angle PQR$ and $\angle POS$.

A2   (*a*) A point P is 2 cm from O, the centre of a circle of radius 6 cm. A chord APB is drawn through P and $AP = 5$ cm. Calculate the length of PB.

(*b*) In the parallelogram $ABCD$, the side $AB = 5$ cm and the side $AD = 8$ cm. The perpendicular distance between AB and DC is 4 cm. Calculate the perpendicular distance between AD and BC.

365

(c) $ABCD$ is a quadrilateral in which AB is the longest side and CD is the shortest side. Prove that

(i) $\angle BDA$ is greater than $\angle ABD$,

(ii) $\angle ADC$ is greater than $\angle ABC$.

A3 (a) Prove that the perpendicular to a chord from the centre of a circle bisects the chord.

(b) In the triangle ABC the middle point of BC is D and the middle point of AD is E. The line BE is produced to meet AC at F, and G is the point on AC such that DG is parallel to BF. Prove that $AF = FG = GC$.

A4 Using ruler and compasses only, construct in a single diagram

(i) the equilateral triangle PQR in which $PQ = 7$ cm,

(ii) the triangle PQS equal in area to the triangle PQR so that S lies on the same side of PQ as R and $\angle PQS = 135°$,

(iii) the point X on PS which is equidistant from QR and QS. Measure QX.

Section B *Answer four questions from this section.*

B5 The acute-angled triangle ABC, in which AB is greater than AC, is inscribed in a circle. The chord BD is parallel to the tangent at A. The line DC produced and the tangent at A meet at E. Prove that

(i) the triangles ABC and EAC are similar,

(ii) $AD = AB$.

B6 Prove that the angle subtended at the centre of a circle by an arc is twice the angle subtended by the same arc at any point on the remaining part of the circumference. The isosceles triangle ABC with $AB = AC$ is inscribed in a circle. With A as centre and AB as radius a second circle is drawn, and a line through B cuts the first circle at L and the second circle at M, the point L lying on the minor arc AC. Prove that

(i) $\angle BLC = 2 \angle BMC$,

(ii) $LC = LM$.

B7 Prove that in any triangle the sum of the squares on any two sides is equal to twice the square on half the third side together with twice the square on the median which bisects the third side.

In the triangle ABC, the middle points of BC and CA are D and E respectively and $AD = BE$. Prove that

$$BC^2 - CA^2 = 2(AE^2 + DC^2)$$

and deduce that the triangle ABC is isosceles.

B8 Using ruler and compasses only, construct on one diagram

(i) the triangle ABC in which $AB = 7$ cm, $BC = 6$ cm, $AC = 11$ cm,

(ii) the circle with centre on AC which passes through A and B, this circle cutting AC at D,

(iii) the tangent from C to touch the circle at T which lies on the opposite side of AC from B,

(iv) the cyclic quadrilateral $ADTE$ in which AT bisects angle DAE.

B9 The point Q is on the side DC of a parallelogram $ABCD$. The lines AQ and BC are produced to meet at R, and DB and AQ intersect at P. Prove that

(i) the triangles APD and BQP are equal in area,

(ii) the triangles APD and RPB are similar,

and deduce that $\dfrac{PQ}{PR} = \dfrac{PD^2}{PB^2}$.

Given that $3PQ = QR$ and $PD = 3$ cm, calculate PB.

B10 Prove that if two circles touch externally, the line joining their centres passes through the point of contact.

Two equal circles touch externally at X. The first circle has its centre at A and the second circle has its centre at B.

A line through X meets the first circle again at C and the second circle again at D. Prove that

(i) AC is parallel to BD,

(ii) $XC = XD$.

367

E

SOUTHERN UNIVERSITIES' JOINT BOARD

GENERAL CERTIFICATE OF EDUCATION
(Ordinary Level)

MATHEMATICS II

Thursday a.m.   27 June 1968

Time: 2½ hours

Answer **all** *questions in Section I and any* **five** *in Section II.*

All essential working must be shown. Any result obtained by slide-rule should be followed by the letters S.R.
[Marks: Section I: 35; Section II: 65.]

SECTION I

1. (*a*) Ordinary matches are 5d per box of 90. Safety matches are 2d per box of 35 or 1s 8d per dozen such boxes. Which is the dearest of these three ways of buying matches and which the cheapest?

(*b*) When a garment is soaked it absorbs a weight of water equal to its own weight. Wringing disposes of three-quarters of the water. Compare the weights of the garment dry (*D*), soaked (*S*) and wrung (*W*), giving the answer in ratio form *D*:*S*:*W* with the least whole numbers.

2. (*a*) Solve as accurately as four-figure tables allow
$$5x^2 + x - 2 = 0.$$

(*b*) $x = (2n - 4)/n$; express n in terms of x. To which topic in geometry does this formula refer?

3. The tangents to a circle at X and Y meet at T. Angle $XTY = 38°$. XZ is the chord parallel to TY. The tangent at Z meets TX produced at W. Calculate angles TYX, XZY, TYZ, XWZ.

4. In triangle ABC, $AB = 13$ cm, $BC = 15$ cm, $AC = 14$ cm. Calculate (i) cos C and hence sin C, (ii) the area of the triangle, (iii) angle A.

368

SECTION II

Answer any **five** *questions.*

5. A Briton holds £240 in an Australian company. Find the amount of dividend at $2\frac{1}{2}$%.

15% of this dividend is withheld as Australian tax. Find the amount of dividend sent to Britain.

From this remaining dividend British income tax at 8s 3d in the £ is deducted. Find to the nearest penny how much the holder actually receives.

To reclaim the Australian tax it must be expressed in Australian dollars (\$A); do this, the rate being \$A1.00 = 8s 0d.

6. A tent has a cylindrical wall of height $2\frac{1}{2}$ ft and a conical roof of vertical height $10\frac{1}{2}$ ft **above** this. Both have radius 7 ft. Calculate how far from the wall a person of height $5\frac{1}{2}$ ft can stand without touching the roof with his head.

Find the volume of air to each of 12 persons in the tent, and the total area of canvas making the walls and roof; (π should be taken as 22/7).

7. (*a*) Verify that 2/3 is a solution of the equation

$$\frac{x + 2}{2x} - \frac{7x - 2}{x - 2} = 4.$$

Find the other solution.

(*b*) A nurseryman buys 6 plants and at the end of a year divides each one into 4. Every year he does the same with all the plants he has then. Use tables to find how many plants he has when they have been divided 7 times, giving the answer to the nearest hundred.

8. Draw the graph of $y = 1 - x - \frac{1}{10} x^3$ for values of x from -2 to $+3$ taking 1 in as unit on both axes.

Find the one real solution of the equation $x^3 + 10x - 10 = 0$.

Find the gradient (vertical/horizontal) of the graph where it crosses $0y$.

9. In triangle ABC, $AB = 8$ cm, $BC = 10$ cm and A is a right angle. The internal bisector of angle B meets AC at X. The internal bisector of angle C meets AB at Y and its external bisector meets BA produced at Z. Find by calculation without trigonometry, or by drawing with ruler and compasses only, the lengths of AC, AY, XY, AZ.

10. Two chords AB, CD intersect at a point X inside a circle. Prove the theorem $AX \cdot XB = CX \cdot XD$.

If X is the midpoint of AB and the perpendicular to CD at X meets the circle on diameter CD at E and F, prove (i) $XE^2 = XB^2$, (ii) a circle can be drawn with centre X to pass through A, E, B, F.

11. A map is on a scale of $1:1,000,000$. Find how many miles are represented by 1 in, giving the answer to one place of decimals.

The north–south distance between Lizard Point and the northern tip of the Shetland Islands is 760 miles. Find the height of a map sheet containing both these.

The area of Great Britain is 89,000 square miles; find in square inches the area it occupies on the map.

12. An electric chandelier hangs by a chain CA. In the elevation shown, B is one of six bulbs which are the vertices of a horizontal regular hexagon with centre A. What is the distance between a pair of adjacent bulbs?

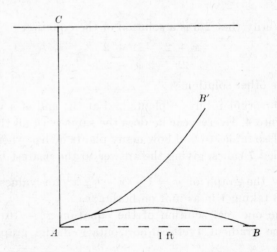

The tube AB is an arc of a circle and makes an angle of $20°$ with AB at A. Find by calculation or drawing the radius of this circle to the nearest inch.

The tube is removed from its socket at A and replaced in the position AB' in the same vertical plane and still at an angle of $20°$ to AB at A. Find the horizontal distance from B' to AC.

F

SOUTHERN UNIVERSITIES' JOINT BOARD

GENERAL CERTIFICATE OF EDUCATION

(*Ordinary Level*)

MATHEMATICS AND STATISTICS FOR COMMERCE II

Thursday a.m. 27 June 1971

Time: $2\frac{1}{2}$ hours

Answer **all** questions in Section A and Section B and **two** questions in Section C. All essential working must be shown. Mathematical tables should be used where appropriate.

Any results obtained by slide-rule should be followed by the letters S.R.

[Marks: Section A: 32; Section B: 34; Section C: 34 – each question 17. Total 100]

SECTION A

1. The following is an extract from a report on Holidays, published by the British Tourist Authority:

'Holiday traffic abroad rose from 1 500 000 people in 1951 to 5 750 000 in 1969. This latter figure represents 15% of the total number of people who took their holidays in Britain. Of those whose holidays were spent in Britain, 62% were taken in July and August.'

(*a*) Calculate the 1969 figure for holidays abroad as a percentage of the corresponding 1951 figure. Give your answer correct to two significant figures.

(*b*) How many people took holidays in Britain in 1969?

(*c*) How many people spent their holidays in Britain in 1969 in months other than July and August?

[Give your answers to (*b*) and (*c*) in the form $A \times 10^n$ where $1 < A < 10$ and n is a positive integer.]

2. If $x = 4500$ and $y = 3400$ and if there is a possible 1% error in both x and y, copy and complete the following:

	Greatest possible value:	Least possible value:
$x + y$		
$x - y$		

3. The following table gives the daily wages and number of men employed in four different grades.

	Grade 1	Grade 2	Grade 3	Grade 4
Daily wage	£3.20	£3.70	£4.20	£5.00
No. of men	300	300	1200	900

Calculate the weighted average daily wage.

4. Southern Rovers football team have to play two matches. They estimate that the probability of winning the first match is 1/4 and the probability of winning the second match is 1/6. Find (i) the probability that they will win both matches; (ii) the probability that they will win the first match but not win the second; (iii) the probability that they will not win at least one match.

SECTION B

5. (a) If $y = x(30 - x)/50$ copy and complete the following table:

x	0	5	10	15	20	25	30
y		2.5			4		

(b) Using a suitable scale on each axis, draw the graph of

$$y = \frac{x}{50} (30 - x).$$

(c) Use your graph to find

(i) the values of x when $y = 3.5$;

(ii) the range of values of x for which $0 < y < 2$.

6. The following table gives the quarterly visible trade balances (in million pounds) for the period 1 July 1969 to 31 December 1970.

	1969		1970			
Quarter	3	4	1	2	3	4
Visible trade balance	+30	+28	+57	−57	−69	+81

(*a*) Calculate the total visible trade balance for the given period.

(*b*) If the total visible trade balance for the whole of 1969 was − 80 million pounds, what was the total balance for the first 6 months of 1969?

(*c*) Draw a suitable bar chart to represent the information in the above table.

On your bar chart draw a line suitably placed to represent the average quarterly trade balance over the given period.

Find the greatest and the least deviation from the average.

SECTION C

Answer **two** questions

7. (*a*) Show the following two series of numbers graphically so that any correlation between them is brought out.

A	5	10	7	8	13	14	12	18	16	21	19	25
B	13	17	15	14	14	22	10	4	7	11	8	3

(*b*) Calculate the coefficient of rank correlation between the two series of numbers.

(*c*) Calculate the mean and the standard deviation of series A.

8. The following table gives the cumulative frequencies for a set of candidates in an examination.

Marks	15–24	25–34	35–44	45–54	55–64	65–74	75–84	85–94
Cumulative frequency	20	54	114	170	203	216	228	229

(*a*) (i) How many candidates took the examination?

(ii) How many candidates scored 45 marks or more?

(*b*) Draw the cumulative frequency curve for this distribution. Hence estimate how many candidates scored not more than 50 marks.

(*c*) Which class interval of marks had the greatest number of candidates in it?

9. (*a*) The following table gives the values of a commodity during the period 1957–1967. These values are shown as readings

taken from a graph drawn on ordinary graph paper using a semi-logarithmic scale.

	1957	1959	1961	1963	1965	1967
Reading	2.00	2.30	2.48	2.43	2.63	2.78

(i) By a suitable use of your tables find, correct to two significant figures, the actual value of the commodity in 1957, in 1961 and in 1965. (The units are pounds.)

(ii) During which two-year period was the increase in price the greatest and what was that greatest price increase?

(b) The following table gives the raw jute prices in pounds per ton during the period 1967–1968.

January 1967	January 1968	August 1968	December 1968
140	132	140	159

Using January 1967 = 100, calculate index numbers, correct to three significant figures, to represent the prices in January, August and December 1968.

G

ASSOCIATED EXAMINING BOARD

(*for the General Certificate of Education*)

June Examination, 1969 – Ordinary Level

MATHEMATICS – SYLLABUS C – (Special)

PAPER II

Four hours allowed

Answer ONE question only.

1. The natural numbers can be put into sets according to the number of 'ones' needed when they are written in binary notation. Thus

$$S_1 = \{1, 2, 4, 8, \ldots\}$$
$$S_2 = \{3, 5, 6, 9, 10, 12, \ldots\}$$
$$S_3 = \{7, 11, \ldots\}$$

begins the separation of the natural numbers into their appropriate sets.

Investigate this situation, discussing other rules for finding out to which sets numbers belong.

2. The set C is the set of all ordered pairs $\{(a, b), (c, d), \ldots\}$, where $a, b, c, d, \ldots$ are any numbers. The pairs are added and multiplied according to the following rules.

$$(a, b) + (c, d) = (a + c, b + d)$$
$$(a, b) \times (c, d) = (ac - bd, ad + bc).$$

Investigate.

3.

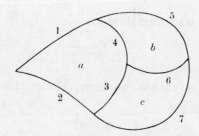

In the network shown in Fig. 1, the arcs are numbered and the regions are labelled a, b and c. The boundary of each region is called a loop; e.g. loop a is formed by arcs 1, 2, 3 and 4.

The three loops may be represented in matrix form thus

$$
\begin{array}{c}
\qquad\qquad\qquad \text{Arcs} \\
\begin{array}{ccccccc}
1 & 2 & 3 & 4 & 5 & 6 & 7
\end{array} \\
\text{Loops }
\begin{array}{c}
a \\ b \\ c
\end{array}
\begin{pmatrix}
1 & 1 & 1 & 1 & 0 & 0 & 0 \\
0 & 0 & 0 & 1 & 1 & 1 & 0 \\
0 & 0 & 1 & 0 & 0 & 1 & 1
\end{pmatrix}
\end{array}
$$

Hence each loop may be represented by a seven-component row-vector, e.g. we may write

$$a = (1, 1, 1, 1, 0, 0, 0).$$

The vectors can be added by adding corresponding elements modulo 2 e.g.

$$a + b = (1, 1, 1, 0, 1, 1, 0).$$

Explain what these sums of vectors mean as far as the network in Fig. 1 is concerned, and investigate this set of vectors under this form of addition.

Investigate the network in Fig. 2 and make general conclusions.

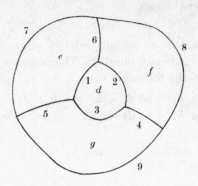

4. Consider the transformation

$$\frac{a}{b} \to \frac{a + 2b}{a + b}$$

where a and b are natural numbers. Under this transformation, starting with $\frac{1}{1}$, we get

$$\frac{1}{1} \to \frac{3}{2} \to \frac{7}{5} \to \frac{17}{12} \to \cdots$$

By writing each fraction in decimal notation, explain what happens if the process is continued.

Investigate the transformation

$$\frac{a}{b} \to \frac{a + nb}{a + b}$$

for various natural numbers n.

5. Write an essay on *Dominoes*.

H

OXFORD AND CAMBRIDGE
SCHOOLS EXAMINATION BOARD

(*on behalf of G.C.E. Examining Boards*)

General Certificate Examination
Advanced Level

SMP 55

SCHOOL MATHEMATICS PROJECT

S.M.P. MATHEMATICS I

Monday 22 June 1970. 3 *hours*

Answer not more than 18 questions.

You are advised to work straight through the paper, omitting questions that you cannot do, and **not** *to spend time reading carefully through the whole paper before you begin.*

1. A particle is projected at 20 m/s at an angle of 50° above the horizontal. Find by drawing and measurement (or by calculation) the velocity and the displacement after 3 s, expressing both results in magnitude and direction. [Take $g = 10$ m/s^2.]

2. Find an expression of the form $a + bh + ch^2$ which approximates to the value of $e^{-x} \sin x$ when $x = \frac{1}{2}\pi + h$, where h is small, evaluating the coefficients a, b, c numerically to two places of decimals.

3. Show that the set of matrices $\begin{pmatrix} \cos\theta & -\sin\theta \\ \sin\theta & \cos\theta \end{pmatrix}$ under multiplication is isomorphic to the set of complex numbers $\cos\theta + \mathrm{j}\sin\theta$ under multiplication. *Name* a set of geometrical transformations, together with an operation, which is isomorphic to each of these sets.

4. A, B, C are points of three-dimensional space with coordinates $(1, 0, 3)$, $(-1, -4, 7)$ and $(3, 3, -3)$ respectively.
Find:

  (i) the magnitude of the angle BAC;

  (ii) expressions for the coordinates of any point of the plane ABC in terms of two parameters s and t.

377

5. Give values of a and b which make the second line of this working correct, and deduce for the expression a simple polynomial in n not involving $\sum$

$$\sum_0^n [r^2(r+1) - (r+2)^2(r+3)] = \sum_0^n r^2(r+1) - \sum_0^n (r+2)^2(r+3)$$
$$= \sum_0^n r^2(r+1) - \sum_a^b r^2(r+1).$$

6. Use integration by parts to evaluate

$$\int_0^\pi x \cos x \, dx.$$

Explain how you could have known before you started the calculation that the value of the integral is negative.

7. Two unbiased dice are rolled and the greater score (or either if they are the same) is recorded. State the set of possible scores and the probabilities associated with these scores. Find the expected value (i.e. theoretical mean) of the recorded scores.

8. A pair of resistors with resistances r, s are connected across a battery of negligible resistance, first in series and then in parallel. In which case will the power output from the battery be greater? Give either a mathematical or a physical reason for your answer.

Find an expression, in as simple an algebraic form as possible, for the ratio of the power output in the second case to that in the first.

9. In a certain book the frequency function for the number of words per page may be taken as approximately Normal with mean 800 and standard deviation 50. If I choose three pages at random, what is the probability that none of them has between 830 and 845 words each?

10. Using tables or a rough graph, find an approximation not more than 0.1 in error to the solution of the equation

$$x + 6 \ln x = 10.$$

Find a closer approximation to the solution of this equation, showing the details of your calculation and giving three decimal places in your answer.

(*Note.* $\ln x$ is an alternative notation for $\log_e x$.)

11. In a routine test for early warning of an unsuspected disease, 8% of those tested react (but do not all have the disease) while 7% in fact have the disease (but do not all react). The probability

378

that a person neither has the disease nor reacts is 0.9. What is the probability that a person with the disease will be detected?

12. Express $8 \cos \theta + 6 \sin \theta$ in the form $R \cos (\theta - \alpha)$, and hence find the set of angles θ between $0°$ and $360°$ for which

$$8 \cos \theta + 6 \sin \theta > 5.$$

13. a, b are elements of a group. Prove that

$$ab = ba \Rightarrow ba^2 = a^2b.$$

Show that the converse is false by taking a to be an element of order 2 in the group of symmetries of an equilateral triangle.

14. Show that the acceleration dv/dt of a particle moving in a straight line can be written, in terms of its velocity v and its displacement x from a point of the line, in the form $v \, dv/dx$.

At a distance x km from the centre of the earth the gravitational acceleration in km/s^2 is given by the formula c/x^2 where $c = 4 \times 10^5$. If a lunar vehicle 10 000 km from the centre of the earth is moving directly away from it at a speed of 10 km/s, at what distance will its speed be half that value?

15. A warship at rest has total mass M, which includes a torpedo of mass m. The torpedo is fired with speed v relative to the torpedo tube which is pointing horizontally at an angle of $45°$ to the length of the ship. Assuming zero resistance to forward or backward motion of the ship and infinite resistance sideways, find an expression for the speed of the ship.

16. Points X, Y, Z have position vectors denoted by $\mathbf{x}$, $\mathbf{y}$, $\mathbf{z}$ relative to an origin O. Write the expression $\mathbf{p} = \frac{1}{2}\mathbf{x} + \frac{2}{5}\mathbf{y} + \frac{1}{10}\mathbf{z}$ in the form $a\mathbf{x} + b(c\mathbf{y} + d\mathbf{z})$, where $a + b = 1$. Hence describe how the point P, with position vector $\mathbf{p}$, is related to the points X, Y, Z.

By writing $\mathbf{p}$ in terms of $\mathbf{x}$, $\mathbf{y}$, $\mathbf{z}$ in another way, give a different description of the relation of P to X, Y, Z.

17. (i) Given that $2 + j$ is a zero of $2x^6 - 9x^4 + 32x^2 + 75$, write down three other zeros.

(ii) The graphs of two polynomials, each with real coefficients, are plotted with the same perpendicular axes and scales. If one polynomial is of degree 5 and the other of degree 4, state the greatest and the least numbers of points in which the graphs could meet.

18. Functions f_1, f_2, f_3 have for domains the set of real numbers. The diagram shows the graph of each of these functions over the interval $0 \leqslant x \leqslant a$. You are given that

379

(i)f_1 is an even function;

(ii) f_2 is a periodic function with period a;

(iii) f_3 is an odd periodic function with period $2a$.

Sketch as much of the graphs of these functions as this information allows.

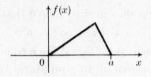

19. Express

$$f(x) = \frac{4x - 1}{(2x - 3)(x + 1)}$$

in partial fractions and hence find $f'(x)$.

Show that $f(\frac{1}{2} - x) = -f(x)$. What does this imply about the symmetry of the graph of f?

20. Express the following complex numbers in the form $r(\cos \theta + j \sin \theta)$, where $r \geqslant 0$ and $-\pi < \theta \leqslant \pi$:

(i) $-3j$;   (ii) $-2 + 2j$;

(iii) $\cos \alpha - j \sin \alpha$ (where $0 < \alpha < \pi$);

(iv) $1 + \cos \beta + j \sin \beta$ (where $0 < \beta < \pi$).

21. Complete the flow-diagram for evaluating

$$a_0 x^n + a_1 x^{n-1} y + \cdots + a_n y^n,$$

where the numbers $n, x, y, a_0, a_1, \ldots, a_n$, in this order, are on a tape:

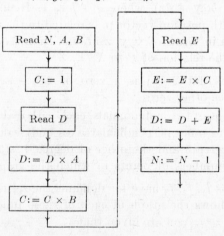

22. Use integration to prove that the volume of a cone of height h having a base (of any shape) with area A is $\frac{1}{3}Ah$.

23. A soldier swings on a rope fixed at its upper end A. The rope, initially inclined at $60°$ to the downward vertical, remains taut and may be considered as inextensible and of negligible mass. The soldier can be idealised as a particle at the end of the rope. If (a) his mass is 70 kg; (b) the rope is 10 m long; (c) he lets go when the rope is vertical; (d) A is 12 m above the ground; find his landing speed. (Neglect air resistance, and take $g = 10$ m/s^2.)

Which of (a), (b), (c), (d) can be varied, one at a time, without affecting this speed?

24. An electric circuit is made up of components – batteries and resistors – whose terminals are joined together in a certain configuration by short wires. With such a circuit we associate a 'mathematical model' in which numbers representing current, E.M.F., potential difference and resistance are assigned to various components or pairs of terminals. Write down the laws on which calculations within this model are based.

25. A vertical light spring has two equal masses fixed to it at its ends, as shown. When the lower mass lies on a table, the spring is compressed a distance d by the weight of the upper mass. If the upper mass is now pushed down a further distance $2.5d$ and then released, examine whether the lower mass will at some time leave the table.

[You may assume that if the lower mass were fixed, the upper mass would oscillate symmetrically about its equilibrium position.]

26. Estimate the value of

$$\int_1^{49} \frac{dx}{1 + \sqrt{x}}$$

by applying Simpson's rule with three ordinates.

SOME SPECIMEN EXAMINATION PAPERS

Show by an appropriate substitution that this integral is equal to

$$\int_1^7 \frac{2u}{1+u}\,du.$$

Hence, using the fact that $u = (1 + u) - 1$ or otherwise, deduce an exact expression for the integral.

27. For the equations:

$$x - 2y - 11z = 28,$$
$$2x + 2y - z = 5,$$
$$3x + 4y + 2z = -2$$

investigate the possibility of deriving the left side of the third equation by adding suitable multiples of the left sides of the first two equations. Explain the significance of your result in terms of the solution set of the equations.

I

OXFORD AND CAMBRIDGE
SCHOOLS EXAMINATION BOARD
(on behalf of G.C.E. Examining Boards)

General Certificate Examination
Advanced Level

MEI 72

M.E.I. SCHOOLS PROJECT

APPLIED MATHEMATICS II

Monday 26 June 1972. $2\frac{1}{2}$ hours

*Candidates should attempt not more than **seven** questions, of which at least **one** must be taken from Section A.*

The paper is divided into Sections A, B and C. Answers to questions in Section C must be handed in separately from those in Sections A and B.

Mathematical and statistical tables and calculating machines may be used.

SOME SPECIMEN EXAMINATION PAPERS

Section A

Probability

A1. A product is manufactured so that $2\frac{1}{2}\%$ of the production is larger than 5.020 units and 5% is less than 4.984 units and the distribution of size is Normal. The selling price per item in the range 5.000 ± 0.015 units is 10p, and in the ranges 4.975 to 4.985 and 5.015 to 5.025 the price is 2p. Find the average selling price, if items outside the range 5.000 ± 0.025 are not saleable.

A2. The average proportion of patients cured of a specific disease by the administration of x units of drug is given by

$$p = \frac{1}{2}\left\{1 + \tanh\left(\frac{x - \mu}{\sigma}\right)\right\},$$

where

$$\tanh y = \frac{e^y - e^{-y}}{e^y + e^{-y}}$$

and μ and σ are constants. When $x = 29$, $p = 0.10$ and when $x = 37$, $p = 0.35$. Prove that $\mu = 40$ and $\sigma = 10$. What dose on average will cure 95% of patients with this disease?

If 100 patients are given this dose what is the probability that at least one will not be cured?

Section B

Further Probability, Applied Calculus and Statistics

B3. N different letters are placed at random into N differently-addressed envelopes. What is the probability that all the letters are in the correct envelopes?

$$X_r = 1 \quad \text{if the } r\text{th letter is in the correct envelope,}$$
$$\quad = 0 \quad \text{otherwise,}$$

for $r = 1, \ldots, N$.
Find $E(X_r)$, $\text{var}(X_r)$ and show that

$$\text{cov}(X_r, X_s) = \frac{1}{N^2(N - 1)} \quad \text{for} \quad r \neq s.$$

Hence show that the expected number of letters in correct envelopes is 1 and find the variance of this number.

383

B4. Use the method of least squares to estimate the constants α, β, γ in the model

$$Y_i = \alpha + \beta \sin x_i + \gamma \cos x_i + \epsilon_i \quad (i = 1, \ldots, 7)$$

using the following data. The errors $\{\epsilon_i\}$ are independent with zero expectation and constant variance.

x_i	0	$\pi/3$	$2\pi/3$	π	$4\pi/3$	$5\pi/3$	2π
y_i	2.6	4.2	2.1	-1.4	-3.0	-1.0	2.7

B5. Twenty-five independent observations $\{X_i\}$ are taken from a Normal distribution with mean μ_1 and variance 9. Another twenty-five independent observations $\{Y_j\}$ are taken from a separate Normal distribution with mean μ_2 and variance 16. Construct 95% confidence intervals for μ_1 and for μ_2 when the mean $\bar{x}$ from the first sample is 3.5 and the mean $\bar{y}$ from the second sample is 5.0.

Show that:

(i) the null hypothesis $\mu_1 = 5$ is rejected by a significance test at the 5% level,

(ii) the null hypothesis $\mu_1 = \mu_2$ is accepted by a significance test at the 5% level.

B6. $\{X_i\}$ are independently distributed random variables with mean μ and variance σ^2 for $i = 1, \ldots, n$. Which of the following expressions give an unbiassed estimate of σ^2?

(i) $nS_2 - S_1^2$,   (ii) $\dfrac{1}{n} S_3$,   (iii) $\dfrac{1}{n-1} S_3$,

(iv) $S_3 - \dfrac{1}{n} S_4^2$,   (v) $\dfrac{1}{n-1} S_5 - \dfrac{1}{n(n-1)} S_4^2$,

where

$$S_1 = \sum_{i=1}^{n} x_i, \qquad S_2 = \sum_{i=1}^{n} x_i^2, \qquad S_3 = \sum_{i=1}^{n} (x_i - \bar{x})^2,$$

$$S_4 = \sum_{i=1}^{n} (x_i - \lambda), \qquad S_5 = \sum_{i=1}^{n} (x_i - \lambda)^2, \qquad \bar{x} = \frac{1}{n} S_1,$$

and λ is a constant.

Prove any one of your statements in answer to the first part and hence derive proofs of the remaining statements.

B7. A sampling inspection scheme is operated by taking a random sample of size 10 from each large batch of product. The batch is rejected if more than 2 defectives are found and otherwise the batch is accepted. Plot the operating characteristic of the given plan.

Explain the application of operating characteristics when choosing a suitable plan.

p	0.05	0.10	0.20	0.30	0.40	0.50
$(1 - p)^{10}$	0.5987	0.3487	0.1074	0.0282	0.0060	0.0010

B8. The structure of a game of tennis is unaltered if the scoring is taken to be 1 for each win of a particular play. With this scoring a player wins if he scores 4 before his opponent scores 3, but if the score is tied at $3 - 3$ a subsequent win requires a joint score of 5–3 or 6–4 or 7–5, etc. Let a player have a constant probability p of winning each play and his opponent a probability $1-p$ of winning each play. Show that the probability that the first player wins with a joint score of 4–1 is

$$\binom{4}{3} p^4 (1 - p)$$

and that the probability that he wins after a tie is

$$\frac{\binom{6}{3} p^5 (1 - p)^3}{1 - 2p + 2p^2}.$$

Hence find the probability that a player wins when his probability of winning a single play is $\frac{2}{3}$.

B9. X takes values 1, 0, -1 with probabilities p_1, p_2, p_3 respectively with $p_1 + p_2 + p_3 = 1$. In n trials the observed frequencies are r_1, r_2, r_3 respectively, $r_1 + r_2 + r_3 = n$. The probability of the event (r_1, r_2, r_3) is

$$\frac{n!}{(r_1)!(r_2)!(r_3)!} p_1^{r_1} p_2^{r_2} p_3^{r_3}, \ (r_1, r_2, r_3 = 0, \ldots, n; \ r_1 + r_2 + r_3 = n).$$

Find $E(r_1)$ and $E(r_3)$; prove that $\operatorname{var}(r_1) = np_1(1 - p_1)$, $\operatorname{var}(r_3) = np_3(1 - p_3)$ and $\operatorname{cov}(r_1, r_3) = -np_1p_3$.

Show how an estimate of the variance of the distribution of X may be derived from $(r_1 + r_3)$ in the case when $p_1 = p_3$.

B10. A probability density function $f(x)$ satisfies the equation

$$\frac{df}{dx} = \frac{3x - a_1 - 2a_2}{(x - a_1)(x - a_2)} f \quad \text{for } a_1 < x < a_2,$$

$$f(x) = 0 \text{ for } x \leqslant a_1, x \geqslant a_2.$$

Find an explicit form for $f(x)$. Give the mode of the distribution and sketch the graph of the p.d.f.

Section C

Applied Calculus and Mathematical Physics

C11. (i) If $z = y \sec x$, find dz/dx in terms of x, y and dy/dx. Hence or otherwise solve

$$\frac{dy}{dx} + y \tan x = \sin^3 x$$

given that $y = 1$ when $x = 0$.

(ii) Given that $y = z\,e^{2x}$ and

$$\frac{d^2y}{dx^2} - 4\frac{dy}{dx} + 4y = x\,e^{2x},$$

find a differential equation for z and hence find the general solution to the original equation.

C12. Sketch the curve $a^2y^2 = x^3(a - x)$.

Find the volume generated by rotating that part of the area contained by the curve which lies in the positive quadrant about

(i) the x-axis; find also the moment of inertia about OX of a uniform mass M occupying this volume;

(ii) the y-axis, using the substitution $x = a \sin^2 \theta$ or otherwise. You may assume that

$$I_{m,n} = \int_0^{\pi/2} \sin^m \theta \cos^n \theta \, d\theta$$

satisfies

$$I_{m,n} = \frac{m - 1}{m + n} I_{m-2,n}.$$

C13. A smuggler's boat with a maximum speed of 12 knots is at a position A. Information is received that a patrol boat, 20 nautical miles due South of A, is travelling at a constant 30 knots due North. In what direction should the smugglers' boat steer to keep as far as possible out of the way of the patrol boat?

What is then the least distance apart of the two boats in the subsequent motion?

On another occasion a similar problem arises with the same boats, but this time the patrol boat is observed to be 20 nautical miles 30° West of South from A. Find the corresponding direction and least distance apart.

C14. A light rod AB of length 1.5 m has a small mass of 2 kg at B and it rotates in a vertical circle about an axis at A. In the first place the rod is constrained by gearing to a motor to rotate with a constant angular velocity of 3 revolutions a second. Find the horizontal and vertical components of the reaction of the mass on the rod when AB makes an angle θ with the upward vertical as it descends.

The constraints are now removed so that the rod can rotate freely about the axis and so that its angular velocity when B is vertically above A is 1 revolution a second. Show that the horizontal component of the reaction of the mass on the rod in the corresponding position is now

$$2 \sin \theta \{6\pi^2 + g(2 - 3 \cos \theta)\} \text{ newton,}$$

and find the vertical component.

C15. A light spring AB of natural length a metre is doubled in length by a force of λg newton. It hangs vertically from A, carrying at B a tray of mass m kg in equilibrium. A particle, also of mass m kg, is dropped from A on to the tray and does not bounce.

Find:

(i) the distance fallen by the particle before hitting the tray.

(ii) the combined velocity of tray and particle after impact.

Show that the tray will descend to a depth (below A) of $a + x_1$, where x_1 is the greater root of

$$\lambda^2 x^2 - 4am\lambda x + ma^2(2m - \lambda) = 0.$$

By considering the significance of the other root of this equation, show that the particle will leave the tray in the subsequent motion if $\lambda > 2m$.

C16. In one stage of building a bridge a uniform section of length l metre and mass m tonne lies on supports at A and B. The next (equal) section BC is pushed out from AB to extend to C with joints made at B. Find the vertical force and the couple which the joints at B have to supply to support the section BC.

A tower of height h metre is then erected on the support at B, as shown in the diagram. A cable passes from A over a pulley and it is attached at C to BC. The cable is then tightened to a tension equal to the weight of T tonne; what will the couple at B be then? Find what value T should have for a zero couple at B.

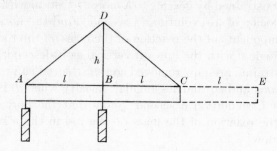

If the construction has been within safe limits so far, discuss whether it is safe to push out a further equal section from C to E. Should T be increased or not? If T is increased what further precautions might be necessary? Consider both B and C joints.

C17. A particle moves in a plane, attached to a fixed point O in the plane by an elastic string which remains taut during the motion. Show that the work done by the tension in the string in any interval of time depends only on the extensions of the string at the beginning and end of the interval and not on the direction either of the string or of the motion of the particle.

A heavy particle hangs from a light elastic string. Its period for small side to side oscillations is twice its period for small vertical oscillations. Show that it would take a particle three times as heavy to extend the string to twice its natural length.

C18. On a straight, two-lane road a car B is travelling 30 metre behind another car A, both going at 90 km per hour. B wishes to pass A; he can accelerate up to 110 km per hour in 10 seconds at a constant rate but must not exceed that speed. Until he reaches a position 30 metre in front of A he will be partly or wholly occupying the other lane. How long will it take B to pass A completely? How far will B travel in that time? If a car C is coming the other way at 110 km per hour on this other lane, how far away from B must C be at the start for the passing to be safe? (It may be assumed that the lengths of the cars need not be considered.)

What difference is made to these answers if B is able to come up to 30 m behind A at 110 km per hour and overtake at this constant speed?

J

UNIVERSITY OF LONDON

General Certificate of Education Examination

Advanced Level

This specimen paper illustrates the new syllabus for Advanced Level Pure Mathematics Paper 3 (Numerical Methods with Computing Applications) which will be examined for the first time in Summer 1972 as an alternative to Pure Mathematics Paper 2.

The new syllabus is contained in the Regulations and Syllabuses for Summer 1972 and January 1973.

PURE MATHEMATICS 3

Three hours

Answer EIGHT questions.

---

1. Find the sum of the first n terms of the series in which the rth term is $1/r(r+1)$, and hence show that the sum
$$1 + 1/2^2 + 1/3^2 + \cdots + 1/n^2$$
is never greater than 2.

   Write a program to find the sum of the first N terms of the series in which the nth term is $1/n^\lambda$, for $N = 10^2, 10^4, 10^6$ and $\lambda = 0.8, 0.9, 1, 1.1, 1.2$.

2. Show that if the roots of the equation
$$x^3 + 3ax^2 + 3bx + c = 0$$
are in arithmetic progression, then
$$2a^3 - 3ab + c = 0.$$

   If this relation holds between the coefficients show that the roots are $-a, -a \pm \sqrt{(3a^2 - 3b)}$.

   Write a program to print, together with their roots, all cubic equations of the above form which have real unequal roots in arithmetic progression, with a and b positive integers and a not greater than 10.

3. Sketch the graphs of $y = \exp x$ and $y = 1 + x^3$, and find the number of real roots of the equation $\exp x = 1 + x^3$.

Write a program to compute the non-zero roots of the equation

$$\exp (kx) = 1 + x^3$$

for $k = 1(0.1)2$.

4. (i) Find the matrix $\mathbf{P}$ such that $\mathbf{AP} = \mathbf{BP} = \mathbf{C}$, where

$$\mathbf{A} = \begin{bmatrix} 3 & 1 & -2 \\ 0 & 4 & -3 \end{bmatrix}, \qquad \mathbf{B} = \begin{bmatrix} 2 & -1 & 0 \\ 2 & 0 & -1 \end{bmatrix},$$

$$\mathbf{C} = \begin{bmatrix} 1 & 2 \\ 0 & 0 \end{bmatrix},$$

(ii) Find the three values of λ for which a non-zero vector $\mathbf{x}$ can be found such that

$$\begin{bmatrix} 2 & -2 & 3 \\ 1 & 1 & 1 \\ 1 & 3 & -1 \end{bmatrix} \mathbf{x} = \lambda \mathbf{x},$$

and find the vector of unit modulus corresponding to each value of λ.

5. Draw up a flow chart for the solution of a well-conditioned system of three simultaneous linear equations in three unknowns by Gaussian elimination. Explain how the residuals corresponding to an approximate solution can be used to provide a closer approximation.

6. Define the functions sinh x, cosh x, tanh x in terms of exp x, and show that

$$\text{artanh } x = \tfrac{1}{2} \ln \{(1 + x)/(1 - x)\}.$$

Write a program to evaluate the expression

$$5 \sinh (2x + y) + 4 \text{ artanh } (2x - y)$$

for the values of x and y in the ranges $x = -1(0.1)1$, $y = -1(0.1)1$ for which the expression is real and finite.

7. If $y = \exp (2x)$, use the expansion of y in powers of x to find the leading term in the error in the approximation $y_{n+1} = y_n + \tfrac{1}{2}h(y'_n + y'_{n+1})$, where y_n and y'_n denote the values of y and dy/dx at $x = nh$. Draw up a flow chart using this approximation to evaluate y for $x = 0(0.02)1$ given that

$$dy/dx = 2y + x^2y^2, \qquad y(0) = 1.$$

State the expected accuracy of the values of y.

8. (i) Differentiate arctan $\{x/\sqrt{(1 - x^2)}\}$ with respect to x.

  (ii) Evaluate (a) $\displaystyle\int_0^{\pi/2} x \sin (2x)\, \mathrm{d}x$, $\quad (b)$ $\displaystyle\int_1^2 \frac{1}{x(1 + x^2)}\, \mathrm{d}x$,

9. By means of the substitution $y = \frac{1}{2}\pi x^2$, show that

$$\int_0^{\pi/2} \frac{\cos y}{\sqrt{(2\pi y)}}\, \mathrm{d}y = \int_0^1 \cos\left(\tfrac{1}{2}\pi x^2\right) \mathrm{d}x.$$

  Evaluate this integral approximately, using (a) the trapezium rule, (b) Simpson's rule, given the table of values below. State the sources of error in your results.

x	0	0.25	0.5	0.75	1
$\cos\left(\tfrac{1}{2}\pi x^2\right)$	1	0.9952	0.9239	0.6344	0

10. If A, B, C are elements in Boolean algebra, define the union $A + B$, the intersection $A \cdot B$ and the inverse of C, C'.

  Show that $(A + B) \cdot (A + B' + C) = A + (B \cdot C)$.

  Simplify the expression

$$(A + B) \cdot (A + C) \cdot (B + C) \cdot C',$$

  and construct a truth table to check your answer.

K

CAMBRIDGE COLLEGES JOINT EXAMINATION
(*Awards and Entrance*)

FURTHER MATHEMATICS

PAPER 504: MATHEMATICS

Thurs. 25 Nov. 71–Morning. 3 hours

Begin each answer on a separate sheet and write
on one side of the paper only.

Aim at answering whole questions.

*Answers to questions marked A and B must be tied up
separately with a separate cover-sheet.*

**The attention of candidates is drawn to the fact
that these questions are intended to cover a
variety of syllabuses, and no candidate should
expect to be familiar with all the topics that
occur.**

1A Let G be a group with identity element e. Prove that the number of solutions of the equation $x^2 = e$ in G is either 1, ∞ or even. [Suppose $a \neq e$ is one solution and consider first the solutions satisfying $ax = xa$.]

2A Let n, p, q be integers with p, q prime, such that q divides $n^p - 1$ but not $n - 1$. Let a relation $\sim$ on the set $\{1, 2, \ldots, q - 1\}$ be defined by writing $x \sim y$ if q divides $y - n^r x$ for some r. Prove that

  (i) $\sim$ is an equivalence relation,
  (ii) each equivalence class has p elements,
  (iii) p divides $q - 1$.

3A Let a, b, c be integers and let $f(x, y) = ax^2 + 2bxy + cy^2$. Show that there are integers, p, q, r, s such that $ps - qr = 1$ and $f(x, y) = 2(px + qy)(rx + sy)$ if and only if a and c are even and $b^2 - ac = 1$.

4A Σ is a conic, and ABC, $A'B'C'$ are triangles such that the lines $B'C'$, $C'A'$, $A'B'$ are the polars with respect to $\sum$ of A, B, C respectively. Show that AA', BB', CC' are concurrent.

5A If A, B are points in the plane, the part of the line AB between A and B is called the *segment* AB. Points P_1, P_2, ..., P_6 in the plane are such that no three are collinear and no three segments P_iP_j, P_kP_l, P_mP_n are concurrent. A *crossing* is a point common to two distinct segments P_iP_j, P_kP_l. Prove that P_1, P_2, ..., P_6 always have at least three crossings, and find six points with exactly three crossings.

6A Prove that, for any four points A, B, C, D in a plane,

$$\begin{vmatrix} 2AB^2 & AB^2+AC^2-BC^2 & AB^2+AD^2-BD^2 \\ AC^2+AB^2-CB^2 & 2AC^2 & AC^2+AD^2-CD^2 \\ AD^2+AB^2-DB^2 & AD^2+AC^2-DC^2 & 2AD^2 \end{vmatrix} = 0.$$

7A Let l_1, l_2, l_3, l_4 be lines in the plane and let C_i be the circumcircle of the triangle obtained by omitting l_i. Prove that

  (i) C_1, C_2, C_3, C_4 have a point O in common.
  (ii) The feet of the perpendiculars from O to l_1, l_2, l_3, l_4 lie on a line l_0.
  (iii) l_0, l_1, l_2, l_3, l_4 touch a parabola with focus O and vertex on l_0.

8A Let

$$f(x) = \sum_{n=1}^{\infty} \frac{x}{n(n+x)}$$

for real positive x. Prove that

$$2f(2x) - f(c) - f(x + \tfrac{1}{2}) = 2\log 2 - 1/(x + \tfrac{1}{2}).$$

9B Find the most general solution of the 'differential equation'

$$f'(x) = \lambda f(1-x),$$

where λ is a real constant.

10B A fair coin is tossed successively until either two heads occur in a row or three tails occur in a row. What is the probability that the sequence ends with two heads?

11B My house lies between two bus stops, one of which lies 90 yards to the right and one 270 yards to the left. If I catch a bus at the left-hand stop it costs me 6 pence. If I catch it at the right-hand stop it will cost me 7 pence and if I miss the bus I must take a taxi which will cost 20 pence. The bus comes from the right and comes into sight at a point 90 yards further away from my house than the bus stop. I reckon to walk at 2 yards a second until I see the bus and then to run at 6 yards a second. The bus travels at 15 yards a second until it reaches the first stop where it waits for 3 seconds and then travels at 15 yards a second until it reaches the second stop where again it waits for 3 seconds and then goes round a corner out of sight.

When I leave the house there is no bus in sight, and I reckon that it does not matter which stop I go to. How frequent are the buses?

12B A farmer wishes to provide his cattle with three nutrients A, B and C, for which he has minimum requirements of 21, 9 and 12 units respectively. Two animal foods F_1 and F_2 are available; their content for unit cost are given in the following table.

	A	B	C
F_1	7	10	20
F_2	30	6	3

How can the farmer most cheaply satisfy his needs?

13B A heavy horizontal carriageway of uniform weight w per unit length is suspended from a heavy flexible wire attached to two pillars a distance $2d$ apart. The weight of the wire per unit length at any point is chosen to be k times the tension it has to sustain. Assuming that the carriageway acts as a continuous vertical load on the wire, and that $2kd < \pi$, show that the vertical load on each pillar is given by $T_0\beta \tan \beta kd$ where T_0 is the minimum tension in the wire and $\beta^2 = (w + T_0 k)/T_0 k$.

14B The bank of a river whose surface lies in the (x, y)-plane is given by $y = 0$. The surface current is in the x-direction and is given by ky. A man who swims steadily at speed V starts from the point $(0, y_0)$ wishing to reach the point $(0, 0)$. Assuming that $V > ky_0$, calculate the time it takes him to reach his destination

(a) if he arranges to swim so that his path is a straight line;
(b) if he swims towards the bank until he reaches it and then swims along the bank;
(c) if he always points himself towards his destination.

Show that each time can be written in the form $T = k^{-1}F(ky_0 V^{-1})$. By means of a series expansion show that, for $ky_0 V^{-1} \ll 1$, method (c) is fastest.

$$[\sin^{-1}u = u + \tfrac{1}{6}u^3 + \tfrac{3}{40}u^5 + \cdots]$$

15B A block of mass M rests on a rough horizontal table, and is attached to one end of an unstretched spring of length l and modulus λ. The other end is suddenly put into motion with uniform velocity V away from the block. The limiting coefficient of static friction μ_s is larger than the coefficient of dynamic friction μ_d. Show that the motion of the block repeats itself every

$$2\left\{\frac{(\mu_s - \mu_d)g}{\alpha^2 V} + \frac{1}{\alpha}\left[\pi - \tan^{-1}\frac{(\mu_s - \mu_d)g}{\alpha V}\right]\right\}$$

units of time, where $\alpha^2 = \lambda/Ml$. (It may be assumed that the tension in the spring is always positive.)

16B A uniform sphere of radius a and mass M moves under gravity in a vertical plane on the inside of a circular cylinder of radius $2a$ and mass M, which is pivoted freely about its own fixed horizontal axis. The centre of the sphere moves in a plane perpendicular to this axis. The centre of gravity of the cylinder (which is not of uniform density) is a distance a from its axis and its radius of gyration about its axis is $2a$. Let ϕ be the angle by which the cylinder departs from its equilibrium position and θ the angle made with the vertical by a line drawn through the centre of the sphere perpendicular to the axis of the cylinder.

In terms of θ and ϕ, what are the equations of motion when the sphere and cylinder are (a) perfectly smooth; (b) perfectly rough?

Show that the motion in (a) must, for small disturbances about equilibrium, be periodic with period either $2\pi\sqrt{(a/g)}$ or $4\pi\sqrt{(a/g)}$, interpreting the result physically. Explain, by reference to the equations of motion, how periodic motions can arise in case (b).

Bibliography

A. JOURNALS

There are many journals which publish papers and articles on mathematical education. Of those published in English, the reader's attention is particularly drawn to:

Educational Studies in Mathematics (Reidel, Netherlands), abbreviated in the Bibliography to *ESM*
International Journal of Mathematical Education in Science and Technology (John Wiley)
The Mathematical Gazette (Bell, for the Mathematical Association)
Mathematics in School (Longman, for the Mathematical Association)
Mathematics Teaching (Association of Teachers of Mathematics)
The Bulletin of the Institute of Mathematics and its Applications
The American Mathematical Monthly (Mathematical Association of America, abbreviated to MAA)
The Mathematics Teacher (National Council of Teachers of Mathematics)

B. PROJECTS

We have not listed the textbooks published by the various projects in the general bibliography, although some SMP Handbooks and SMSG Studies can be found there entered under the name of the person(s) responsible for that volume. Below, however, we give details of some of the leading projects in the UK and the USA, together with their publishers where appropriate. Addresses of Directors etc. can be found in the annual reports of the International Clearinghouse on Science and Mathematics Curricular Developments, University of Maryland.

(i) *United Kingdom*

CSM (Contemporary School Mathematics – St Dunstan's Syllabus), Edward Arnold
MMG (Manchester Mathematics Group), Rupert Hart-Davis
MEI (Mathematics in Education and Industry)
MME (Midlands Mathematical Experiment), Harrap
National Council for Educational Technology Continuing Mathematics Project
Nuffield Mathematics Teaching Project, John Murray and W. & R. Chambers (John Wiley in the USA)
Psychology and Mathematics Project, University of London Press
SMP (School Mathematics Project), Cambridge University Press (Cuisenaire in the USA)
Schools Council Mathematics for the Majority Project, Chatto & Windus
Schools Council Sixth Form Project, Heinemann Educational
SMG (Scottish Mathematics Group), Blackie and W. & R. Chambers
Shropshire Mathematics Experiment, Penguin

(ii) *United States*

CCSM (Cambridge Conference on School Mathematics)
  [1] *Goals for School Mathematics*, Houghton Mifflin, 1963
  [2] *Goals for Mathematical Education of Elementary School Teachers*, Houghton Mifflin, 1967
CEMREL–CSMP (Central Midwestern Regional Educational Laboratory – Comprehensive School Mathematics Program)
Computer-Based Mathematics Instruction at the Stanford-Based Laboratory for Learning and Teaching (Suppes Project)
CUPM (Committee on the Undergraduate Program in Mathematics), MAA
ECCP (Engineering Concepts Curriculum Project), McGraw-Hill
Madison Project, Addison-Wesley
Minnemast (Minnesota Mathematics and Science Teaching Project – producers of Minnemath reports)
SEED (Special Elementary Education for the Disadvantaged)
SMSG (School Mathematics Study Group), L. W. Singer, Random House and Stanford
SSMCIS (Secondary School Mathematics Curriculum Improvement Study), Teachers College Press, Columbia
Stanford–Brentwood Computer Assisted Instruction Laboratory (Suppes Project)
University of Illinois Arithmetic Project, Macmillan
UICSM (University of Illinois Committee on School Mathematics), Heath

C. ASSOCIATIONS AND ORGANISATIONS

ATCDE (Association of Teachers in Colleges and Departments of Education)
  [1] *Teaching Mathematics*, ATCDE, 1967
  [2] *The Development of the B.Ed. Degree in Mathematics*, ATCDE, 1970
ATM (Association of Teachers of Mathematics)
  [1] *Some Lessons in Mathematics*, Cambridge University Press, 1964
  [2] *Notes on Mathematics in Primary Schools*, Cambridge University Press, 1967
  [3] *The Development of Mathematical Activity in Children – The Place of the Problem in this Development*, ATM, 1966
  [4] *Examinations and Assessment*, ATM, 1968
HMSO (Her Majesty's Stationery Office)
  [1] *Report of the Commissioners appointed to inquire into the State of Popular Education in England* (Newcastle Commission), 1861
  [2] *Report of Her Majesty's Commissioners appointed to Inquire into the Revenues and Management of Certain Colleges and Schools* (Clarendon Commission), 1864
  [3] *Report of the Schools Inquiry Commission 1868* (Taunton Commission), 1868
  [4] *The Education of the Adolescent* (Hadow Report), 1926
  [5] *Secondary Education with Special Reference to Grammar Schools and Technical High Schools* (Spens Report), 1938
  [6] *Curriculum and Examinations in Secondary Schools* (Norwood Report), 1941
  [7] *Secondary School Examinations other than the GCE* (Beloe Report), 1960
  [8] *Half our Future* (Newsom Report), 1963

BIBLIOGRAPHY

[9] *Report of the Working Party on the Schools' Curriculum and Examinations* (Lockwood Report), 1964

[10] *Children and Their Primary Schools* (Plowden Report), 1967

[11] *Enquiry into the Flow of Candidates in Science and Technology into Higher Education* (Dainton Report), 1968

[12] *Teacher Education and Training* (James Report), 1972

[13] *Examinations in Secondary Schools*, 1911

[14] *Special Reports on the Teaching of Mathematics in the United Kingdom* (2 vols), 1912

[15] *Mathematics in Primary Schools*, 1931

[16] *Teaching Mathematics in Secondary Schools*, 1958

[17] *Mathematics in Primary Schools* (Curriculum Bulletin No. 1), 1965

[18] *Computers and the Schools* (Curriculum Papers No. 6), 1969

[19] *Statistics of Education* – annual

[20] *Statistics of Education* Special Series, No. 4, Part 2, *The Curriculum*, 1971

ICMI (International Commission on Mathematical Instruction)
> *Developments in Mathematical Education* (Proc. 2nd ICME) (ed. Howson, A. G.) Cambridge University Press, 1973

Maryland, University of, *Annual Reports of the International Clearing House on Science and Mathematics Curricular Developments* (annually from 1963)

Mathematical Association

[1] *The Teaching of Geometry in Schools*, Bell, 1923

[2] *A Second Report on the Teaching of Geometry in Schools*, Bell, 1939

[3] *Computers and the Teaching of Numerical Mathematics in the Upper Secondary School*, Bell, 1971

[4] Newsletter, June 1971

[5] *Applications of Sixth Form Mathematics*, Bell, 1967

[6] *Primary Mathematics: a further report*, Bell, 1970

[7] *Mathematics Projects in British Secondary Schools*, Bell, 1967

[8] *Mathematical Laboratories in Schools*, Bell, 1968

MAA (Mathematical Association of America)

[1] *Recommendations for the Training of Teachers of Mathematics*, MAA, 1961 (Revised, 1971)

[2] *The Mathematical Association of America: Its First Fifty Years* (Ed. May, K. O.), MAA, 1972

[3] *Studies in Mathematics*, 7 vols, various editors and subjects (for example, Modern Analysis, Modern Algebra, Applied Mathematics), Prentice Hall

Ministry of Education
> *Report of Cambridge Conference 1963* (duplicated notes)

NCET (National Council for Educational Technology)

[1] *Towards More Effective Learning*, Councils and Education Press, 1969

[2] *Computers for Education*, Councils and Education Press, 1969

NCTM (National Council of Teachers of Mathematics)

[1] *Experiences in Mathematics*

[2] *A History of Mathematical Education in the United States and Canada* (32nd Yearbook), 1970

[3] *Multi-Sensory Aids in the Teaching of Mathematics* (18th Yearbook), 1945

[4] *Growth of Mathematical Ideas* (24th Yearbook), 1959

[5] *Evaluation in Mathematics* (26th Yearbook), 1961

[6] *Enrichment Mathematics for High School* (28th Yearbook), 1963

BIBLIOGRAPHY

NSSE (National Society for the Study of Education)
Mathematics Education (69th Yearbook), University of Chicago Press, 1970

Nuffield Foundation, *The Use of Mathematics in the Electrical Industry*, 1961

Nuffield Foundation – CEDO (Centre for Educational Development Overseas)
[1] *Mathematics: the First Three Years*, Chambers and Murray, 1970
[2] *Mathematics: the Later Primary Years*, Chambers, Murray and Wiley, 1972

OECD (Organisation for Economic Cooperation and Development)
[1] *Curriculum Improvement and Educational Development*, 1966
[2] *OECD Observer*, October 1968
[3] *Supply, Recruitment and Training of Science and Mathematics Teachers* (undated)

(CERI – Centre for Educational Research and Innovation)
[4] *Equal Educational Opportunity* 1, 1971
[5] *Educational Technology: The Design and Implementation of Learning Systems*, 1971
[6] *Computer Science in Secondary Education*, 1971

OEEC (Organisation for European Economic Cooperation)
[1] *New Thinking in School Mathematics*, 1961
[2] *Synopses for Modern Secondary School Mathematics*, 1961

OISE (Ontario Institute for Studies in Education)
Geometry: Kindergarten to Grade 13, OISE, 1967

Schools Council (and SSEC)
[1] Working Paper 5, *Sixth Form Curriculum and Examinations*, HMSO, 1966
[2] Working Paper 12, *The Educational Implications of Social and Economic Change*, HMSO, 1967
[3] Working Paper 14, *Mathematics for the Majority*, HMSO, 1967
[4] Examinations Bulletin 1, *Some Suggestions for Teachers and Examiners* (SSEC), HMSO, 1963
[5] Examinations Bulletin 2, *Experimental Examinations: Mathematics* (SSEC), HMSO, 1964
[6] Examinations Bulletin 3, *An Introduction to Some Techniques of Examining* (SSEC), HMSO, 1964
[7] Examinations Bulletin 4, *An Introduction to Objective-Type Examinations* (SSEC), HMSO, 1964
[8] Examinations Bulletin 7, *Experimental Examinations: Mathematics II* HMSO, 1965
[9] Examinations Bulletin 23, *A Common System of examining at 16+*, Evans/Methuen Educational, 1971
[10] Examinations Bulletin 25, *CSE: Mode 1 Examinations in Mathematics*, Evans/Methuen Educational, 1972
[11] *The Schools Council, The First Three Years, 1964–7*, HMSO, 1968
[12] *Curriculum Innovation in Practice*, HMSO, 1968
[13] *Enquiry 1: Young School Leavers*, HMSO, 1968
[14] Working Paper 47, *Preparation for Degree Courses*, Evans/Methuen Educational, 1973

UNESCO (United Nations Educational, Scientific and Cultural Organisation)
[1] *Statistical Yearbook*, 1966
[2] Bulletin of the Regional Office for Education in Asia, 1966

BIBLIOGRAPHY

[3] *The Further Training of Mathematics Teachers at Secondary Level*, 1969

[4] *Modernization of Mathematics Teaching in European Countries*, Editions Didactiques et Pedagogiques, Bucarest, 1968

[5] *New Trends in Mathematics Teaching*, **1** (1966), **2** (1970), **3** (1972)

D. INDIVIDUALS

Abraham, R. *Foundations of Mechanics*, Benjamin, 1967

Ahlfors, L. V. [1] *et al.* 'On the mathematics curriculum of the high school', *Maths Teacher* **55** (1962), 191–4; *Amer. Math. Monthly* **69** (1962), 189–93

[2] *Complex Analysis*, McGraw-Hill, 1966

Albert, A. A. and Sandler, R. S. *An Introduction to Finite Projective Planes*, Holt, Rinehart and Winston, 1968

Anderson, C. A. and Bowman, M. J. (Eds) *Education and Economic Development*, Cass, 1966

Armitage, J. V. and Griffiths, H. B. *A Companion to Advanced Mathematics 1* (SMP), Cambridge University Press, 1969

Armytage, W. H. G. *Four Hundred Years of English Education*, Cambridge University Press, 1965

Atkinson, R. C. and Wilson, H. A. *Computer-Assisted Instruction*, Academic Press, 1969

Austwick, K. 'Towards a technology of instruction', in *Educational Administration and the Social Sciences* (Ed. Baron, G. and Taylor, W.), Athlone Press, 1969

Avital, S. M. and Shettleworth, S. J. *Objectives for Mathematics Learning*, Ontario Institute for Studies in Education, 1968

Baker, J. E. *et al.* 'A pedagogic approach to morphisms', *ESM* **4** (1971), 252–63

Ball, W. W. R. [1] *A History of the Study of Mathematics in Cambridge*, Cambridge University Press, 1889

[2] *Mathematical Essays and Recreations*, 11th Edn (revised by Coxeter, H. S. M.), Macmillan, 1939

Banks, B. 'The "disaster kit"', *Math. Gazette* **55** (1971), 17–22

Barnard, H. C. *A Short History of English Education*, University of London Press, 1947

Barnes, D. *et al. Language, the Learner and the School*, Penguin, 1971

Barnett, C. 'The education of military elites', *Journal of Contemporary History*, July 1967

Beberman, M. *An Emerging Program of Secondary School Mathematics*, Harvard University Press, 1958

Becher, R. A. *et al. New Patterns and Problems in Educational Publishing and Manufacturing*, CERI, Paris, 1971

Beeby, C. E. [1] 'Curriculum planning', in Howson [4]

[2] *The Quality of Education in Developing Countries*, Harvard University Press, 1966

[3] 'Educational aims and content of instruction', in *Essays on World Education* (Ed. Bereday, G. Z.), Oxford University Press, 1969

Begle, E. G. [1] 'The role of research in the improvement of mathematical education', *ESM* **2** (1969) 100–12

[2] Paper on geometry teaching circulated at the Ditchley Conference 1966 (see Thwaites [3])

Bell, E. T. [1] Letter in *Amer. Math. Monthly*, 1935

400

[2] *Men of Mathematics*, Penguin, 1953
[3] *Development of Mathematics*, McGraw-Hill, 1945
Bell, M. S. (Ed.) *Some Uses of Mathematics*, SMSG, Stanford, 1967
Benacerraf, P. and Putnam, H. *Philosophy of Mathematics*, Blackwell, 1964
Beth, E. W. and Piaget, J. *Mathematical Epistemology and Psychology*, Reidel, 1966
Biggs, E. E. [1] and Maclean, J. R. *Freedom to Learn*, Addison-Wesley, 1969
[2] *Mathematics for Older Children*, Macmillan, 1972
Birkhoff, G. D. [1] 'A set of postulates for plane geometry', *Annals of Maths* **33** (1932), 329–45
[2] and Beatley, R. *Basic Geometry*, Scott, Foresman, 1940
Blackie, J. 'The character and aims of British Primary education', in Howson [3]
Blandford, B. *A Study of Mathematical Education*, Oxford University Press, 1908
Blaug, M. (Ed.) [1] *Economics of Education* (2 vols.), Penguin, 1968–9
[2] *An Introduction to the Economics of Education*, Allen Lane, 1970
Bloom, B. S. *et al.* [1] *Handbook on Formative and Summative Evaluation of Student Learning*, McGraw-Hill, 1971
[2] *Taxonomy of Educational Objectives: Handbook 1, Cognitive Domain*, Longman, 1956
[3] *Taxonomy of Educational Objectives: Handbook 2, Affective Domain*, Longman, 1964
Blumenthal, L. M. *A Modern View of Geometry*, Freeman, 1961
Boehm, G. (Ed.) *The Math Sciences: a collection of essays* (COSRIMS), MIT, 1969
Bolt, A. B. and Wardle, M. E. *Communicating with a Computer*, Cambridge University Press, 1970
Bourbaki, N. *Eléments d'histoire des mathématiques*, Hermann, 1960
Bowen, J. *A History of Western Education*, vol. 1, *The Ancient World: Orient and Mediterranean, 2000 BC–AD 1054*, Methuen, 1972
Brauer, F. 'The non-linear simple pendulum', *Amer. Math. Monthly* **79** (1972) 348–54
Brown, J. W. *et al. AV Instruction: Media and Methods*, 3rd Edn., McGraw-Hill, 1969
Brown, M. '"Real" problems for mathematics teachers', *Int. J. Math. Educ. Sci. Technol.* **3** (1972) 223–6
Bruce, G. *Secondary School Examinations*, Pergamon, 1969
Bruner, J. *Toward a Theory for Instruction*, Harvard University Press, 1966
Burgess, T. *Inside Comprehensive Schools*, HMSO, 1970
Butler, Lord *The Art of the Possible*, Hamish Hamilton, 1971
Cane, B. *In-Service Training*, NFER, 1969
Cartwright, Dame Mary. Obituary of Grace Chisholm Young, *J. Lond. Math. Soc.* **19** (1944), 185–92.
Castle, E. B. *The Teacher*, Oxford University Press, 1970
Choquet, G. *Geometry in a Modern Setting*, Kershaw, 1969
Clark, Lord. *Civilization*, John Murray with BBC Publications, 1969
Clegg, Sir Alec (Ed.) *The Changing Primary School*, Chatto & Windus, 1972
Coombs, P. H. *The World Educational Crisis*, Oxford University Press, 1968
Courant, R. [1] and Robbins, H. *What is Mathematics?*, Oxford University Press, 1941

[2] and Hilbert, D. *Methods of Mathematical Physics*, Interscience, 1943

Cox, C. B. and Dyson, A. E. [1]–[3] *The Black Papers*, The Critical Quarterly Society, 1969–71

Coxeter, H. S. M. *Introduction to Geometry*, Wiley, 1961

Crank, J. *The Differential Analyser*, Longman, 1947

Crowell, R. H. *Knots and Wheels* in NCTM [6]

Curle, A. *Educational Strategy for Developing Societies*, 2nd Edn, Tavistock, 1970

Curtis, M. *Oxford and Cambridge in Transition, 1558–1642*, Oxford University Press, 1959

Curtis, S. J. *History of Education in Great Britain*, 7th Edn, University Tutorial Press, 1967

Davis, H. T. *Introduction to Non-Linear Differential and Integral Equations*, Dover, 1960

Davis, P. J. 'Fidelity in mathematical discourse', *Amer. Math. Monthly* **79** (1972), 252–62

Davis, R. B. [1] *The Changing Curriculum: Mathematics*, Association for Supervision and Curriculum Development, 1967

[2] *Explorations in Mathematics*, Addison-Wesley, 1967

Demott, B. 'The math. wars' in *New Curricula* (Ed. Heath, R. W.), Harper & Row, 1964

Dienes, Z. P. [1] *Building up Mathematics*, 4th Edn, Hutchinson Educational, 1971

[2] *An Experimental Study of Mathematics-Learning*, Hutchinson, 1963

[3] *Mathematics in the Primary School*, Macmillan, 1964

Dieudonné, J. A. [1] *Linear Algebra and Geometry*, Kershaw, 1969

[2] 'New Thinking in School Mathematics', in OEEC [1]

[3] 'The work of Nicholas Bourbaki', *Amer. Math. Monthly* **77** (1970), 134–45

[4] 'Should we teach "modern" mathematics?', *American Scientist* **61** (1973), 16–19

Dorn, W. S. and Greenberg, H. J. *Mathematics and Computing*, Wiley, 1967

Dubbey, J. M. [1] *Development of Modern Mathematics*, Butterworth, 1970

[2] 'Charles Babbage and his computer', *Bull. IMA* **9** (1973) 62–6

Dubisch, R. [1] *The Teaching of Mathematics*, Wiley, 1963

[2] *The Nature of Number*, Ronald, 1952

Durell, C. V. *The Teaching of Elementary Algebra*, Bell, 1931

Durran, J. H. *Statistics and Probability*, Cambridge University Press, 1970

Einstein, A. 'Geometrie und Erfahrung', *Berl. Akad. S. Ber.* **85** (1921)

Ellul, J. *The Technological Society*, Cape, 1965

Engel, A. [1] 'Geometrical activities for the upper elementary school', *ESM* **3** (1971), 353–94

[2] 'Systematic use of applications in mathematics teaching', *ESM* **1** (1968), 202–21

Eraut, M. R. *In-service Education for Innovation* (NCET), Councils and Education Press, 1972

Félix, L. *The Modern Aspect of Mathematics*, Basic Books, 1960

Fielker, D. and Stephens (née Mold,) J. *Topics from Mathematics*, various booklets, Cambridge University Press

Fletcher, T. J. [1] 'The teaching of geometry', *ESM* **3** (1971), 395–412

[2] *Linear Algebra through its Applications*, Van Nostrand, 1972

402

[3] 'Thinking with arrows', *Maths Teaching* **57** (1971) 2–5

Freudenthal, H. [1] 'Geometry between the devil and the deep sea', *ESM* **3** (1971), 413–35

[2] 'What groups mean in mathematics and what they should mean in mathematical education', in ICMI [1].

Gale, D. 'The jeep once more, or jeeper by the dozen', *Amer. Math. Monthly* **77** (1970), 493–501. Correction, **78** (1971), 644–5

Gattegno, C. *For the Teaching of Mathematics*, I (3 vols). Educational Explorers, 1963

George, T. W. 'Mathematics in the training of geologists', in UNESCO [5]

Gillings, R. J. *Mathematics in the Time of the Pharaohs*, MIT, 1972

Glaser, R. (Ed.) *Teaching Machines and Programmed Learning II*, Nat. Educ. Assn, USA, 1965

Glaymann, M. 'Initiation to vector spaces', *ESM* **2** (1969), 69–79

Godfrey, C. 'Geometry teaching: the next step', *Math. Gazette* **10** (1920), 20–4

Goodstein, R. L. 'The definition of number' *Math. Gazette* **41** (1957), 180–6

Gosden, P. H. J. H. *The Evolution of a Profession*, Blackwell, 1972

Griffiths, H. B. [1] and Hilton, P. J. *A Comprehensive Textbook of Classical Mathematics*, Van Nostrand, 1970

[2] 'Mathematical insight and mathematical curricula', *Maths in School* **1** (3) (1972), 3–7

[3] '1871: our mathematical ignorance', *Amer. Math. Monthly* **78** (1971), 1067–85

Gruenberg, K. W. and Weir, A. J. *Linear Geometry*, Van Nostrand, 1967

Haggett, P. and Chorley, R. D. *Network Analysis in Human Geography*, Arnold, 1969

Haldane, J. B. S. 'On being the right size', *and* 'The mathematics of natural selection', in Newman [1]

Halls, W. D. and Humphreys, D. *European Curriculum Studies 1: Mathematics*, Council for Cultural Cooperation, Strasbourg, 1968

Halmos, P. R. 'Nicholas Bourbaki', in Kline [1], pp. 77–81

Halsted, G. B. *Rational Geometry*, Wiley, 1904

Hammersley, J. M. 'On the enfeeblement of mathematical skills by the teaching of modern mathematics and similar intellectual trash in schools and universities', *Bull IMA* **4** (1968), 68–85

Hardy, G. H. [1] *A Mathematician's Apology*, Cambridge University Press, 1940 (reprinted 1967 with foreword by C. P. Snow)

[2] 'The case against the Tripos', *Math. Gazette* **62** (1948), 134–45 (reprint)

[3] *Pure Mathematics*, Cambridge University Press, 1908 (10th Edn 1952)

Hastad, M. 'Mathematics and engineers', *ESM* **1** (1968), 93–7

Heading, J. [1] *The Present Position of Applied Mathematics in the UK*, University of Wales, 1969

[2] 'Revival in applied mathematics', *Bull. IMA* **7** (1971), 262–8

Heath, Sir Thomas. *A Manual of Greek Mathematics*, Oxford University Press, 1931

Herstein, I. N. *Topics in Algebra*, Blaisdell, 1964

Hilbert, D. [1] *The Foundations of Geometry*, 3rd Edn, Open Court, 1938

[2] and Cohn-Vossen, S. *Geometry and the Imagination*, Chelsea, 1956

Hilgard, E. R. [1] *Theories of Learning*, 2nd Edn, Methuen, 1958

[2] and Bower, G. H. *Theories of Learning*, 3rd Edn, Appleton-Century-Crofts 1966

Hirst, K. E. and Rhodes, F. *Conceptual Models in Mathematics*, Allen & Unwin, 1971

Holt, J. *How Children Fail*, Pitman, 1964

Holt, M. and Marjoram, D. T. E. *Mathematics in a Changing World*, Heinemann Educational, 1973

Hooper, R. (Ed.) *The Curriculum*, Oliver & Boyd, 1971

Howson, A. G. [1] 'Milestone or millstone?', *Math. Gazette* **57** (1973), 258–66

[2] and Eraut, M. R. *Continuing Mathematics*, Councils and Education Press, 1969

[3] (Ed.) *Children at School*, Heinemann Educational, 1969

[4] (Ed.) *Developing a New Curriculum*, Heinemann Educational, 1970

[5] *The International Transfer of Learning Systems*, Educational Developments International, **1** (1973), 143–7

[6] 'Function Boxes: a model for differentiation', *Math. Gazette* **58** (1974)

[7] 'Charles Godfrey (1873–1924) and the reform of mathematical education', *ESM* **5** (1973), 157–80

Husén, T. (Ed.) *International Study of Achievement in Mathematics* (2 vols), Wiley, 1967

Jardine, J. *Physics is Fun: Introductory Course for Secondary Schools* (4 vols), Heinemann Educational, 1966

Jeger, M. [1] *Transformation Geometry*, Allen & Unwin, 1966

[2] and Rueff, M. *Sets and Boolean Algebra*, Allen & Unwin, 1970

Kant, I. *On Education*, 1803

Kaufman, B. A. and Steiner, H. G. 'The CSMP approach to a content-oriented, highly individualised mathematics education', *ESM* **1** (1969), 312–36

Kemeny, J. G. and Snell, J. L. [1] *Mathematical Models in the Social Sciences*, Ginn, 1962

[2] and Thompson, G. L. *Introduction to Finite Mathematics*, Prentice-Hall, 1959

Kerr, J. F. 'Curriculum change in emergent countries' in Howson [4]

Khinchin, A. M. *The Teaching of Mathematics*, Pergamon, 1968

Klein, F. *Elementary Mathematics from an Advanced Viewpoint* (2 vols.), Dover, 1939

Kline, M. [1] (Ed.) *Mathematics in the Modern World*, Freeman, 1968

[2] 'The ancients versus the moderns: a new Battle of the Books', *Maths. Teacher* **51** (1958), 418–27

[3] 'Math teaching assailed as peril to US scientific progress', *New York University Alumni News*, October 1961

[4] *Mathematical Thought from Ancient to Modern Times*, Oxford University Press, 1973

[5] *Mathematics in Western Culture*, Penguin, 1972

Knott, C. G. *The Life and Scientific Work of P. G. Tait*, Cambridge University Press, 1910

Lanchester, F. W. 'Mathematics in warfare', in Newman [1], ɔp. 2138–57

Lawrence, E. *The Origins and Growth of Modern Education*, Penguin, 1970

Leith, G. D. M. *Second Thoughts on Programmed Learning* (NCET), Councils and Education, 1969

Levi, H. 'Geometric algebra for the high school program', *ESM* **3** (1971), 490–500

BIBLIOGRAPHY

Lien Pu (Sung Dynasty, *c.* 981). 'Madame D', from *Famous Chinese Short Stories*, John Day

Lighthill, Sir James. 'The art of teaching the art of applying mathematics', *Math. Gazette* **55** (1971), 249–70

Livingstone, Sir Richard, W. *The Future in Education*, Cambridge University Press, 1941

Long, R. S. *et al.* 'Research in mathematical education', *ESM* **2** (1970), 446–68

Low, D. A. *Geometry for Engineers*, in HMSO [14]

Macdonald, I. D. [1] 'Problems about problems', *Amer. Math. Monthly* **72** (1965), 648–51

[2] 'Some more problems about problems', *Austral. Math. Teacher* **23** (1967)

MacLane, S. [1] 'Of course and courses', *Amer. Math. Monthly* **61** (1954), 151–7

[2] and Birkhoff, G. *Algebra*, Macmillan, 1967

[3] *Categories for the Working Mathematician*, Springer, 1971

McLone, R. R. *The Training of Mathematicians – A Survey Report*, SSRC, 1973

Manheim, J. H. *The Genesis of Point Set Topology*, Pergamon, 1964

Mansfield, D. E. and Thompson, D. [1] *Mathematics: A New Approach*, vol. 1, Chatto & Windus, 1962

[2] *Mathematics: A New Approach* (Second Series), vol. 2, Chatto & Windus, 1970

Matthews, G. 'Calculus', in Servais [1], pp. 83–93

Maxwell, E. A. *A Gateway to Abstract Mathematics*, Cambridge University Press, 1965

Maynard Smith, J. *Mathematical Ideas in Biology*, Cambridge University Press, 1968

Mill, J. S. *The Principles of Political Economy*, 1867

Miller, H. P. 'Income in relation to education', *Amer. Economic Review*, December 1960

Moise, E. E. [1] *Elementary Geometry from an Advanced Standpoint*, Addison-Wesley, 1963

[2] *The Number Systems of Elementary Mathematics*, Addison-Wesley, 1966

Montgomery, R. J. *Examinations*, Longman, 1965

Moore, E. H. 'On the foundations of mathematics', *Bull. AMS*, 1903

Morgan, J. B. 'The International Baccalaureate', in ICMI [1]

Needham, J. *Science and Civilization in China*, vol. 3, Cambridge University Press, 1959

Newman, J. R. (Ed.) *The World of Mathematics* (4 vols), Allen & Unwin, 1961

Newsom, C. V. 'The image of the mathematician', *Amer. Math. Monthly* **79** (1972), 878–82

Nickson, M. *Educational Technology*, Ward Lock, 1971

Noble, B. *Applications of Undergraduate Mathematics*, Macmillan (NY) 1967

Nyerere, J. *Education for Self Reliance*, Government Printer, Dar es Salaam, 1967

O'Hara, C. W. and Ward, D. R. *An Introduction to Projective Geometry*, Oxford University Press, 1937

Opie, Iona and Peter, *The Language and Lore of Schoolchildren*, Oxford University Press, 1959

Ore, O. *Graphs and their Uses*, Random House, 1963

Papy, G. [1] Arlon 1 *Elements of Topology*, 1959

Arlon 6 *Document for Teaching Analysis*, 1964
Arlon 7 *Document on Vector Spaces and Scalar Product*, 1965
Arlon 8 *First Lessons in Analysis*, 1966
Arlon 9 *New Lessons in Analysis*, 1967
(All published by the Centre Belge de Pédagogie de la Mathématique)
[2] *Groups*, Macmillan, 1964
[3] *Mathématique moderne* 1–6, Didier, 1963–7
(Some volumes are published in translation by Collier-Macmillan)
[4] *Le premier Enseignement de L'Analyse*, Presses Universitaires, Bruxelles, 1968
Pedley, R. *The Comprehensive School*, Penguin, 1969
Peel, E. A. 'Psychological and educational research bearing on mathematics teaching', in Servais and Varga [1]
Pendlebury, C. *A 'Shilling' Arithmetic*, Bell, 1889
Peterson, A. D. C. *International Baccalaureate*, Harrap, 1972
Philp, H. 'Mathematics in developing countries: some problems of teaching and learning', in ICMI [1]
Phythian, J. E. 'Mathematical Kujitegemea in Tanzania', *ESM* **4** (1971) 187–200
Piaget, J. [1] *The Child's Conception of Number*, Routledge & Kegan Paul, 1952
[2] and Inhelder, B. *The Child's Conception of Space*, Routledge & Kegan Paul, 1956
Pickert, G. 'Axiomatische Begründung der ebener euklidischen Geometrie in vektorieller Darstellung', *Math. Phys. Semesterbericht* **10** (1964)
Pidgeon, D. A. [Ed.] *Achievement in Mathematics*, NFER, 1967
Pollak, H. [1] 'Applications of mathematics in mathematics education', in NSSE [1]
[2] 'How can we teach applications of mathematics?', *ESM* **2** (1969), 261–272
Pólya, G. [1] 'Fundamental ideas and objectives of mathematical education' in *Mathematics in Commonwealth Schools*, Commonwealth Secretariat, 1969
[2] *How to Solve it*, Doubleday Anchor Books, 1957
[3] *Mathematical Methods in Science*, SMSG Studies XI, Stanford, 1963
Prost, A. *L'enseignement en France, 1810–1967*, Colin, 1968
Quadling, D. A. [1] 'The teaching of mathematics in schools in relation to undergraduate courses', *Bull. IMA* **7** (1971), 119–21
[2] *Mathematical Analysis*, Oxford University Press, 1955
Quereshi, A. R. and Richmond, P. E. 'Armstrong's heuristic method', *School Science Review* No. 157, June 1964
Reid, C. *Hilbert*, Springer, 1970
Rhodes, F. 'A geometric duality for two metrics for the coordinate plane', *Math. Gazette* **54** (1970), 19–23
Richardson, L. F. 'Mathematics of war and foreign politics, etc.', in Newman [1], pp. 1240–63
Richmond, W. K. [1] *The Concept of Educational Technology*, Weidenfeld & Nicolson, 1970
[2] *The School Curriculum*, Methuen, 1971
Roach, J. *Public Examinations in England 1850–1900*, Cambridge University Press, 1971

Rollett, A. P. [1] 'Class consciousness', *Math. Gazette* **52** (1968), 219–41
  [2] 'A history of the teaching of modern mathematics in England', *Math. Gazette* **47** (1963), 299–306
Room, T. G. *A Background to Geometry*, Cambridge University Press, 1967
Rosen, R. *Optimality Principles in Biology*, Butterworth, 1967
Rosskopf, M. F., Steffe, L. B. and Taback, S., *Piagetian Cognitive Development Research and Mathematical Education*, Washington DC, NCTM, 1971
Rugg, H. *American Life and the School Curriculum*, Ginn, 1936
Russell, Earl (Bertrand). *Mysticism and Logic*, Allen & Unwin, 1921
Ryser, H. J. *Combinatorial Mathematics*, Wiley, 1963
Sawyer, W. W. *Prelude to Mathematics*, Penguin, 1955.
Schiffer, M. M. 'Applied mathematics in the high school', *SMSG Studies* **10**, Stanford, 1963
Schramm, W. 'The new educational technology' in *Essays on World Education* (Ed. Bereday, G. Z.), Oxford University Press, 1969
Schuster, S. 'On the teaching of geometry', *ESM* **4** (1971), 76–86
Servais, W. and Varga, T. (Eds.) *Teaching School Mathematics*, Penguin (for UNESCO), 1971
Shannon, C. E. and Weaver, W. *The Mathematical Theory of Communication*, University of Illinois, 1949
Shenton, W. F. 'The first English Euclid', *Amer. Math. Monthly*, December 1928
Shibata, T. 'Axioms in contemporary mathematics and in mathematical education' in ICMI [1]
Shulman, L. S. 'Psychology and mathematics education', in NSSE [1]
Siddons, A. W. 'Progress', *Math. Gazette* **20** (1936), 7–26.
Singh, J. *Operations Research*, Penguin, 1971
Skemp, R. R. *The Psychology of Learning Mathematics*, Penguin, 1971
Skinner, B. F. *The Technology of Teaching*, Appleton-Century-Crofts, 1968
Skorov, G. *Integration of Educational and Economic Planning in Tanzania*, UNESCO, IIEP, 1966
Smith, E. R. and Tyler, R. W. *Appraising and Recording Student Progress*, Harper, 1942
Stamper, A. W. *A History of the Teaching of Elementary Geometry*, Columbia Teachers College, 1909
Steiner, H. G. 'Examples of exercises in mathematization on the secondary school level', *ESM* **1** (1968), 181–99
Stewart, W. A. C. [1] and McCann, W. P. *The Educational Innovators, 1750–1880*, Macmillan, 1967
  [2] *The Educational Innovators, 1881–1967*, Macmillan, 1968
Stone, M. H. [1] 'Reform in school mathematics', in OEEC [1]
  [2] Quoted in Demott [1]
  [3] Review of CCSM [1], *Maths. Teacher* **58** (1965), 353–60
Stones, E. (Ed.) *Readings in Educational Psychology*, Methuen, 1970
Strachey, R. *Millicent Garret Fawcett*, Murray, 1931
Strezikozin, V. 'Examinations in the Soviet Union', in *World Year Book of Education*, Evans Bros., 1969
Suppes, P. [1] 'The uses of computers in education', *Scientific American*, September 1966
  [2] *et al. CAI: Stanford's 1965–6 Arithmetic Program*, Academic Press, 1968
Sylvester, D. W. *Educational Documents, 800–1816*, Methuen, 1970

BIBLIOGRAPHY

Tammadge, A. 'How much does it cost to keep a dog?', *Maths. Teaching* **57** (1971), 9–11

Tate, T. *Principles of Geometry, Mensuration, Trigonometry, Land-Surveying and Levelling*, Longman, 1848

Taylor, L. C. *Resources for Learning*, Penguin, 1971

Taylor, W. (Ed.) *Towards a Policy for the Education of Teachers*, Butterworth, 1969

Têng, Ssu-Yü. 'Chinese influence on the western examination system', *Harvard Journal of Asiatic Studies* **7** (1942–3), 267–312

Thom, R. [1] '"Modern" mathematics: an educational and philosophic error?', *American Scientist* **59** (1971), 695–9
  [2] 'Modern mathematics, does it exist?', in ICMI [1]
  [3] *Stabilité structurelle et Morphogénèse*, Benjamin, 1972

Thompson, Sir D'Arcy W. *On Growth and Form* (Abridged Bonner, J. T.), Cambridge University Press, 1961

Thurston, H. A. *The Number System*, Blackie, 1956

Thwaites, B. [1] (Ed.) *On Teaching Mathematics*, Pergamon Press, 1961
  [2] '1984: Mathematics ⇔ computers?' *Bull. IMA* **3**, 1967
  [3] *SMP: The First Ten Years*, Cambridge University Press, 1972

Tibble, J. W. 'The educational effects of examinations in England and Wales,' in *World Year Book of Education*, Evans Bros., 1969

Trevelyan, G. M. *English Social History*, Longman, 1942

Tuller, A. *A Modern Introduction to Geometries*, Van Nostrand, 1967

Twersky, V. 'Calculus and science', *SMSG Studies in Mathematics* **15** (1967)

Vaizey, J. [1] *Education in the Modern World*, World University Library, 1967
  [2] 'Ask a silly question', *Times Educational Supplement*, 2 July 1971
  [3] *The Economics of Education*, Faber, 1962
  [4] and Sheehan, J. *Resources for Education*, Allen & Unwin, 1968
  [5] *et al. The Political Economy of Education*, Duckworth, 1972

Valentine, J. A. 'Selection for education by examination in the USA', in *World Year Book of Education*, Evans Bros., 1969

Vogeli, B. [1] 'Sweep away all cows, ghosts and devils', *ESM* **2** (1970), 496–500
  [2] *Soviet Secondary Schools for the Mathematically Talented*, NCTM, 1968

von Neumann, J. and Morgenstern, O. *The Theory of Games and Economic Behaviour*, Princeton University Press, 1944

Wallis, B. W. 'How is mathematics used?' *Maths. Teaching* **56** (1971), 14–17

Wardle, D. *English Popular Education 1780–1970*, Cambridge University Press, 1970

Weber, M. *Theory of Social and Economic Organization*, Collier-Macmillan, 1966

Weston, J. D. and Godwin, H. J. *Some Exercises in Pure Mathematics*, Cambridge University Press, 1968

Wheeler, D. 'So what?', in *New Approaches to Mathematics Teaching* (Ed. Land, F. W.), Macmillan, 1963

White, H. C. *An Anatomy of Kinship*, Prentice-Hall, 1963

Whitehead, A. N. 'The aims of education', *Math. Gazette* **8** (1916) 191–203

Whitney, H. 'The mathematics of physical quantities', *Amer. Math. Monthly* **75** (1968), 115–38, 227–56

Whittaker, E. T. and Watson, G. N. *Modern Analysis*, Cambridge University Press, 1920

Wilder, R. L. 'History in the mathematics curriculum: its status, quality and function', *Amer. Math. Monthly* **79** (1972), 479–94

Wilkinson, R. H. (Ed.) *Governing Elites: Studies in Training and Selection*, Oxford University Press, 1969

Wilson, J. M. 'The early history of the Association, or the passing of Euclid', *Math. Gazette* **10** (1921), 239–47

Wilson, J. W. [1] *et al. Z-population test batteries* (NLSMA Report 3), Stanford (SMSG), 1968

[2] in Bloom [1], pp. 643–96

Wilson, R. *Equality, Fraternity and Dissent*, quoted in Schools Council [2]

Wittenberg, A. [1] *Priorities and Responsibilities in the Reform of Mathematical Education* (cyclostyled notes), Toronto, 1964

[2] 'Sampling a mathematical sample test', *Amer. Math. Monthly* **70** (1963), 452–9

Wood, R. 'Objectives in the teaching of mathematics', *Educational Research* **10** (1968), 83–98

Woodward, W. H. *Education in the Age of the Renaissance*, Cambridge University Press, 1906

Wrigley, J. 'Assessment of children's progress, and evaluation of programmes' in *Mathematics in Commonwealth Schools*, Commonwealth Secretariat, 1969, pp. 85–93

Yaglom, I. M. *Geometric Transformations*, Random House, 1962

Young, G. C. and Young, W. H. *The First Book of Geometry*, Dent, 1905 (reprinted by Chelsea)

Young, G. S. 'The computer and the calculus', *ESM* **1** (1968), 105–10

Young, M. *The Rise of the Meritocracy*, Penguin, 1961

Zeeman, E. C. [1] 'Mathematics and creative thinking', *Psychiatric Quarterly, New York* (April 1966), reprinted in *Mathematics in School* **1** (2) (1972), 3–5

[2] 'The geometry of catastrophe', *Times Literary Supplement*, 10 December 1971

Name Index

Abbot, E. A. 129
Abelard 212
Ahlfors, L. V. 5
Albert, Prince 12
Anselm 212
Archimedes 8, 9
Armstrong, H. E. 16, 36
Arnold, M. 34, 65
Arnold, T. 13, 127
Ashford, Sir Cyril 160
Austwick, K. 84
Avital, S. M. 327, 331

Babbage, C. 115
Bacon, F. 68
Baker, H. F. 40
Ball, W. W. Rouse 108
Barnes, D. 67
Baron, M. 111
Beatley, R. 238
Beberman, M. 138
Beeby, C. E. 62, 127, 143
Begle, E. G. 139, 242
Bell, E. T. 25
Beltrami, E. 234
Bentham, J. 34
Bers, L. 104
Biggs, E. E. 96, 135
Billingsley, H. 93
Birkhoff, G. D. 238
Blackie, J. 77
Blandford, B. 217
Blaug, M. 74
Bloom, B. S. 45–9, 68, 324, 327ff
Bolyai, W. 104, 234
Boole, G. 103
Boole, M. 78
Bourbaki, N. 106, 110–11, 120, 214, 228
Brauer, R. 139
Bruner, J. 24, 97
Byrne, O. 93

Cantor, G. 253–5
Charp, S. 98

Chisholm, G. *see* Young, G. C.
Choquet, G. 239
Christiansen, B. 202
Clairaut, A. C. 91
Clark, Lord (Kenneth) 212
Cohn-Vossen, C. 108
Courant, R. 108, 121, 231
Coxeter, H. S. N. 235
Crowder, N. A. 89
Crowell, R. H. 293
Curle, A. 53

Dienes, Z. P. 134
Dieudonné, J. A. 236, 239
Dilworth, T. 90
Dirichlet, P. G. L. 103
Durell, C. V. 202

Einstein, A. 105, 114, 243
Eliot, G. 75
Eratosthenes 313
Euclid 8, 9, 15, 90, 93, 128, 226, 233, 237, 297ff
Euler, L. 292–3

Fawcett, P. 15
Félix, L. 93
Fermat, P. de 297ff
Flanders, H. 68
Fletcher, T. J. 95, 238
Floyd, P. J. 190ff
Freudenthal, H. 111, 240
Froebel, F. 78, 301

Gagné, R. 85
Galois, E. 103
Gattegno, C. 94, 129
Gauss, C. F. 103–4, 234, 297
Godfrey, C. 136
Goedel, K. 105
Goodstein, R. L. 140
Grassmann, H. G. 104
Green, G. 102

NAME INDEX

Verhulst, A. 306–7, 317
Villa, M. 236

Wall, W. D. 71
Weaver, W. 113
Weierstrass, K. T. 105
Weil, A. 245
Wells, H. G. 107
Weston, J. D. 245
Wheeler, D. H. 25
Whitehead, A. N. 24

Willsford, T. 90
Wilson, J. M. 128
Wilson, J. W. 328ff
Wilson, R. 58
Wittenberg, A. 67–8, 201
Wood, R. 327–8
Wrigley, J. 333

Young, G. C. 15, 94

Zaccharias, J. 137

Subject Index

Abbey Wood School, London, 333

ability of pupils, differences in, 54, 58, 60–1, 190; examinations to test, 326, 328

Abitur examination (Prussia), 37, 43

abstraction, faculty of, 303

abstractness in mathematics, 106, 120, 121, 140, 240, 282

acceleration, 309, 315

accuracy, 160, 162–3, 182, 278

actuarial mathematics, 102–3

adding machines, 305

addition, 263–4; definitions of, 248, 251, 280; of fractions, 279; of negative numbers, 273–4; of real numbers, 272, 285; of vectors, 283

administrative framework, required for educational reform, 135

aerodynamics, 113

affective domain in learning, 47, 328, 332, 340

affine plane, 240, 241

African Educational Program, 145

A-level, *see* GCE A-level

algebra, 8, 90; in applied mathematics, 305–9; Boolean, 103, 113, 306; in CCTT proposals, 209; modern, 109, 180, 240; in RN College syllabus, 163–167, 176; in SMP syllabus, 180–1

algorithms, 103, 111, 211, 218, 268, 338

American Mathematical Monthly, 129, 139

analogue methods of solving problems, 298

analysis, mathematical, 105, 108–9, 216, 245, 253, 314; global, 107; numerical, 117

angles, 170, 182, 185

anti-derivative, 314

anti-logarithms, 252–3, 257

applied mathematics, 113–15, 287–90, 318–22; algebra in, 305–9; arithmetic in, 304–5; British teaching of, 18, 19, 118–19, 141, 287; calculus, in 314–18; CCTT proposals on, 210; difficulties of

including in curriculum, 321–2; geometry in, 311–14; models in, 290–300; in physics, 288, 295, 296, 309–11; in primary schools, 301–3; versus 'pure' mathematics, 120–3, 292, 296

approximation, 163, 173, 302, 304, 314, 318–19

Arabs, mathematics of, 8

Archimedes, 8; axiom of, 284

area, 166, 169, 197, 199, 301, 306; under graphs, 174, 180, 314

arithmetic, 90, 108, 133, 209, 301; in applied mathematics, 304–5; approximate, 302; cardinal, 215, 253–5, 261; in CCTT proposals, 209; clock (modular), 241n, 302, 334; in elementary schools, 15; Greek neglect of, 8; number theory and, 247, 292; in RN College syllabus, 162–3

Arlon seminars, 140

Association for the Improvement of Geometrical Teaching (AIGT), 15–16, 129, 158, 229

Association of Teachers in Colleges and Departments of Education (ATCDE), 66

Association of Teachers of Mathematics (ATM), 130–1, 134, 242; Sixth-Form Project, of, 131, 334

Associations: *beside the above, see also* Mathematical Association, Mathematical Association of America, National Council of Teachers of Mathematics

associativity, 180, 249, 264, 268, 272, 273, 277

astronomy, mathematics in, 8, 169, 309–310

atomic energy industry, 114, 294

authoritarian methods: in reform, 151; in teaching, 132, 205, 221, 258, 296

automatic guidance systems, 114

axioms, 106, 107, 109, 226–33; categoricity of, 255–7, 259, 278; and the development of geometry, 233–5; in discussion of number systems, 246; pedagogical, 228–30, 235, 238, 253, 255; progressive use of (Papy), 283; in school geometry, 237–8; two attitudes to, 235–6; uses of, outside geometry, 244–5; *see also individual axioms*

Baccalauréat examination (France), 43
Bachelor of Education degree (BEd), 63–4, 219–21
Beeby model for educational system, 62–3, 152, 204
Belgium, 140, 307
belief and understanding, 212
'bicimals', 285
bijection (matching), 219, 253, 254, 257, 260, 261, 263, 283
binary number base, 232, 284
biology, mathematics in, 114, 289, 294, 311, 315
Birmingham, 19n, 128; Hazelwood School at, 126
'black box', 260
Black Papers, 79
Board of Education, 19, 160; *see also* Department of Education and Science, Ministry of Education
books: expenditure per pupil on, 73n; expository, 107–9, 110; of projects, 142, 146; in teaching, 89–94, 128, 151, 203
Boyle's law, 166, 175
Brighton, scholarships to secondary schools in, 44
'British' approach to mathematical teaching, a, 207–9
British Association for the Advancement of Science, 16–17
British Broadcasting Corporation (BBC), Maths To-day project of, 92, 96
British Computer Society, 118
bubble, as model, 298, 300
building, application of geometry, in, 8

calculating machines, 108, 115, 192
calculus, 8, 110, 114, 288; A-level question on, 335–6; in applied mathematics, 314–18; for engineers, 122; Newtonian, 102; in RN College syllabus, 162, 171–176; in school curriculum, 20, 108, 180, 184, 202; of variations, 114
Cambodia, 70
Cambridge (Mass.) Conference on School Mathematics (CCSM, 1963), 204–6, 208, 258, 279; K–6 syllabus of, 186, 187, 206, 209, 210, 301; pamphlet of (*Goals for School Mathematics*), 205

Cambridge (Mass.) Conference on Teacher Training (CCTT, 1966), 66, 204, 206–12; pamphlet of (*Goals for Mathematical Training of Elementary School Teachers*), 206
Cambridge University, 13, 15, 32, 33, 36, 344; entrance examination paper for, 344, 392–5
Canada, 18, 133
cardinal arithmetic, 215, 253–5
cardinal numbers, 197, 199
cardinality, 261
Cartesian product, 266–7
categoricity of axiom systems, 255–7, 259, 278
categories and functors, 104, 210n, 211
Cauchy's integral theorem, 229
China, 8, 31, 78; Cultural Revolution in, 73; examinations for public office in, 28–9
'Chinese-box' nesting property, 282
Christ's Hospital, Mathematical School at, 10
circle, 170, 177–8
Civil Service, examinations for, 33, 34–6
Clarendon Commission (1864), 14, 65
classification theorems, 106
classroom organisation, 23, 95, 132, 136, 191
codes, cracking of, 114, 116
cognitive domain in learning, 47, 243, 340; taxonomic levels of, 48, 327, 328, 329
College Entrance Examinations Board (USA), 43–4
Colleges of Advanced Technology, 19n
Colleges of Education, 63, 66, 132, 219
Columbia University, New York; Teachers College at, 80n
combinatorial mathematics, 108, 114, 116, 241
commerce, pressures of, 99, 206
commercial mathematics, 114; examination paper in, 271–3
common sense in engineering, limitations of, 113
communication, mathematical theory of, 113
commutative diagrams, 307–8
commutative ring, 249–50, 252
commutativity, 180, 249, 264, 268, 272, 273, 277, 280
comparison, 262–3
completeness of real number system, 253, 277, 281–2
complex numbers, 103, 104n, 282
components, approach to trigonometry through, 182

probability, 102, 113; in applied mathematics, 289, 297, 302, 306, 320; axiomatic approach to, 245; flow diagram for work on, 88; in SMP syllabus, 182, 186; theory of, 316
problem-solving: by analogue methods, 298; in geometry, 235, 236, 242; as heart of mathematics, 140, 231–2; models in, 292–3; strategies for, 211
problems, combinatorial, 108
profession, mathematics as a, 45, 68, 192, 212, 322
programmed learning, 84, 87–9, 98
project work, 133, 191–3, 198
projects for curriculum development, 137–8, 145–51; British 140–3; European, 140; life-span of, 146, 151–2; Schools Council and, 144–5; and teacher-training, 143–4; in USA, 138–40; *see also individual projects*
proofs: axioms and, 226, 227, 245; beautiful, 48, 328; deductive, 218, 220; formal, 167, 186, 198, 237, 331; impossibility, 208, 240
Prussia, 3, 15, 16, 37
psychology, 130, 149, 212; of learning, 87, 192; of mathematicians, 235, 322
Psychology and Mathematics Project, 150
psychomotor domain in learning, 48, 243

quadratic equations, 165, 185
quantum theory, 102, 114, 296, 297, 320
quaternions, 103

radio, 113; in education, 94
rate of change, 172
Reckoning Master, profession of, 8
refraction of light, 298
relations, 180, 185
relativity, 102, 114, 296, 306
Renaissance, 8, 32
'Renewal' tendency in mathematics teaching, 213, 218, 220, 221
research: in computer-assisted education, 98; in education, 71, 77, 82–3; in mathematics, 207, 231
research-type activity, at pupil's level, 131
Riemann metric, Riemann surfaces, 104
rigour, mathematical, 120, 221, 245, 315; versus intuition, 218
Roman numerals, 133
routes, matrix description of, 185
Royal Institution, Christmas lectures at, 107
Royal Naval Colleges, syllabus for, 19, 35, 158–78
Royal Society, 10
Royaumont seminars, 140, 239
Russia, 23, 138, 319; examinations in, 43; special schools for the mathematically gifted in, 60

St Dunstan's Project, *see* Contemporary School Mathematics Series
salaries: effect of education on, 75; of teachers, 17, 70, 71, 220
Samoa, 96
Sandhurst Military Academy, 14
scalar product, 239, 242
scholarships: to secondary schools, 44, 54; to universities, 39
School Certificate Examination, 39
school-leaving age, 52
School Mathematics Project (SMP), 111, 141–2, 147, 148, 341; *Additional Mathematics* of, 245, 282; *Advanced Mathematics* of, 314, applied mathematics in, 304, 316, 319, 321; CSE series of, 280, 281, 341; *Further Mathematics* of, 242; GCE O-level syllabus of, 92, 117, 179–86
School Mathematics Study Group (SMSG), 111, 147, 151; curricula of, 121, 139–40, 238, 289–90; and disadvantaged child, 60; research report from, 83
schools: comprehensive, 57–9, 130, 131; dame, 63; Dissenters', 10; elementary, 13, 15, 132; grammar, 9, 11, 14, 15, 54, 55, 56, 131; higher grade, 15; modern, 54, 55, 56; multilateral, 55, 58; pilot, in development of projects, 148; preparatory, 167; primary, 59, 77, 78, 95, 131–5, 218, 229, 279, 301–4; private, 13, 14, 17, 57; public, 11, 13, 14, 65, 127, 131; secondary, 15, 16, 52, 61, 77, 130, 133, 283, 304–5; secondary modern, 41, 130; technical, 15, 55, 56; *see also individual schools*
Schools Council for the Curriculum and Examinations, 42, 54, 72, 144, 326; examination bulletins of, 331; finance of, 70, 145, 146; Mathematics for the Majority Project of, 60, 151, 190–4; Sixth-Form Mathematics Project of, 321
science fiction, 107
science, mathematics and, 21, 111, 119, 160, 161, 319; *see also* applied mathematics, physics, *etc.*
Scotland: educational system of, 3–4, 11–12, 62, 74, 141; Inspectorate in, 65, 142; universities of, 11, 37

SUBJECT INDEX

Scottish Education Department, 118, 130
Scottish Leaving Certificate (now Certificate of Education), 43, 62
Scottish Mathematics Group (SMG), 96, 143, 147, 148, 242, 281
Secondary Schools Examination Council, 39, 40, 42
Secondary Schools Mathematics Curriculum Improvement Study (USA), 147, 148, 240–1, 245, 282, 283, 316
selection, examinations for, 27–31, 56–9, 132
sets: approach to number systems through, 253–5, 279; finite and infinite, 254, 261; theory of, 105, 106, 180, 183, 229, 292
setting of pupils, for ability in different subjects, 58
significance, level of, 300
similarity of figures, 168–9, 275
Simpson's rule, 175
sine and cosine, 170, 171, 176
'situations' in mathematics, 191, 222, 260, 332
Sixth Forms, 55; ATM Project for, 131, 334; Schools Council Mathematics Curriculum Project for, 321; SMP course for, 242
slide rules, 182
Snell's law, 298, 300
social ladder, 29, 39, 53, 57
Southampton University, 19n; questions set in BSc Mathematics examination at, 199–202; and SMP, 142; syllabus for course in mathematics curriculum studies at, 194–9
space: Euclidean, 104, 181, 283; 'real', 105, 297, 298; sense of, 169; vector, 110, 239, 240, 242, 272, 283
Special Elementary Education for the Disadvantaged (SEED), 60
specialisation, premature, 40, 42, 72, 145
speed, 172
Spens Report (1938), 20–1, 39, 40, 54
spherical geometry/trigonometry, 169, 171
spiral approach to mathematics, 205, 238
Sputnik satellite, 138
statics, 170
statistics, 102, 113, 119, 209, 306; computer and, 116; in SMP course, 182, 186
stimulus–response theory of learning, 80
stock control, 114
streaming in schools, 58, 59
structural stability, 319–20

structure in mathematics, 106, 120, 202, 231, 232, 244, 262, 282, 322; algebraic, 103; non-isomorphic, 256; recognition of, 288; in SMP syllabus, 186
subtraction, 264–5; of negative numbers, 274
successor function, 247, 266
sun, 309
surds, 165
surfaces: Riemann, 104; topology of, 108
surveying, mathematics for, 11
Swansea scheme, 245
Sweden, 57, 98, 122
switching circuits, design of, 306
Switzerland, 44, 71, 242
syllabus, 190, 222; of CSM, 186, 188–9; for course in mathematical curriculum studies, 194–9; at RN Colleges, 158–78; of SMP for O-level GCE, 179–86; traditional, 133, 156
symmetry, 104, 181, 186
symmetry groups, different geometries distinguished by, 234
systems: axioms of, 227–8, 234; construction of, and mappings between, 211
systems analysis, 84, 114
systems approach to teaching, 84–5, 98, 156

tangents, 170, 171, 172, 177
Tanzania, 73, 76
tape-recorder, 97
Taunton Commission (1868), 38
taxpayers, and education, 20, 22
teacher-training, 63–4, 65–7, 206–7; conference on, 206–12; in-service, 64–5, 139, 143, 149; for projects, 137, 138, 143–4
teachers of mathematics: anti-garbling philosophies for, 204; 'average', 143, 150, 240n; in Beeby model, 62–3, 152, 204; centres for, 149; and curriculum, 62; and examinations, 39–40, 42; freedom of, 94, 143–4; and projects (England), 142–3, 147; qualities desired in, 67; responsibility of, 341; salaries of, 17, 70, 71, 220; shortage of, 65–6
teaching: computer-aided, 97–8; father- and mother-centred styles of, 132; individualised, 97–9; logical and psychological, 192; mandarin, 49; method of, 156–7, stages of, 158–9; theory of, 82
teaching apparatus, 99, 134, 135, 296
technical colleges, 55

422